Conditional Order Statistics and Conditional Spacings

Zhengcheng Zhang

Conditional Order Statistics and Conditional Spacings

Springer

Zhengcheng Zhang
School of Mathematics and Statistics
Hainan Normal University
Haikou, Hainan, China

ISBN 978-981-96-7150-2 ISBN 978-981-96-7151-9 (eBook)
https://doi.org/10.1007/978-981-96-7151-9

This Springer imprint is published by the registered company Springer Nature Singapore Pte Ltd.
The registered company address is: 152 Beach Road, #21-01/04 Gateway East, Singapore 189721, Singapore

Preface

There are several definitions of reliability. Formally, reliability is defined as the probability that a product, system, or service will perform its intended function adequately for a specified period of time, or will operate in a defined environment without failure. Informally, reliability may be defined as quality over time. Reliability is typically the aspect of quality that impacts customers the most, and it is well known that an unreliable product is not a high-quality product. Manufacturers need to design and build increasingly higher reliability products for rapid advances in technology, development of highly sophisticated products, intense global competition, and great customer expectations etc.. Reliability evaluations present a challenge to both manufacturers and consumers because of the elapsed time between when the product is designed and built, and when the needed reliability information is forthcoming. A manufacturer might, for example, needs to design and build a product that is expected to operate successfully for ten years but has only four or six years to do so in practice. Ensuring high reliability in the design of a modern product is, first and foremost, an engineering challenge.

Comparing the means of two distribution functions is the simplest way of comparing two corresponding distribution functions. However, such a comparison is based on only two single numbers of the means, and so it is often not very informative. Moreover, the means sometimes do not exist at all. In many instances of applications one has more detailed information, for the purpose of comparison of two distribution functions, than just the two means. Then, stochastic orders with inequalities have been used popularly during the past 50 years, at an accelerated rate, in many diverse areas, such as reliability theory, queuing theory, survival analysis, insurance, actuarial science, operations research, management science etc..

The purpose of this book is to collect the new results that authors have obtained about various order properties of conditional coherent systems as well as stochastic properties and statistical ones. Specifically, in Preliminaries part, the aim and scope of the monograph are given, also some concepts such as order statistics and stochastic ordering etc. are introduced, which are frequently used in this book. Chapter 2 focuses on the residual life and inactivity time of conditional coherent systems with independent component lifetimes. Chapter 3 is about the residual and inactivity time of conditional coherent systems with independent component lifetimes under double monitoring. Chapter 4 investigates aging properties and stochastic comparisons of the residual lives of the remaining components for an $(n - k + 1)$-out-of-n system with independent and identical components after the kth failure has been observed. Furthermore, the residual life of surviving components is mainly focused on within coherent systems. In Chap. 5, stochastic behaviors of the number of failed components of coherent systems are investigated. Chapter 6 is for the stochastic properties and parameter estimation within load-standby systems. In Chap. 7, several concepts of conditional spacings are introduced when a coherent system fails at certain time, and their stochastic and statistical properties are researched.

I acknowledge the support of National Natural Science Foundation of China and the Natural Science Foundation of Hainan Province of China. Special thanks to my wife Yonghong Yang and my daughter Yiyang Zhang for long-term support and courage. I thank those with whom I had the pleasure of discussing reliability-related problems: Narayanaswamy Balakrishnan, Xiaojun Zhu, Xiaoxiao Hu, Xiaoliang ling, Longxiang Fang etc.. Also, I thank Yingcai Tang for his help with Tex techniques. Finally, I am indebted to the Springer Nature staff for their editorial work.

Haikou, China Zhengcheng Zhang

Competing Interests The author has no competing interests to declare that are relevant to the content of this manuscript.

Notes

Throughout the whole book, it is implicitly assumed that all random variables under consideration have 0 as the common left end point of their *supports*, and the terms *increasing* and *decreasing* stand for *monotone non-decreasing* and *monotone non-increasing*, respectively. Expectations and variances etc. of random variables are assumed to exist whenever they are written. The inverse of a distribution function (which is not strictly monotone) means the right continuous version of it, unless stated otherwise. Also for any conditional random variable $(A|B)$, we assume that $P(B) > 0$.

IID stands for "independent and identically distributed", while INID stands for "independent but not identically distributed".

Contents

Chapter 1
Prelimilaries

1.1 Aim and Scope

As the title suggests, this monograph is devoted to conditional order statistics and conditional spacings for reliability analysis and other disciplines that employ the notion of the order statistics and spacings. The conditional lifetime is widely used in survival analysis and reliability-related statistics, and the conditional spacing is very important in goodness-of-fit and in Type-II censoring etc.. Although the main focus in the text is on this crucial characteristic, our presentation cannot be restricted to conditional order statistics and conditional spacings analysis alone, other important reliability and corresponding statistic properties are studied as well.

This book provides a comprehensive treatment of different reliability models focused on properties of the residual life and conditional spacing of coherent systems and load-sharing systems. Our presentation is about our research findings of the last fifteen years. The book is mostly targeted at researchers and engineers related to reliability theory, stochastic analysis and other fields. The results obtained can be used as a supplement and extensions and generalizations to existing reliability theory etc. Moreover, most of them such as optimal allocation of a load standby in a system etc., may be regarded as technical support for engineering designers to design better products with higher reliability performance.

1.2 Main Concepts

1.2.1 Order Statistics

Let $X_1, X_2, \ldots, X_n$ be n independent random variables. Denote the corresponding order statistics by $X_{(1)} \leq X_{(2)} \leq \cdots \leq X_{(n)}$. When we want to emphasize the dependence on n, we denote the order statistics by $X_{1,n} \leq X_{2,n} \leq \cdots \leq X_{n,n}$, or $X_{1:n} \leq$

Z. Zhang, *Conditional Order Statistics and Conditional Spacings*,
https://doi.org/10.1007/978-981-96-7151-9_1

$X_{2:n} \leq \cdots \leq X_{n:n}$. And it is clear, $X_{1,n}(X_{1:n}) = \min\{X_1, \ldots, X_n\}$, $X_{n,n}(X_{n:n}) = \max\{X_1, \ldots, X_n\}$.

Let $F(x)$ be the distribution function of X. Denote the distribution function of the kth order statistic $X_{k,n}$ by $F_{k,n}(x)$, then,

$$F_{k,n}(x) = \sum_{j=0}^{k} \frac{n!}{j!(n-j)!}[F(x)]^j[1-F(x)]^{n-j}.$$

If X is a continuous random variable with the density function $f(x)$, then, the density function (denoted by $f_{k,n}(x)$) of the kth order statistic $X_{k,n}$ is

$$f_{k,n}(x) = \frac{n!}{(k-1)!(n-k)!}[F(x)]^{k-1}[1-F(x)]^{n-k}f(x).$$

The joint density function of the $(X_{i,n}, X_{j,n})^T$ $(i < j)$ is represented as

$$\begin{aligned} f_{ij}(x, y) = &\frac{n!}{(i-1)!(j-i-1)!(n-j)!}[F(x)]^{i-1} \cdot [F(y) - F(x)]^{j-i-1} \\ &\cdot [1-F(y)]^{n-j} f(x) f(y), x \leq y. \end{aligned}$$

Specially, the joint density function of $(X_{1,n}, X_{n,n})^T$ is

$$f_{1n}(x, y) = \begin{cases} n(n-1)[F(y) - F(x)]^{n-2} f(x) f(y), & x \leq y \\ 0, & x > y. \end{cases}$$

Moreover, the joint density function of $(X_{1,n}, X_{2,n}, \ldots, X_{n,n})^T$ is

$$f(y_1, y_2, \ldots, y_n) = \begin{cases} n! \prod_{i=1}^{n} f(y_i), & y_1 < y_2 < \cdots < y_n \\ 0, & \text{otherwise.} \end{cases}$$

1.2.2 Coherent Systems

The coherent system plays an quite important role in reliability theory. A system is said to be coherent if each of its components is relevant (that is, the system would not contain any component whose functioning has absolutely no influence on whether or not the system works) and if its structure function is monotone (that is, replacing a failed component by a working component cannot cause a working system to fail). The k-out-of-n system is a very important structure of redundancy, which works if and only if at least k components are operating. Important particular cases of the k-out-of-n system are series and parallel systems when $k = n$ and $k = 1$, respectively.

The order statistic $X_{k,n}$ is just the lifetime of the $(n-k+1)$-out-of-n system, hence the study of order statistic is equivalent to that of k-out-of-n system.

For a coherent system with n independent and identical components having lifetimes $X_1, \ldots, X_n$, its lifetime can be expressed as $T = \tau(X_1, \ldots, X_n)$, where τ is a structural function (see Esary and Marshall (1970) and Barlow and Proschan (1975) for the definitions and properties of coherent life functions). Samaniego (1985) firstly introduced the concept of signature to express the distribution function of the lifetime of a coherent system in terms of those of component lifetimes. Subsequently, Kochar et al. (1999) further observed that the distribution function of a coherent system with component lifetimes $X_1, \ldots, X_n$ may be expressed as a mixture of the distribution functions of the order statistics $X_{1,n}, X_{2,n}, \ldots, X_{n,n}$. Specifically, the signature of a system is defined as a probability vector $\mathbf{p} = (p_1, \ldots, p_n)$ with

$$\begin{aligned} p_i &= P(\tau(X_1, \ldots, X_n) = X_{i:n}) \\ &= \frac{\text{number of orderings for which the } i\text{-th failure causes the system failure}}{n!}. \end{aligned}$$

Of course, p_i is nonnegative and $\sum_{i=1}^{n} p_i = 1$.

The signature of a coherent system is an important tool for investigating the performance of a system structure and comparing different structures. It has been proven to be a useful metric for a system design, as it is a distribution-free measure that efficiently captures the precise features of a system performance. For example, often, a poor system design with good components will outperform a good system design with poor components. If, however, two different systems have common component lifetimes, then any difference between them must be attributable to the system design.

1.2.3 Standby Systems

Generally, there are three types of standby redundancies, viz., cold standby, hot standby, and warm standby. In a cold standby redundancy, the redundant component is inactive, that is, it has zero failure rate and hence never fails in standby state; in a system with hot standby redundancy, the redundant component also works in the usual environment and thus may fail when it is in standby state. That is, the active component and the redundant one form a parallel system. The warm standby redundancy works in a milder environment in the standby state and hence the failure rate is nonzero and smaller than that in the usual environment; it is just an intermediate stage between the cold and the hot ones.

1.2.4 Load-Sharing Systems

In practice, the components of many systems share a given load, functioning dependently on each other. This is because one component failure may affect the failure rates of the remaining surviving components of the system. CPU of a multiprocessor computer, generators in a power plant, cables in a suspension bridge, valves or pumps in a hydraulic system, kidneys in a human body, etc. are some of the real examples of load-sharing system models. The mechanism of these models is such that after the failure of one component, the surviving components have to shoulder extra load and hence are prone to failure at an earlier time than what is expected under the original model. The key factor that determines the changes in the failure patterns of the operative components is the rule that states how the load is to be distributed among the working components when any of the components within the system fails. Some commonly used load-sharing rules are Equal, Local and Monotone load-sharing. For more details about load-sharing rules, see Singh and Misra (1994).

1.2.5 Stochastic Orders

In this section, some stochastic orders are introduced that compare the location or the magnitude of random variables, all of them are closely related to those main results to be presented in this book.

Let X and Y be the lifetimes of two components, with their distribution functions $F(x)$ and $G(x)$, and their probability density functions $f(x)$ and $g(x)$, respectively. Denote their corresponding survival functions by $\bar{F}(x) = 1 - F(x)$ and $\bar{G}(x) = 1 - G(x)$.

Definition 1.2.1 The random variable X is said to be smaller than Y in the:

(a) usual stochastic order (denoted by $X \leq_{st} Y$) if $\bar{F}(x) \leq \bar{G}(x)$ for all x.
(b) hazard rate order (denoted by $X \leq_{hr} Y$) if $\bar{F}(x)/\bar{G}(x)$ is *decreasing* in x.
(c) reversed hazard rate order (denoted by $X \leq_{rh} Y$) if $F(x)/G(x)$ is *decreasing* in x.
(d) likelihood ratio order (denoted by $X \leq_{lr} Y$) if $f(x)/g(x)$ is *decreasing* in the union of their *supports*.
(e) increasing convex order (denoted by $X \leq_{icx} Y$) if $\mathrm{E}h(X) \leq \mathrm{E}h(Y) < \infty$ for all increasing and convex functions h, or equivalently

$$\int_t^\infty \bar{F}(x)dx \leq \int_t^\infty \bar{G}(x)dx \quad \text{for all} \quad t \geq 0;$$

(f) upshifted (downshifted) likelihood ratio order (denoted by $X \leq_{lr\uparrow} Y$)(denoted by $X \leq_{lr\downarrow} Y$) if $f(x+y)/g(x)$ is *decreasing (increasing)* in x for $y \geq 0$.

Definition 1.2.2 For two discrete distributions $\mathbf{p} = (p_1, \ldots, p_n)$ and $\mathbf{q} = (q_1, \ldots, q_n)$, $\mathbf{p}$ is said to be smaller than $\mathbf{q}$ in the

(a) usual stochastic order (denoted by $\mathbf{p} \leq_{st} \mathbf{q}$) if $\sum_{j=i}^{n} p_j \leq \sum_{j=i}^{n} q_j$ for all $i = 1, 2, \ldots, n$.
(b) hazard rate order (denoted by $\mathbf{p} \leq_{hr} \mathbf{q}$) if $\sum_{j=i}^{n} p_j / \sum_{j=i}^{n} q_j$ is *decreasing* in i.
(c) reversed hazard rate order (denoted by $\mathbf{p} \leq_{rh} \mathbf{q}$) if $\sum_{j=1}^{i} p_j / \sum_{j=1}^{i} q_j$ is *decreasing* in i.
(d) likelihood ratio order (denoted by $\mathbf{p} \leq_{lr} \mathbf{q}$) if p_i/q_i is *decreasing* in i.

The following relation of implications can easily be established:

$$X \leq_{lr} Y \Longrightarrow X \leq_{hr} (\leq_{rh})\, Y \Longrightarrow X \leq_{st} Y \Longrightarrow X \leq_{icx} (\leq_{icv}) Y. \tag{1.2.1}$$

Recall that a random vector $\mathbf{X} = (X_1, \ldots, X_n)^T$ is said to be smaller than another random vector $\mathbf{Y} = (Y_1, \ldots, Y_n)^T$ in the upper orthant order; (denoted by $\mathbf{X} \leq_{uo} \mathbf{Y}$) if $\bar{F}_X(x_1, \ldots, x_n) \leq \bar{G}_Y(x_1, \ldots, x_n)$ for all $\mathbf{x} = (x_1, \ldots, x_n)$, where $\bar{F}_X(x_1, \ldots, x_n)$, $\bar{G}_Y(x_1, \ldots, x_n)$ are the survival functions of $\mathbf{X}$ and $\mathbf{Y}$, respectively.

For more details of those stochastic orderings and comprehensive discussions on properties, readers may refer to Shaked and Shanthikumar (1994).

1.2.6 Life Distribution Classes

In this part, some lifetime distribution classes are introduced based on the failure rate and residual life distribution or other concepts. In many practical situations, we may not need or have full knowledge for distributions of random variables. For example, in reliability maintenance we may be only interested in the mean residual life to determine the maintenance schedule. In insurance mathematics, we may be only interested in the mean stop-and-loss claims. Since the notions are originated from reliability, so their main properties will be discussed from reliability terminology.

Definition 1.2.3 The random variable X is of:

(a) DLR (decreasing likelihood ratio) if its density function f is log-convex; ILR (increasing likelihood ratio) if the density function f is log-concave;
(b) DFR (decreasing failure rate) if X_t stochastically *increases* in $t \geq 0$; IFR (decreasing failure rate) if X_t stochastically *decreases* in $t \geq 0$;
(c) IRHR (increasing reversed failure rate) if $F(t+x)/F(x)$ is *increasing* in $x \geq 0$; DRHR (decreasing reversed hazard rate) if $F(t+x)/F(x)$ is *decreasing* in $x \geq 0$.
(d) NBU (new better than used) if $X_t \leq_{st} X$ for all $t \geq 0$; NWU (new worse than used) if $X_t \geq_{st} X$ for all $t \geq 0$;
(e) NBUC (new better than used in convex order) if $X_t \leq_{icx} X$ for all $t \geq 0$, or equivalently

$$\int_t^{\infty} \bar{F}(y+x)dx \leq \bar{F}(y) \int_t^{\infty} \bar{F}(x)dx \quad \text{for all} \quad t,\ y \geq 0;$$

NWUC (new worsen than used in convex order) if $X_t \geq_{icx} X$;

(f) NBU(2) (new better than used of second-order stochastic dominance) if $X_t \leq_{icv} X$ for all $t \geq 0$, or equivalently

$$\int_0^{t} \bar{F}(y+x)dx \leq \bar{F}(y) \int_0^{t} \bar{F}(x)dx \quad \text{for all} \quad t,\ y \geq 0;$$

NWU(2) (new worse than used of second-order stochastic dominance) if $X_t \geq_{icv} X$;

(g) NBUE (new better than used in expectation) if $\int_0^{\infty} \bar{F}(y+x)dx \leq \bar{F}(y)E(X)$ for all $y \geq 0$; NWUE (new worse than used in expectation) if $\int_0^{\infty} \bar{F}(y+x)dx \geq \bar{F}(y)E(X)$.

According to Samaniego (1985), log-convex f implies log-convex $\bar{F}$. That is, for any $x \geq 0$,

$$\frac{f(x+t)}{f(t)} \quad \text{is increasing in} \quad t \geq 0 \tag{1.2.2}$$

implies that

$$\frac{\bar{F}(x+t)}{\bar{F}(t)} \quad \text{is increasing in} \quad t \geq 0. \tag{1.2.3}$$

For more details on the results to be summarized here, please refer to Samaniego (1985), and Barlow and Proschan (1975).

1.2.7 Residual Life and Inactivity Time

Assume that T is the lifetime of a system with reliability function F and density function f. Then, the residual life of the system at time t ($t > 0$) is defined as $(T - t \mid T > t)$, with corresponding reliability function and density function

$$\bar{F}_t(x) = \frac{\bar{F}(t+x)}{\bar{F}(t)}, \qquad f_t(x) = \frac{f(t+x)}{\bar{F}(t)}.$$

As a dual notion to the residual life, inactivity time, which measures the time elapsed since the system failure, has a close connection with the so-called autopsy data, i.e., information obtained by examining the component states of a failed system. The inactivity time of the system at time t is defined as $(t - T \mid T \leq t)$, with the reliability function and density function

$$\bar{F}_{(t)}(x) = \frac{F(t-x)}{F(t)}, \qquad f_{(t)}(x) = \frac{f(t-x)}{F(t)}.$$

1.2.8 TP2 and RR2

In the part, we recall the following conception of dependence which will be repeatedly utilized in the monograph. Dependence relations between two variables are studied extensively in probability and statistics.

Definition 1.2.4 A bivariate function $f(x, y) \geq 0$ is said to be *totally positive of order two*, abbreviated as TP2, if

$$f(x, y) f(x', y') \geq f(x', y) f(x, y'), \quad \text{for all } x \leq x', y \leq y'.$$

As the inequality above is reversed, $f(x, y)$ is said to be *reverse regular of order two*, abbreviated as RR2. The concept of bivariate dependence can be easily extended to the multivariate case. A function $h : R^n \rightarrow [0, \infty)$ is said to be multivariate total positive of order two (denoted by MTP_2) if, for every $\mathbf{x}$ and $\mathbf{y}$ in R^n,

$$h(\mathbf{x})h(\mathbf{y}) \leq h(\mathbf{x} \wedge \mathbf{y})h(\mathbf{x} \vee \mathbf{y}),$$

where $\mathbf{x} \wedge \mathbf{y} = (\min(x_1, y_1), \ldots, \min(x_n, y_n))$ and $\mathbf{x} \vee \mathbf{y} = (\max(x_1, y_1), \ldots, \max(x_n, y_n))$; For more details about TP2 and RR2, please see Karlin (1968).

Chapter 2
The Residual Life and Inactivity Time of Coherent Systems

This chapter mainly investigates the residual life and inactivity time of coherent systems with independent or weakly dependent components. Some aging properties and stochastic comparisons of the residual life and inactivity time of $(n-k+1)$-out-of-n systems are given. Moreover, stochastic comparisons on the *residual life* and the *inactivity time* of two general coherent systems with stochastically ordered *signature* are conducted as well.

2.1 On Coherent Systems with IID Components

This section investigates ordered properties of the residual lives and inactivity time of coherent systems with independent and identically distributed components.

2.1.1 The Residual Life of $(n-k+1)$*-Out-of-n Systems*

In this subsection, we introduce a new residual life of $(n-k+1)$-out-of-n systems consisting of n independent and identically distributed components, given that the total number of failures of components is not less than l at time t, and the system is still working. Some stochastic comparisons of residual lifetimes are conducted, and behaviors of DLR and ILR are given.

2.1.1.1 Introduction

In the literature, many authors paid their attentions to the residual life and its mean value of the $(n-k+1)$-out-of-n system. For example, Langberg et al. (1980) obtained some properties of residual life

Z. Zhang, *Conditional Order Statistics and Conditional Spacings*,
https://doi.org/10.1007/978-981-96-7151-9_2

$$(X_{k:n} - X_{k-1:n} | X_{k-1:n} = t)$$

for $t \geq 0$. Afterwards, Belzunce et al. (1999), and Li and Zuo (2002)subsequently provided additional results on the conditional random variable. Asadi and Bairamov (2005) studied the mean residual life $\mathsf{E}(X_{n:n} - t | X_{k:n} > t)$, which extended the results in Bairamov et al. (2002), there they studied the mean residual life $\mathsf{E}(X_{n:n} - t | X_{1:n} > t)$. For more details, readers may refer to Li and Chen (2004), Asadi and Bairamov (2005), Navarro and Shaked (2006), Navarro and Eryilmaz (2007), Hu et al. (2007a), Misra et al. (2008), Gurler and Bairamov (2009) etc.

Li and Zhao (2006), Khaledi and Shaked (2007) respectively obtained some stochastic comparisons and aging properties of general residual life

$$\mathrm{GRL}_{l,k,n}(t) = (X_{k:n} - t | X_{l:n} > t), \ 1 \leq l \leq k \leq n.$$

Khaledi and Shaked (2007), Poursaeed and Nematollahi (2008) researched the mean residual life

$$\mathsf{E}(X_{n:n} - t_2 | X_{l:n} < t_1 < X_{l+1:n}, X_{n:n} > t_2),$$

given that before time t_1, $l\ (< n)$ of components fail, and at time t_2 the system is still working.

In many practical situations, it is possible for us to miss some failure components of a functioning system for a variety of reasons. The only information we can observe at time t is that the number of the component's failures is not less a certain value. Suppose that the $(n - k + 1)$-out-of-n system is equipped with a warning light that comes up when at least l components have failed, and the system is still functioning with probability 1. The operator may now consider some maintenance, or consider replacing the whole system. Therefore it may be useful for operators to know some properties of the residual life of the system, which is defined by

$$\mathrm{GRL}(2)_{l,k,n}(t) = (X_{k:n} - t | X_{l:n} < t < X_{k:n}),$$

or to stochastically compare the residual lives of such systems with different collections of components. In what follows at this part, some ordered properties are researched.

At first, several lemmas are given, which will be useful for some important results in next section.

Lemma 2.1.1 (Misra and Meulen 2003) *Let Θ be a subset of the real line $\Re$, and let X be a nonnegative random variable having a distribution belonging to the family $\mathcal{P} = \{H(\cdot|\theta),\ \theta \in \Theta\}$ which satisfies that*

$$H(\cdot|\theta_2) \geq_{st} H(\cdot|\theta_1) \quad \textit{whenever } \theta_1, \theta_2 \in \Theta, \textit{ and } \theta_2 > \theta_1 > 0.$$

Let $\Psi(x, \theta)$ be a real valued function defined on $\Re \times \Theta$, and be measurable in x for each θ such that $\mathsf{E}_\theta[\Psi(x, \theta)]$ exists, then $\mathsf{E}_\theta[\Psi(X, \theta)]$ is increasing in θ if $\Psi(x, \theta)$ is increasing in θ and increasing in x.

Lemma 2.1.2 *Let Z_1 and Z_2 be two binomial random variables with parameters (n, p_1) and (n, p_2), respectively. If $p_1 \leq p_2$, then, for $l = 1, \ldots, k-1$,*

$$(Z_1 \mid l \leq Z_1 \leq (k-1)) \leq_{lr} (Z_2 \mid l \leq Z_2 \leq (k-1)).$$

The following lemma gives the survival function of $\mathrm{GRL}(2)_{l,k,n}(t)$.

Lemma 2.1.3 *For any $x \geq 0$ and $0 \leq l < k < n$,*

$$P(X_{k:n} - t > x | X_{l:n} < t < X_{k:n}) = \frac{\sum_{i=l}^{k-1} P(N(t) = i) P(X^t_{k-i:n-i} > x)}{P(X_{l:n} < t < X_{k:n})},$$

where $X^t_{k-i:n-i}$ denotes the $(k-i)$th order statistic among the independent and identically distributed variables with the same distribution as $(X - t \mid X > t)$, and $N(t)$ represents the total number of failed components up to time t.

Proof For any $x \geq 0$,

$$\begin{aligned}
&P(X_{k:n} > x + t, X_{l:n} < t < X_{k:n}) \\
&= \sum_{i=l}^{k-1} \binom{n}{i} F^i(t) \sum_{j=n-k+1}^{n-i} \binom{n-i}{j} \bar{F}^j(t+x)[\bar{F}(t) - \bar{F}(t+x)]^{n-i-j} \\
&= \sum_{i=l}^{k-1} \binom{n}{i} F^i(t) \bar{F}^{n-i}(t) \sum_{j=n-k+1}^{n-i} \binom{n-i}{j} \bar{F}_t^j(x)[1 - \bar{F}_t(x)]^{n-i-j} \\
&= \sum_{i=l}^{k-1} P(N(t) = i) P(X^t_{k-i:n-i} > x),
\end{aligned}$$

hence the desired result follows directly.

2.1.1.2 Aging Properties and Stochastic Comparisons

The first theorem shows that the $\mathrm{GRL}(2)_{l,k,n}(t)$ is *decreasing* in l in the sense of the likelihood ratio order.

Theorem 2.1.4 *For any $0 \leq l < k < n$ and $t \geq 0$,*

$$\mathrm{GRL}(2)_{l-1,k,n}(t) \geq_{lr} \mathrm{GRL}(2)_{l,k,n}(t).$$

Proof Denote by $h_{l,k,n,t}(x)$ the corresponding density function of $\mathrm{GRL}(2)_{l,k,n}(t)$. Then, by Lemma 2.1.3, the density functions of $(X_{k:n} - t | X_{l-1:n} < t < X_{k:n})$ and $(X_{k:n} - t | X_{l:n} < t < X_{k:n})$ are, respectively

$$h_{l-1,k,n,t}(x) = \frac{\sum_{i=l-1}^{k-1} a_i \bar{F}_t^{n-k}(x)[1-\bar{F}_t(x)]^{k-i-1} f_t(x)}{P(X_{l-1:n} < t < X_{k:n})},$$

and

$$h_{l,k,n,t}(x) = \frac{\sum_{i=l}^{k-1} a_i \bar{F}_t^{n-k}(x)[1-\bar{F}_t(x)]^{k-i-1} f_t(x)}{P(X_{l:n} < t < X_{k:n})},$$

where $a_i = P(N(t) = i)(n-k+1)\binom{n-i}{n-k+1}$ $(l-1 \le i \le k-1)$. Hence, their ratio can be represented as

$$\begin{aligned}
\frac{h_{l-1,k,n,t}(x)}{h_{l,k,n,t}(x)} &= \left(1 + \frac{a_{l-1}\bar{F}_t^{n-k}(x)[1-\bar{F}_t(x)]^{k-l} f_t(x)}{\sum_{i=l}^{k-1} a_i \bar{F}_t^{n-k}(x)[1-\bar{F}_t(x)]^{k-i-1} f_t(x)}\right) \cdot \frac{P(X_{l:n} < t < X_{k:n})}{P(X_{l-1:n} < t < X_{k:n})} \\
&= \left(1 + \frac{a_{l-1}[1-\bar{F}_t(x)]^{-l}}{\sum_{i=l}^{k-1} a_i [1-\bar{F}_t(x)]^{-i-1}}\right) \cdot \frac{P(X_{l:n} < t < X_{k:n})}{P(X_{l-1:n} < t < X_{k:n})} \\
&= \left(1 + \frac{a_{l-1}}{\sum_{i=l}^{k-1} a_i [1-\bar{F}_t(x)]^{l-i-1}}\right) \cdot \frac{P(X_{l:n} < t < X_{k:n})}{P(X_{l-1:n} < t < X_{k:n})}.
\end{aligned}$$

Noting that $[1-\bar{F}_t(x)]^{l-i-1}$ is decreasing in x for all $l \le i \le k-1$, thus $h_{l-1,k,n,t}(x)/h_{l,k,n,t}(x)$ is increasing in $x \ge 0$, and hence the required result holds.

Theorem 2.1.5 *For any* $0 \le l < k < n$ *and* $t \ge 0$,

$$\mathrm{GRL}(2)_{l-1,k-1,n}(t) \le_{lr} \mathrm{GRL}(2)_{l,k,n}(t).$$

Proof For any random variable V, denote by f_V the density function of V. For $z \ge 0$, let $X_{k-l:n-l}^z$ be the $(k-l)$th order statistics of a sample of size $n-l$ from the distribution as $(X - z \mid X > z)$, then, from Keseling (1999), $(X_{k:n} - X_{l:n} \mid X_{l:n} = y) \stackrel{st}{=} X_{k-l:n-l}^y$, here $\stackrel{st}{=}$ means equal in distribution. Therefore, for any $t \ge 0$,

$$\begin{aligned}
&P(X_{k:n} - t > x, X_{l:n} < t < X_{k:n}) \\
&= \int_0^{+\infty} P(X_{k:n} - t > x, X_{l:n} < t < X_{k:n} \mid X_{l:n} = z) dF_{X_{l:n}(z)} \\
&= \int_0^t P(X_{k:n} - t > x \mid X_{l:n} = z) dF_{X_{l:n}(z)} \\
&= \int_0^t P(X_{k-l:n-l}^z + z > x + t) f_{X_{l:n}}(z) dz.
\end{aligned}$$

In order to prove the result, it is enough to verify that

$$\begin{aligned}
\frac{h_{l,k,n,t}(x)}{h_{l-1,k-1,n,t}(x)} &= \frac{\int_0^t f_{X_{k-l:n-l}^z + z}(x+t) f_{X_{l:n}}(z) dz}{\int_0^t f_{X_{k-1:n-l+1}^z + z}(x+t) f_{X_{l-1:n}}(z) dz} \times \frac{P(X_{l-1:n} < t < X_{k-1:n})}{P(X_{l:n} < t < X_{k:n})} \\
&\propto \mathrm{E}_x[\psi(x, Z)]
\end{aligned}$$

is increasing in x, where

$$\psi(x, z) = \frac{f_{X^z_{k-l:n-l}+z}(x+t) f_{X_{l:n}}(z)}{f_{X^z_{k-1:n-l+1}+z}(x+t) f_{X_{l-1:n}}(z)} \quad \text{for } 0 \le z \le t,$$

and the random variable Z has a distribution function belonging to the family $\mathcal{P} = \{H(\cdot \mid x), x \in \Re^+\}$ with corresponding densities given by

$$h(z \mid x) = c(x) f_{X^z_{k-l:n-l+1}+z}(x+t) f_{X_{l-1:n}}(z) \quad \text{for } 0 \le z \le t,$$

here $c(x)$ is the normalizing constant.

Note that $X^{z_1}_{k-l:n-l+1} + z_1$ ($X^{z_2}_{k-l:n-l+1} + z_2$) is the $(k-l)$th order statistics based on the distribution of $(X \mid X > z_1)$ $((X \mid X > z_2))$. From Shaked and Shanthikumar (2007),

$$(X \mid X > z_1) \le_{lr} (X \mid X > z_2) \quad \text{for } z_1 \le z_2 < t,$$

hence from Hu et al. (2007a),

$$X^{z_1}_{k-l:n-l+1} + z_1 \le_{lr} X^{z_2}_{k-l:n-l+1} + z_2,$$

which implies that

$$\frac{h(z_2 \mid x)}{h(z_1 \mid x)} = \frac{f_{X^{z_2}_{k-l:n-l+1}+z_2}(x+t)}{f_{X^{z_1}_{k-l:n-l+1}+z_1}(x+t)}$$

is increasing in x, whenever $0 \le z_1 \le z_2$, that is $H(\cdot \mid x_1) \le_{lr} H(\cdot \mid x_2)$, and hence, $H(\cdot \mid x_1) \le_{st} H(\cdot \mid x_2)$ whenever $0 \le x_1 \le x_2$. On the other hand,

$$\psi(x, z) \propto \frac{\left[1 - \frac{\bar{F}(x+t)}{\bar{F}(z)}\right]^{k-l-1} \left[\frac{\bar{F}(x+t)}{\bar{F}(z)}\right]^{n-k} \frac{f(x+t)}{\bar{F}(z)}}{\left[1 - \frac{\bar{F}(x+t)}{\bar{F}(z)}\right]^{k-l-1} \left[\frac{\bar{F}(x+t)}{\bar{F}(z)}\right]^{n-k+1} \frac{f(x+t)}{\bar{F}(z)}} \times \frac{F^{l-1}(z) \bar{F}^{n-l}(z) f(z)}{F^{l-2}(z) \bar{F}^{n-l+1}(z) f(z)}$$
$$= \frac{F(z)}{\bar{F}(x+t)}$$

is increasing in x and z. Thus, from Lemma 2.1.1, the desired result follows.

The theorem below shows that the $\text{GRL}(2)_{l,k,n}(t)$ is *decreasing* in n in the sense of the likelihood ratio order. Its proof is similar as that of Theorem 2.1.5, and therefore is omitted.

Theorem 2.1.6 *For any* $0 \le l < k < n$ *and* $t \ge 0$,

$$\text{GRL}(2)_{l,k,n-1}(t) \ge_{lr} \text{GRL}(2)_{l,k,n}(t).$$

It can be easily seen that, from Theorems 2.1.5 and 2.1.6, the following corollary holds.

Corollary 2.1.7 *For any* $0 \le l < k < n$ *and* $t \ge 0$,

$$\mathrm{GRL}(2)_{l,k-1,n}(t) \le_{lr} \mathrm{GRL}(2)_{l,k,n}(t).$$

The following theorem conducts a stochastic comparison of two $(n-k+1)$-out-of-n systems based on their hazard rate ordering.

Theorem 2.1.8 *For a* $(n-k+1)$*-out-of-n system with independent random units* $X_1, X_2, \ldots, X_n$*, which are identically distributed as* X*. For another* $(n-k+1)$*-out-of-n system with independent random units* $Y_1, Y_2, \ldots, Y_n$*, which have common distribution* G *as* Y*. If* $X \ge_{hr} Y$*, then, for any* $t \ge 0$,

$$\mathrm{GRL}(2)^X_{l,k,n}(t) \ge_{st} \mathrm{GRL}(2)^Y_{l,k,n}(t),$$

where $\mathrm{GRL}(2)^X_{l,k,n}(t) = (X_{k:n} - t | X_{l:n} < t < X_{k:n})$, $\mathrm{GRL}(2)^Y_{l,k,n}(t) = (Y_{k:n} - t | Y_{l:n} < t < Y_{k:n})$.

Proof For fixed $l < k \le n$ and $t \ge 0$,

$$\begin{aligned} P(X_{k:n} - t > x | X_{l:n} < t < X_{k:n}) &= \frac{\sum_{i=l}^{k-1} P(N(t) = i) P(X^t_{k-i:n-i} > x)}{P(X_{l:n} < t < X_{k:n})} \\ &= \sum_{i=l}^{k-1} \frac{\binom{n}{i} F^i(t) \bar{F}^{n-i}(t) P(X^t_{k-i:n-i} > x)}{\sum_{m=l}^{k-1} \binom{n}{m} F^m(t) \bar{F}^{n-m}(t)}. \end{aligned}$$

Let Z_1 denote a binomial random variable with parameters $(n, F(t))$, and let $p_i = P(Z_1 = i \mid l \le Z_1 \le (k-1))$, then

$$P(X_{k:n} - t > x | X_{l:n} < t < X_{k:n}) = \sum_{i=l}^{k-1} p_i P(X^t_{k-i:n-i} > x), \quad x \ge 0. \tag{2.1}$$

Similarly, the distribution function of $(Y_{k:n} - t \mid Y_{l:n} < t < Y_{k:n})$ can be written as

$$P(Y_{k:n} - t > x | Y_{l:n} < t < Y_{k:n}) = \sum_{i=l}^{k-1} q_i P(Y^t_{k-i:n-i} > x), \quad x \ge 0,$$

where $q_i = P(Z_2 = i \mid l \le Z_2 \le (k-1))$ and Z_2 is a binomial random variable with parameters $(n, G(t))$, $Y^t_{k-i:n-i}$ denotes the $(k-i)$th order statistic among the independent and identically variables with the same distribution as $(Y - t \mid Y > t)$.

For all $t \ge 0$, $X \ge_{hr} Y$ implies $F(t) \le G(t)$ therefore, by Lemma 2.1.2, it holds that

$$(Z_1 \mid l \le Z_1 \le (k-1)) \le_{st} (Z_2 \mid l \le Z_2 \le (k-1)).$$

Also $X \ge_{hr} Y$ if and only if $(X - t \mid X > t) \ge_{st} (Y - t \mid Y > t)$, hence

$$P(X^t_{k-i:n-i} > x) \geq P(Y^t_{k-i:n-i} > x), \quad x \geq 0. \tag{2.2}$$

According to Mi and Shaked (2002),

$$P(Y^t_{k-i:n-i} > x) \downarrow \quad \text{in} \quad i = l, l+1, \ldots, k-1,$$

thus for $x \geq 0$, we have

$$\begin{aligned} P(X_{k:n} - t > x | X_{l:n} < t < X_{k:n}) &= \sum_{i=l}^{k-1} p_i P(X^t_{k-i:n-i} > x) \\ &\geq \sum_{i=l}^{k-1} p_i P(Y^t_{k-i:n-i} > x) \\ &\geq \sum_{i=l}^{k-1} q_i P(Y^t_{k-i:n-i} > x) \\ &= P(Y_{k:n} - t > x | Y_{l:n} < t < Y_{k:n}), \end{aligned}$$

where the first inequality follows from (2.2), and the second inequality is obtained from the characterization of the order $\leq_{st}$ given in Shaked and Shanthikumar (2007). This validates the proof.

Example 2.1.9 Let $X_1, X_2, \ldots, X_n$ be a collection of independent exponential random variables with common hazard rate λ. For any $t \geq 0$, let Z denote a binomial random variable with parameters $(n, 1 - e^{-\lambda t})$. Then, for $l < k \leq n$, we have, from the memoryless property of exponential distribution,

$$\mathsf{E}(X_{k:n} - t | X_{l:n} < t < X_{k:n}) = \sum_{i=l}^{k-1} p_i \mathsf{E} X_{k-i:n-i}$$

where

$$p_i = \frac{\binom{n}{i}(1 - e^{-\lambda t})^i e^{-(n-i)\lambda t}}{\sum_{m=l}^{k-1} \binom{n}{m}(1 - e^{-\lambda t})^m e^{-(n-m)\lambda t}}, \quad i = l, l+1, \ldots, k-1.$$

Therefore, by Barlow and Proschan (1975),

$$\mathsf{E} X_{k-i:n-i} = \frac{1}{\lambda(n-i)} + \frac{1}{\lambda(n-i-1)} + \cdots + \frac{1}{\lambda(n-k+1)}, \quad i = l, l+1, \ldots, k-1.$$

Hence, we get

$$\mathsf{E}(X_{k:n} - t | X_{l:n} < t < X_{k:n}) = \frac{1}{\lambda} \sum_{i=l}^{k-1} p_i \sum_{m=n-k+1}^{n-i} \frac{1}{m}. \tag{2.3}$$

Remark 2.1.10 It is obvious, from (2.3), that $\mathsf{E}(X_{k:n} - t | X_{l:n} < t < X_{k:n})$ is decreasing in l for $l < k \le n$ and $t \ge 0$, and increasing in k; Furthermore, if $X \ge_{hr} Y$, then $\mathsf{E}(X_{k:n} - t | X_{l:n} < t < X_{k:n}) \ge \mathsf{E}(Y_{k:n} - t | Y_{l:n} < t < Y_{k:n})$.

In order to investigate how the failure time of a component affects the lifetime of a $(n - k + 1)$-out-of-n system, Lanberg et al. (1980) presented the characterizations for the IFR (DFR), NBU (NWU) classes of life distributions in terms of the monotonicity of random variable $(X_{k:n} - X_{k-1:n} | X_{k-1:n} = t)$. Li and Zuo (2002) obtained the characterizations of DMRL , NBU(2) and NBUC.

It is easy to see, $\mathrm{GRL}(2)_{k-1,k,n}(t) \stackrel{st}{=} (X_{k:n} - X_{k-1:n} | X_{k-1:n} = t)$. In the following, we give the characterizations of DLR and ILR.

Theorem 2.1.11 *For any integers k such that $1 < k \le n$ and $0 \le t' \le t < \infty$, X is $DLR(ILR)$ if and only if*

$$\mathrm{GRL}(2)_{k-1,k,n}(t) \ge_{lr} (\le_{lr})\mathrm{GRL}(2)_{k-1,k,n}(t').$$

Proof We only give the proof for the case of DLR, the ILR property of X can be derived in a similar way.

We first prove the 'only if' part of the theorem. Note that, for all $t \ge 0$ and $x \ge 0$,

$$\bar{H}_{k-1,k,n,t}(x) = \bar{F}_t^{n-k+1}(x),$$

it follows that, for all $t \ge 0$,

$$h_{k-1,k,n,t}(x) = (n - k + 1)\bar{F}_t^{n-k}(x) f_t(x).$$

X is DLR implies X is DFR, hence, we have

$$\frac{h_{k-1,k,n,t}(x)}{h_{k-1,k,n,t'}(x)} = \left[\frac{\bar{F}(t+x)}{\bar{F}(t'+x)}\right]^{n-k} \left[\frac{\bar{F}(t')}{\bar{F}(t)}\right]^{n-k+1} \cdot \frac{f(t+x)}{f(t'+x)}$$

is increasing in $x \ge 0$;

The 'if' part of the theorem is easily obtained by letting $k = n$ in $h_{k-1,k,n,t}(x)/h_{k-1,k,n,t'}(x)$.

2.1.2 Residual Life and Inactivity Time of Coherent Systems Under Component Setting

This section investigates the residual life (inactivity time) of a kind of general coherent systems with independent and identical components under the condition that the

r-th component survives time t (the r-th component has failed at time t). Stochastic comparison on the *residual life* and the *inactivity time* of two systems with stochastically ordered *signature*'s is conducted.

2.1.2.1 Introduction

The coherent systems play an quite important role in reliability theory. Some authors devoted themselves to stochastically comparing the lifetimes of coherent systems during the past decade. For example, Nanda et al. (1998) studied the preservation properties of some partial orderings of coherent systems with IID components. Subsequently, Belzunce et al. (2001) investigated partial orderings between coherent systems with different structures. Afterward, Navarro et al. (2005) developed comparison of systems with dependent components by using signatures, and hence extended some ordering properties on systems with IID components due to Kochar et al. (1999). Navarro and Shaked (2006) also obtained some meaningful results of the hazard rate function of the lifetimes of various coherent systems.

For some coherent systems of order n, the system is still in functioning state with probability one when less than $r(< n)$ components fail at time $t > 0$, and an operator may do some maintenance, replace the system, or compare performances of competing systems in terms of residual life.

In the literature, many authors pay their attentions to the residual life and inactivity time of some coherent systems, especially k-out-of-n systems. For example, Li and Lu (2003) established some stochastic comparisons on the residual lives and inactivity time of series and parallel systems, Asadi (2004) studied the mean inactivity time $\mathsf{E}(t - X_{n,n} \mid X_{n,n} \leq t)$ of a parallel system given that the system's failure occurred at or before time $t > 0$. Recently, Khaledi and Shaked (2007), Li and Zhao (2006) respectively studied the aging properties and stochastic comparisons on the conditional random variable $\mathrm{GRL}_{r,k,n}(t) = (X_{k,n} - t \mid X_{r,n} > t)$, the general residual life of a $(n - k + 1)$-out-of-n system at time $t \geq 0$. Khaledi and Shaked (2007) was among the first to investigate the residual life and inactivity time of coherent systems. They showed that the reliability functions of the residual life, and of the inactivity time of a coherent system with n IID components can be expressed as mixtures of those of the residual lives, and of those of the inactivity times of k-out-of-n systems, respectively. Also, they obtained the usual stochastic orders on $(T_X - t|X_{r:n} > t)$ and $(t - T_X|X_{r:n} \leq t)$, respectively, here $T_X = \tau(X_1, \cdots, X_n)$ is the structure function of a coherent system with IID components.

If two coherent systems have a common set of IID components, then any remaining differences of these two systems must lie only in their structure functions. This motivates us to compare such systems in terms of the residual life and inactivity time based on some properties of systems' signatures, and this kind of studies is useful for an operator to select better structure of the system in the early design period with the reference to residual life and inactivity time.

Khaledi and Shaked (2007) further pointed out that the signature of a coherent system, which functions with probability 1 as long as at least $n - s + 1$ $(1 \leq s \leq n)$

components function, must be of the form $(0, \ldots, 0, p_s, \ldots, p_n)$. It was also showed there that the reliability function of the residual life of a coherent system with IID components can be written as

$$P(T_X - t > x \mid X_{r,n} > t) = \sum_{i=s}^{n} p_i \, P(X_{i,n} - t > x \mid X_{r,n} > t), \qquad r \le s. \quad (2.4)$$

Denote $h_{r,i,n,t}(x)$, $h_{i,r,n,(t)}(x)$ the density functions of random variables $GRL_{r,i,n}(t) = (X_{i,n} - t \mid X_{r,n} > t)$ and $(t - X_{i,n} \mid X_{r,n} \le t)$, respectively. The following lemma is useful in next sections.

Lemma 2.1.12 *(i) For $r \le i \le n$, $h_{r,i,n,t}(x)$ is $TP2$ in $(x, i) \in (0, \infty) \times \{s, \cdots, n\}$;*
(ii) For $i \le r \le n$, $h_{i,r,n,(t)}(x)$ is RR_2 *in $(x, i) \in (0, \infty) \times \{1, \cdots, s\}$.*

2.1.2.2 Stochastic Comparison of Residual Life

Suppose $X_1, \ldots, X_n$ are lifetimes of mutually independent and identically distributed components and $T_1 = \tau_1(X_1, \ldots, X_n)$ and $T_2 = \tau_2(X_1, \ldots, X_n)$ are two coherent systems with their respective signatures $\mathbf{p} = (0, 0, \ldots, p_s, p_{s+1}, \ldots, p_n)$ and $\mathbf{q} = (0, 0, \ldots, q_s, q_{s+1}, \ldots, q_n)$ for some $r \le s \le n$.

Before proceeding to discussing how the variation of the system's signature has effect on its residual life, we introduce a useful lemma below.

Lemma 2.1.13 (Shaked and Shanthikumar (2007)) Let Θ_i be a random variable with support $\mathcal{X} \subset \mathbb{R}$ and distribution function F_i, $i = 1, 2$, $X(\theta)$ has the distribution belonging to the family $\{G_\theta,\ \theta \in \mathcal{X}\}$. Suppose $X(\Theta_i)$ has the distribution function

$$H_i(y) = \int_{\mathcal{X}} G_\theta(y) \, \mathrm{d}F_i(\theta), \quad y \in \mathbb{R},\ i = 1, 2.$$

If $X(\theta) \le_{hr} (\le_{lr}) X(\theta')$ whenever $\theta \le \theta'$ and $\Theta_1 \le_{hr} (\le_{lr}) \Theta_2$, then $X(\Theta_1) \le_{hr} (\le_{lr}) X(\Theta_2)$.

Theorem 2.1.14 *If $\boldsymbol{p} \le_{hr} \boldsymbol{q}$, then, for $r \le s$,*

$$(T_1 - t \mid X_{r,n} > t) \le_{hr} (T_2 - t \mid X_{r,n} > t).$$

Proof Denote $\bar{H}_{r,i,n,t}(x)$, the reliability function of the general residual life $GRL_{r,i,n}(t) = (X_{i,n} - t \mid X_{r,n} > t)$. By (2.4), for $r \le s \le n$,

$$P(T_1 - t > x \mid X_{r,n} > t) = \sum_{i=s}^{n} p_i \, \bar{H}_{r,i,n,t}(x),$$

$$P(T_2 - t > x \mid X_{r,n} > t) = \sum_{i=s}^{n} q_i \bar{H}_{r,i,n,t}(x).$$

According to Lemma 2.1.12(i) and (2.4), $\mathrm{GRL}_{r,i,n}(t) \le_{hr} \mathrm{GRL}_{r,j,n}(t)$, for all $s \le i \le j \le n$. By the hazard rate order $\mathbf{p} \le_{hr} \mathbf{q}$ and Lemma 2.1.13, the desired result follows immediately.

Theorem 2.1.15 *If* $\mathbf{p} \le_{lr} \mathbf{q}$, *then, for* $r \le s$,

$$(T_1 - t \mid X_{r,n} > t) \le_{lr} (T_2 - t \mid X_{r,n} > t).$$

Proof For $r \le s \le n$, the density functions of $(T_1 - t \mid X_{r,n} > t)$ and $(T_2 - t \mid X_{r,n} > t)$ are respectively

$$h_1(x) = \sum_{i=s}^{n} p_i h_{r,i,n,t}(x)$$

and

$$h_2(x) = \sum_{i=s}^{n} q_i h_{r,i,n,t}(x).$$

By Lemma 2.1.12(i), $\mathrm{GRL}_{r,i,n}(t) \le_{lr} \mathrm{GRL}_{r,j,n}(t)$, for all $s \le i \le j \le n$. By the likelihood ratio order $\mathbf{p} \le_{lr} \mathbf{q}$ and Lemma 2.1.13 again, the desired result follows immediately.

Example 2.1.16 Consider the two systems of order 4 depicted in Fig. 2.1, it can be evaluated that the signature of the left one is $\mathbf{q} = \left(0, 0, \frac{1}{2}, \frac{1}{2}\right)$ and that of the right one is $\mathbf{p} = \left(0, \frac{1}{6}, \frac{7}{12}, \frac{1}{4}\right)$. It can be verified that $\mathbf{p} \le_{lr} \mathbf{q}$ and hence $\mathbf{p} \le_{hr} \mathbf{q}$. By the two theorems above the system in the left side is better in the sense that it has stochastically longer general residual life.

2.1.2.3 Stochastic Comparison of Inactivity Time

Suppose, for some $s \le r \le n$, that the signatures of τ_1 and τ_2 are $\mathbf{p} = (p_1, \ldots, p_s, 0, \cdots, 0)$ and $\mathbf{q} = (q_1, \cdots, q_s, 0, \ldots, 0)$. It can be shown that, for

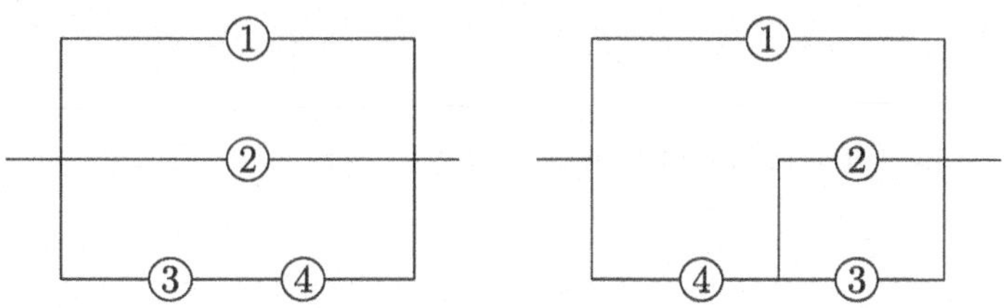

Fig. 2.1 Two coherent systems with hazard rate ordered signatures

$s \le r \le n$, the reliability function of the inactivity time of a coherent system with IID components is

$$P(t - T_1 > x \mid X_{r,n} \le t) = \sum_{i=1}^{s} p_i P(t - X_{i,n} > x \mid X_{r,n} \le t). \tag{2.5}$$

The following lemma is useful in proving the main conclusions in this section.

Lemma 2.1.17 (Shaked and Shanthikumar, 2007) *Let α and β be real valued functions such that α is nonnegative and β/α and α are non-increasing. If X_i has distribution F_i, $i = 1, 2$ and $X_1 \le_{rh} X_2$, then*

$$\frac{\int_{-\infty}^{\infty} \beta(x)\, \mathrm{d}F_1(x)}{\int_{-\infty}^{\infty} \alpha(x)\, \mathrm{d}F_1(x)} \ge \frac{\int_{-\infty}^{\infty} \beta(x)\, \mathrm{d}F_2(x)}{\int_{-\infty}^{\infty} \alpha(x)\, \mathrm{d}F_2(x)}.$$

Theorem 2.1.18 *If $\boldsymbol{p} \le_{rh} \boldsymbol{q}$, then, for $r \ge s$,*

$$(t - T_1 \mid X_{r,n} \le t) \ge_{hr} (t - T_2 \mid X_{r,n} \le t).$$

Proof Let $\bar{H}_{i,r,n,(t)}(x)$ be the reliability function of the random variable $(t - X_{i,n} \mid X_{r,n} \le t)$, $1 \le i \le r \le n$. Then, by (2.5),

$$P(t - T_1 > x \mid X_{r,n} \le t) = \sum_{i=1}^{s} p_i \bar{H}_{i,r,n,(t)}(x),$$

$$P(t - T_2 > x \mid X_{r,n} \le t) = \sum_{i=1}^{s} q_i \bar{H}_{i,r,n,(t)}(x).$$

As a result, it is sufficient to prove that

$$\frac{P(t - T_1 > x \mid X_{r,n} \le t)}{P(t - T_2 > x \mid X_{r,n} \le t)} = \frac{\sum_{i=1}^{s} p_i \bar{H}_{i,r,n,(t)}(x)}{\sum_{i=1}^{s} q_i \bar{H}_{i,r,n,(t)}(x)}$$

is increasing in x, that is, for all $x_2 \ge x_1 \ge 0$,

$$\frac{\sum_{i=s}^{n} p_i \bar{H}_{i,r,n,(t)}(x_2)}{\sum_{i=s}^{n} p_i \bar{H}_{i,r,n,(t)}(x_1)} \geq \frac{\sum_{i=s}^{n} q_i \bar{H}_{i,r,n,(t)}(x_2)}{\sum_{i=s}^{n} q_i \bar{H}_{i,r,n,(t)}(x_1)}.$$

Let $\alpha(i) = \bar{H}_{i,r,n,(t)}(x_1)$ and $\beta(i) = \bar{H}_{i,r,n,(t)}(x_2)$. It is clear that $\alpha(i)$ is decreasing in $i = s, \cdots, n$, and by Lemma 2.1.12(ii), $\beta(i)/\alpha(i)$ is decreasing in i for all $t \geq 0$. From Lemma 2.1.17, the desired result follows.

Theorem 2.1.19 *If* $\mathbf{p} \leq_{lr} \mathbf{q}$*, then, for* $r \geq s$*,*

$$(t - T_1 \mid X_{r,n} \leq t) \geq_{lr} (t - T_2 \mid X_{r,n} \leq t).$$

Proof Note that, for $s \leq r \leq n$, the density functions of $(t - T_1 \mid X_{r,n} > t)$ and $(t - T_2 \mid X_{r,n} > t)$ are $\sum_{i=1}^{s} p_i h_{i,r,n,(t)}(x)$ and $\sum_{i=1}^{s} q_i h_{i,r,n,(t)}(x)$, respectively, we only need to prove that

$$\frac{\sum_{i=1}^{s} p_i h_{i,r,n,(t)}(x)}{\sum_{i=1}^{s} q_i h_{i,r,n,(t)}(x)}$$

is increasing in $x \geq 0$. Obviously, this is equivalent to, for all $x_2 \geq x_1 \geq 0$,

$$g(x_1, x_2) = \sum_{i=1}^{s} \sum_{j=1}^{s} p_i q_j \left[h_{i,r,n,(t)}(x_2) h_{j,r,n,(t)}(x_1) - h_{i,r,n,(t)}(x_1) h_{j,r,n,(t)}(x_2) \right] \geq 0.$$

Since $\mathbf{p} \leq_{lr} \mathbf{q}$ and $h_{i,r,n,(t)}(x)$ is RR2 in (i, x), we have

$$\begin{aligned}
g(x_1, x_2) = & \sum_{i=1}^{s} \sum_{j=1}^{s} p_i q_j \left[h_{i,r,n,(t)}(x_2) h_{j,r,n,(t)}(x_1) - h_{i,r,n,(t)}(x_1) h_{j,r,n,(t)}(x_2) \right] \\
= & \sum_{i=1}^{s} \sum_{j=i}^{s} p_i q_j \left[h_{i,r,n,(t)}(x_2) h_{j,r,n,(t)}(x_1) - h_{i,r,n,(t)}(x_1) h_{j,r,n,(t)}(x_2) \right] \\
& + \sum_{i=1}^{s} \sum_{j=1}^{i} p_i q_j \left[h_{i,r,n,(t)}(x_2) h_{j,r,n,(t)}(x_1) - h_{i,r,n,(t)}(x_1) h_{j,r,n,(t)}(x_2) \right] \\
= & \sum_{i=1}^{s} \sum_{j=i}^{s} p_i q_j \left[h_{i,r,n,(t)}(x_2) h_{j,r,n,(t)}(x_1) - h_{i,r,n,(t)}(x_1) h_{j,r,n,(t)}(x_2) \right] \\
& + \sum_{i=1}^{s} \sum_{j=i}^{s} p_j q_i \left[h_{j,r,n,(t)}(x_2) h_{i,r,n,(t)}(x_1) - h_{j,r,n,(t)}(x_1) h_{i,r,n,(t)}(x_2) \right]
\end{aligned}$$

$$= \sum_{i=1}^{s}\sum_{j=i}^{s} p_i p_j \left(\frac{q_j}{p_j} - \frac{q_i}{p_i}\right) \left[h_{i,r,n,(t)}(x_2)h_{j,r,n,(t)}(x_1) - h_{i,r,n,(t)}(x_1)h_{j,r,n,(t)}(x_2)\right]$$

$$\geq 0.$$

This validates the desired result.

Example 2.1.20 Consider the two systems of order 4 displayed in Fig. 2.2, it can easily be computed that the signature of the left one $\mathbf{p} = (\frac{1}{2}, \frac{1}{2}, 0, 0)$ and that of the right one is $\mathbf{q} = (\frac{1}{4}, \frac{1}{4}, \frac{1}{2}, 0)$, respectively. It is clear that $\mathbf{p} \leq_{lr} \mathbf{q}$, thus $\mathbf{p} \leq_{rh} \mathbf{q}$. Therefore, from the two theorems above, the system in the right side is better than the one in the left side in the sense of inactivity time in hazard rate order and likelihood ratio order.

The results above indicate that variation of the signature of a coherent system has a direct influence on its lifetime distribution of both the residual life and the inactivity time. To end this note, we present an example to illustrate systems depicted in Figs. 2.3 and 2.4 which satisfy $\mathbf{p} \leq_{hr} (\leq_{rh})\mathbf{q}$ but $\mathbf{p} \nleq_{lr} \mathbf{q}$.

Example 2.1.21 Consider the two systems of order 5 depicted in Fig. 2.3, it can be computed that the signature of the left one is $\mathbf{p} = \left(\frac{1}{5}, \frac{1}{5}, \frac{2}{5}, \frac{1}{5}, 0\right)$ and that of the right one is $\mathbf{q} = \left(0, \frac{3}{10}, \frac{1}{2}, \frac{1}{5}, 0\right)$. It can be verified that $\mathbf{p} \leq_{rh} \mathbf{q}$, but $\mathbf{p} \nleq_{lr} \mathbf{q}$. Consider the two systems of order 5 displayed in Fig. 2.4, it can be evaluated that the signature of the left one is $\mathbf{p} = \left(0, \frac{1}{5}, \frac{2}{5}, \frac{1}{5}, \frac{1}{5}\right)$ and that of the right one is $\mathbf{q} = \left(0, \frac{1}{5}, \frac{1}{2}, \frac{3}{10}, 0\right)$. It can be checked that $\mathbf{p} \leq_{hr} \mathbf{q}$, but $\mathbf{p} \nleq_{lr} \mathbf{q}$.

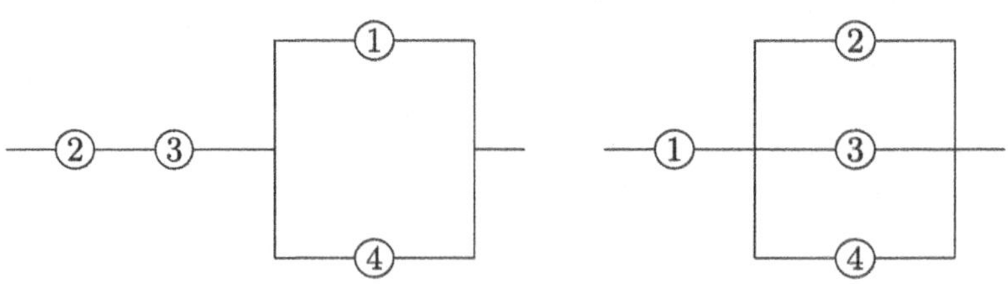

Fig. 2.2 Two coherent systems with reverse hazard rate ordered signatures

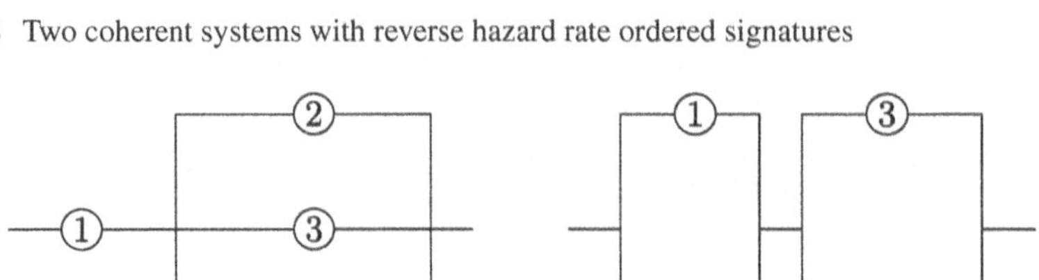

Fig. 2.3 Systems having only the reversed hazard rate order between their signatures

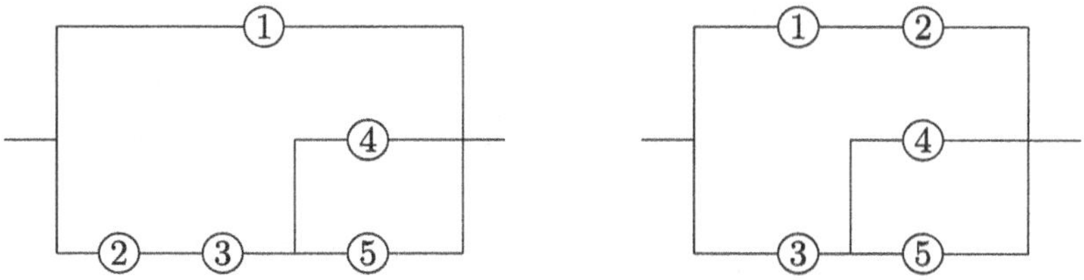

Fig. 2.4 Systems having only the hazard rate order between their signatures

2.1.3 *Inactivity Time of General Coherent Systems Under System Setting*

This section investigates the inactivity time of general coherent systems with independent and identical components under the condition that the system has failed at time t. Some ordered or stochastic properties are conducted.

2.1.3.1 Introduction

In a real life situation, one might be interested in getting inference more about the history of the system. Suppose that if a coherent system with lifetime T has failed at or some time before $t(>0)$, then an operator may measure the time elapsed since the system failure, that is, he or she may consider the conditional random variable $(t - T \mid T \leq t)$, the inactivity time, which usually has a close connection with the so-called autopsy data, i.e., information obtained by examining the component states of a failed system. For more details on this, readers may refer to Meilijson (1981) among others.

This part investigates the inactivity time $(t - T \mid T \leq t)$ of a coherent system with IID component lifetimes. We obtain several useful mixture representations of the reliability function of the inactivity time of a coherent system under the condition that the system has failed at time $t \geq 0$ in terms of the reliability functions of inactivity times of order statistics. Some ordering properties of the *inactivity times* of coherent systems with independent and identically distributed components are obtained, based on the stochastically ordered coefficient vectors between systems.

2.1.3.2 Ordering Properties of the Inactivity Times

Let T be the lifetime of a coherent system with n IID component lifetimes $X_1, \ldots, X_n$, and $\mathbf{p} = (p_1, \ldots, p_n)$ be its signature. At first, we establish a representation for the reliability function of the inactivity time of a failed system at time $t > 0$.

Theorem 2.1.22 *For* $t \geq x \geq 0$,

$$P(t - T > x \mid T \leq t) = \sum_{i=1}^{n} p_i(t) P(X_{i,n} \leq t - x \mid X_{i,n} \leq t). \tag{2.6}$$

Proof It is noted that, for $t \geq x \geq 0$, and $i = 1, 2, \ldots, n$,

$$P(X_{i,n} \leq t - x \mid X_{i,n} \leq t) P(X_{i,n} \leq t) = P(X_{i,n} \leq t - x),$$

therefore,

$$\begin{aligned} P(t - T > x \mid T \leq t) &= \frac{P(T \leq t - x, T \leq t)}{P(T \leq t)} \\ &= \frac{\sum_{i=1}^{n} p_i P(X_{i,n} \leq t - x)}{\sum_{i=1}^{n} p_i P(X_{i,n} \leq t)} \\ &= \frac{\sum_{i=1}^{n} p_i P(X_{i,n} \leq t - x \mid X_{i,n} \leq t) P(X_{i,n} \leq t)}{\sum_{i=1}^{n} p_i P(X_{i,n} \leq t)} \\ &= \sum_{i=1}^{n} p_i(t) P(X_{i,n} \leq t - x \mid X_{i,n} \leq t), \end{aligned}$$

where the function $p_i(t) = p_i P(X_{i,n} \leq t)/F_T(t)$ for $i = 1, 2, \ldots, n$.

Remark 2.1.23 It can be seen that $p_i(t)$ ($i = 1, 2, \ldots, n$) depends on the structure function of the system and F, and may be regarded as the probability of conditional event $\{T = X_{i,n} \mid T \leq t\}$ as follows:

$$\begin{aligned} P(T = X_{i,n} \mid T \leq t) &= \frac{P(T = X_{i,n}, T \leq t)}{P(T \leq t)} \\ &= \frac{P(T = X_{i,n}) P(T \leq t \mid T = X_{i,n})}{P(T \leq t)} \\ &= \frac{P(T = X_{i,n}) P(X_{i,n} \leq t \mid T = X_{i,n})}{P(T \leq t)} \\ &= \frac{p_i P(X_{i,n} \leq t)}{F_T(t)} \\ &= p_i(t). \end{aligned}$$

Remark 2.1.24 Expression (2.6) indicates that the inactivity time $(t - T \mid T \leq t)$ of the failed system at time t is a mixture of the inactivity time $(t - X_{i,n} \mid X_{i,n} \leq t)$ of the order statistics at time t with coefficients $p_i(t)$ for $i = 1, 2, \ldots, n$, that is, $(t - T \mid T \leq t)$ is equivalent in distribution to a mixed system of the inactivity times of i-out-of-n systems. For instance, considering the system with lifetime $T = \max\{X_1, \min\{X_2, X_3\}\}$, whose signature is $\mathbf{p} = (0, \frac{2}{3}, \frac{1}{3})$, and that we know it is not working at time t, then by some computations, the coefficient vector is given by

Table 2.1 Vectors of coefficients in (2.6) with $n = 3$ for coherent systems with 1-3 IID components.

System	$T = \phi(X_1, X_2, X_3)$	$\mathbf{p}(t)$
1	X_1	$(1 - F(t) + \frac{1}{3}F^2(t), F(t) - \frac{2}{3}F^2(t), \frac{1}{3}F^2(t))$
2	$\min\{X_1, X_2\}$	$\left(\frac{6-6F(t)+2F^2(t)}{6-3F(t)}, \frac{3F(t)-2F^2(t)}{6-3F(t)}, 0\right)$
3	$\max\{X_1, X_2\}$	$(0, 1 - \frac{2}{3}F(t), \frac{2}{3}F(t))$
4	$\min\{X_1, X_2, X_3\}$	$(1, 0, 0)$
5	$\min\{X_1, \max\{X_2, X_3\}\}$	$\left(\frac{3-3F(t)+F^2(t)}{3(1+F(t)-F^2(t))}, \frac{2F(t)(3-2F(t))}{3(1+F(t)-F^2(t))}, 0\right)$
6	$X_{2,3}$(2-out-of-3)	$(0, 1, 0)$
7	$\max\{X_1, \min\{X_2, X_3\}\}$	$\left(0, \frac{6-4F(t)}{6-3F(t)}, \frac{F(t)}{6-3F(t)}\right)$
8	$\max\{X_1, X_2, X_3\}$	$(0, 0, 1)$

$$\mathbf{p}(t) = \left(0, \frac{6 - 4F(t)}{6 - 3F(t)}, \frac{F(t)}{6 - 3F(t)}\right).$$

Hence, from Theorem 2.1.22, we have

$$\begin{aligned} &P(t - T > x \mid T \leq t) \\ &= \frac{6 - 4F(t)}{6 - 3F(t)} P(X_{2,3} \leq t - x \mid X_{2,3} \leq t) + \frac{F(t)}{6 - 3F(t)} P(X_{3,3} \leq t - x \mid X_{3,3} \leq t). \end{aligned}$$

Note that $\lim_{t\to\infty}\mathbf{p}(t) = (0, \frac{2}{3}, \frac{1}{3})$, and $\mathbf{p}(0) = (0, 1, 0)$, that is, when t closes to infinity, the coefficient vector equal the signature $\mathbf{p}$, and at the origin point, the inactivity time $(t - T \mid T \leq t)$ of the system is equivalent to the inactivity time $(t - X_{2,3} \mid X_{2,3} \leq t)$ of the 2-out-of-3 system with probability one.

Remark 2.1.25 Given that the component lifetimes are IID, the representation in (2.6) can be extended to mixed systems even with fewer than n components. For example, the system with one component $X_{1,1}$ equals in distribution the mixed system with signature $\mathbf{p} = (\frac{1}{3}, \frac{1}{3}, \frac{1}{3})$, hence, under the condition that it is not working at time t, (2.6) holds with vector

$$\mathbf{p}(t) = \left(1 - F(t) + \frac{1}{3}F^2(t), F(t) - \frac{2}{3}F^2(t), \frac{1}{3}F^2(t)\right).$$

In Table 2.1, we have presented the vectors of coefficients order 3 in (2.6) for coherent systems with 1–3 IID components.

It is easy to show that, for $t \geq 0$ and $i = 1, 2, \ldots, n - 1$,

$$(t - X_{i,n} \mid X_{i,n} \leq t) \geq_{lr} (t - X_{i+1,n} \mid X_{i+1,n} \leq t), \tag{2.7}$$

and hence,

$$(t - X_{i,n} \mid X_{i,n} \leq t) \geq_{hr} (\geq_{st})(t - X_{i+1,n} \mid X_{i+1,n} \leq t). \tag{2.8}$$

For a fixed $t > 0$, let $\mathbf{p}(t)$, $\mathbf{q}(t)$ be the vectors of coefficients in (2.6) of two mixed systems having lifetimes T_1 and T_2, with common n IID component lifetimes $X_1, X_2, \ldots, X_n$, according to the continuous distribution F. The following three theorems give implications of the usual stochastic ordering, reversed hazard rate ordering and likelihood ratio ordering of two coefficient vectors, respectively.

Theorem 2.1.26 *If $\mathbf{p}(t) \leq_{st} \mathbf{q}(t)$, then $(t - T_1 \mid T_1 \leq t) \geq_{st} (t - T_2 \mid T_2 \leq t)$.*

Proof Denote $\bar{H}_{i,n,(t)}(x)$ the reliability function of $(t - X_{i,n} \mid X_{i,n} \leq t)$ for $t \geq x \geq 0$. By (2.6),

$$P(t - T_1 > x \mid T_1 \leq t) = \sum_{i=1}^{n} p_i(t)\bar{H}_{i,n,(t)}(x),$$

$$P(t - T_2 > x \mid T_2 \leq t) = \sum_{i=1}^{n} q_i(t)\bar{H}_{i,n,(t)}(x).$$

According to expression (2.8), $\bar{H}_{i,n,(t)}(x)$ is decreasing in $i = 1, 2, \ldots, n$, hence, by the condition $\mathbf{p}(t) \leq_{st} \mathbf{q}(t)$ and Shaked and Shanthikumar (2007), the desired result holds.

Theorem 2.1.27 *If $\mathbf{p}(t) \leq_{rh} \mathbf{q}(t)$, then $(t - T_1 \mid T_1 \leq t) \geq_{hr} (t - T_2 \mid T_2 \leq t)$.*

Proof By (2.6), it is enough to prove that

$$\frac{P(t - T_1 > x \mid T_1 \leq t)}{P(t - T_2 > x \mid T_2 \leq t)} = \frac{\sum_{i=1}^{n} p_i(t)\bar{H}_{i,n,(t)}(x)}{\sum_{i=1}^{n} q_i(t)\bar{H}_{i,n,(t)}(x)}$$

is increasing in $x \geq 0$, that is, for all $x_2 \geq x_1 \geq 0$,

$$\frac{\sum_{i=1}^{n} p_i(t)\bar{H}_{i,n,(t)}(x_2)}{\sum_{i=1}^{n} p_i(t)\bar{H}_{i,n,(t)}(x_1)} \geq \frac{\sum_{i=1}^{n} q_i(t)\bar{H}_{i,n,(t)}(x_2)}{\sum_{i=1}^{s} q_i(t)\bar{H}_{i,n,(t)}(x_1)}.$$

Let $\alpha(i) = \bar{H}_{i,n,(t)}(x_1)$ and $\beta(i) = \bar{H}_{i,n,(t)}(x_2)$. By (2.8), $\alpha(i)$ is decreasing in $i = 1, \ldots, n$ and $\beta(i)/\alpha(i)$ is decreasing in i for all $t \geq 0$. Hence, from Theorem 1.B.50 of Shaked and Shanthikumar (2007), the desired result follows.

Theorem 2.1.28 *If $\mathbf{p}(t) \leq_{lr} \mathbf{q}(t)$, then $(t - T_1 \mid T_1 \leq t) \geq_{lr} (t - T_2 \mid T_2 \leq t)$.*

Proof Denote $h_{i,n,(t)}(x)$ the density function of $(t - X_{i,n} \mid X_{i,n} \leq t)$ for $t \geq x \geq 0$. Then the density functions of $(t - T_1 \mid T_1 \leq t)$ and $(t - T_2 \mid T_2 \leq t)$ are $h_1(x) = \sum_{i=1}^{n} p_i(t)h_{i,n,(t)}(x)$ and $h_2(x) = \sum_{i=1}^{n} q_i(t)h_{i,n,(t)}(x)$, respectively. For $c > 0$, let us consider the function

$$h(x) = h_2(x) - ch_1(x) = \sum_{i=1}^{n}(q_i(t) - cp_i(t))h_{i,n,(t)}(x).$$

The expression (2.8) implies that $h_{i,n,(t)}(x)$ is RR2 for $x \in \Re$ and $i = 1, \ldots, n$. Since $\mathbf{p}(t) \leq_{lr} \mathbf{q}(t)$, $q_i(t)/p_i(t)$ is increasing in $i = 1, \ldots, n$, and hence the sequence $q_i(t) - cp_i(t)$ has at most one change of sign from negative to positive as i from 1 to n. According to Karlin (1968), we have $h(x)$ has at most one change of sign from positive to negative as x increases. This validates the result.

For example, considering the system with lifetime $T_1 = X_1$, and another system with lifetime $T_2 = \max\{X_1, X_2\}$. From Table 2.1, for any $t \geq 0$, their coefficient vectors are

$$\mathbf{p}(t) = \left(1 - F(t) + \frac{1}{3}F^2(t), F(t) - \frac{2}{3}F^2(t), \frac{1}{3}F^2(t)\right), \mathbf{q}(t) = \left(0, 1 - \frac{2}{3}F(t), \frac{2}{3}F(t)\right),$$

respectively. It is easy to check that $\mathbf{p}(t) \leq_{lr} \mathbf{q}(t)$, hence according to the theorem above, if F is absolutely continuous, then, for all $t \geq 0$,

$$(t - X_1 \mid X_1 \leq t) \geq_{lr} (t - \max\{X_1, X_2\} \mid \max\{X_1, X_2\} \leq t).$$

Let us take a look at the relative behavior of pairs of systems that are non-comparable via the stochastic ordering of lifetimes. If a system is better than another up to the system s intended mission time, whether or not it has uniformly superior reliability performance for all time t, the next example gives a negative answer.

Example 2.1.29 Considering a system with lifetime $T_1 = X_{1,2}$ and another one with lifetime $T_2 = \min\{X_{2,3}, X_4\}$, whose signatures of order 4 are $\mathbf{p} = (\frac{1}{2}, \frac{1}{3}, \frac{1}{6}, 0)$ and $\mathbf{q} = (\frac{1}{4}, \frac{3}{4}, 0, 0)$, respectively. If they have failed at time t, then by straightforward computations, the corresponding coefficient vectors are

$$\mathbf{p}(t) = \left(1 - F(t) + \frac{1}{2}F^2(t), \frac{6F(t) - 8F^2(t) + 3F^3(t)}{6 - 3F(t)}, \frac{4F^2(t) - 3F^3(t)}{12 - 6F(t)}, 0\right),$$

and

$$\mathbf{q}(t) = \left(\frac{(2 - F(t))(2 - 2F(t) + F^2(t))}{4(2 - F^3(t) - 5F^2(t) + 3F(t) + 1)}, \frac{3F(t)(6 - 8F(t) + 3F^2(t))}{4(2 - F^3(t) - 5F^2(t) + 3F(t) + 1)}, 0, 0\right),$$

respectively. It can be checked that $\mathbf{p}(t) \not\geq_{st} \mathbf{q}(t)$ at any time, but $\mathbf{p}(t) \geq_{st} \mathbf{q}(t)$ for the set $\{t : F(t) \in [0.73, 1)\}$. Hence, from Theorem 3.1.5, $(t - X_{1,2} \mid X_{1,2} \leq t) \leq_{st} (t - \min\{X_{2,3}, X_4\} \mid \min\{X_{2,3}, X_4\} \leq t)$ whenever $t \in \{t : F(t) \in [0.73, 1)\}$.

The following theorem shows that the coefficient vector, the conditional distribution of the event $\{T = X_{i,n} \mid T \leq t\}$ is stochastically decreasing in $t \geq 0$.

Theorem 2.1.30 *Assume $\mathbf{p}(t)$ is the vector of coefficients in* (2.6) *of a mixed system with n IID components. Then, for all* $0 \leq t_1 \leq t_2$, $\mathbf{p}(t_1) \geq_{st} \mathbf{p}(t_2)$.

Proof It is enough to verify that, for all $k = 1, 2, \ldots, n$, and $0 \leq t_1 \leq t_2$,

$$\sum_{i=k}^{n} p_i P(X_{i,n} \leq t_1)/F_T(t_1) \geq \sum_{i=k}^{n} p_i P(X_{i,n} \leq t_2)/F_T(t_2).$$

It is equivalent to

$$\sum_{i=k}^{n}\sum_{l=1}^{n} p_i p_l [F_{X_{i,n}}(t_1) F_{X_{l,n}}(t_2) - F_{X_{i,n}}(t_2) F_{X_{l,n}}(t_1)] \geq 0.$$

Since

$$\sum_{i=k}^{n}\sum_{l=k}^{n} p_i p_l [F_{X_{i,n}}(t_1) F_{X_{l,n}}(t_2) - F_{X_{i,n}}(t_2) F_{X_{l,n}}(t_1)] = 0,$$

we only prove that

$$\sum_{i=k}^{n}\sum_{l=1}^{k-1} p_i p_l [F_{X_{i,n}}(t_1) F_{X_{l,n}}(t_2) - F_{X_{i,n}}(t_2) F_{X_{l,n}}(t_1)] \geq 0.$$

It is well known that $X_{l,n} \leq_{rh} X_{i,n}$ for $l \leq i$, that is, $\frac{F_{X_{i,n}(t)}}{F_{X_{l,n}(t)}}$ is increasing in $t \geq 0$, hence, for all $0 \leq t_1 \leq t_2$, $F_{X_{i,n}}(t_1) F_{X_{l,n}}(t_2) - F_{X_{i,n}}(t_2) F_{X_{l,n}}(t_1) \geq 0$, which validates the result.

The result below indicates that any coherent system has its tail stochastic behaviour similar to that of a i-out-of-n system. Its proof is simple and hence is omitted.

Theorem 2.1.31 *Assume that a coherent system with signature* $\mathbf{p} = (0, \ldots, 0, p_j, p_{j+1} \ldots, p_n)$ *and* $p_j > 0$ *for* $j \in \{1, 2, \ldots, n\}$, *then* $\lim_{t \to 0} \mathbf{p}(t) = (\underbrace{0, 0, \ldots, 0}_{j-1 \text{ times}}, 1, \underbrace{0, 0, \ldots, 0}_{n-j \text{ times}})$.

By Navarro et al. (2008b), if $X_1, X_2, \ldots, X_n$ are exchangeable component lifetimes, then

$$P(T \le t) = \sum_{i=1}^{n} \alpha_i P(X_{1,i} \le t), \tag{2.9}$$

where the coefficients vector $\boldsymbol{\alpha} = (\alpha_1, \alpha, \ldots, \alpha_n)$ is called as the domination vector or the minimal signature, the component of which may be negative with $\sum_{i=1}^{n} \alpha_i = 1$. In the following, we give another mixture representation of the inactivity time in terms of expression (2.9). Its proof procedure has the same way as that of Theorem 2.1.22, and hence is omitted.

Theorem 2.1.32 *For any $t \ge x \ge 0$,*

$$P(t - T > x \mid T \le t) = \sum_{i=1}^{n} \alpha_i(t) P(X_{1,i} \le t - x \mid X_{1,i} \le t) \tag{2.10}$$

where $\alpha_i(t) = \frac{\alpha_i F_{1,i}(t)}{F_T(t)}$, $i = 1, 2, \ldots, n$, and some of these coefficients may be negative. The coefficient vector $\boldsymbol{\alpha}(t) = (\alpha_1(t), \alpha_2(t), \ldots, \alpha_n(t))$ may be called the conditional domination vector given $T \le t$.

Expression (2.10) indicates that the inactivity time $(t - T \mid T \le t)$ of the system at time t is a mixture of the inactivity time $(t - X_{1,i} \mid X_{1,i} \le t)$ of the order statistics at time t with coefficients $a_i(t)$ for $i = 1, 2, \ldots, n$, that is, $(t - T \mid T \le t)$ is equivalent in distribution to a mixed system of the inactivity times of series systems. For example, considering the system with lifetime $T = X_{2,3}$, whose minimal signature is $\boldsymbol{\alpha} = (0, 3, -2)$, and that we know it is not working at time t, then by some computations, the coefficient vector is given by

$$\boldsymbol{\alpha}(t) = \left(0, \frac{6 - 3F(t)}{F(t)(3 - 2F(t))}, \frac{6F(t) - 6 - 2F^2(t)}{F(t)(3 - 2F(t))}\right).$$

Hence from Theorem 2.1.32, we have

$$\begin{aligned} &P(t - T > x \mid T \le t) \\ &= \frac{6 - 3F(t)}{F(t)(3 - 2F(t))} P(X_{1,2} \le t - x \mid X_{1,2} \le t) + \frac{6F(t) - 6 - 2F^2(t)}{F(t)(3 - 2F(t))} P(X_{1,3} \le t - x \mid X_{1,3} \le t). \end{aligned}$$

In Table 2.2, we have presented the vectors of coefficients order 3 in (2.10) for coherent systems of 3 IID components. It is noted that from Table 2.2, $\lim_{t \to \infty} \boldsymbol{\alpha}(t) = \boldsymbol{\alpha}$, that is, when t closes to infinity, the coefficient vector equals the minimal signature $\boldsymbol{\alpha}$.

By Navarro et al. (2008b),

$$P(T \le t) = \sum_{i=1}^{n} \beta_i P(X_{i,i} \le t), \tag{2.11}$$

Table 2.2 Vectors of coefficients in (2.10) with $n = 3$ for coherent systems with 3 IID components.

	$T = \phi(X_1, X_2, X_3)$	$\alpha(t)$
1	$\min\{X_1, X_2, X_3\}$	$(0, 0, 1)$
2	$\min\{X_1, \max\{X_2, X_3\}\}$	$\left(0, \frac{4-2F(t)}{(1+F(t)-F^2(t))}, \frac{3F(t)-3-F^2(t)}{(1+F(t)-F^2(t))}\right)$
3	$X_{2,3}$(2-out-of-3)	$\left(0, \frac{6-3F(t)}{F(t)(3-2F(t))}, \frac{6F(t)-6-2F^2(t)}{F(t)(3-2F(t))}\right)$
4	$\max\{X_1, \min\{X_2, X_3\}\}$	$\left(\frac{1}{F(t)(2-F(t))}, \frac{1}{F(t)}, \frac{3F(t)-3-F^2(t)}{F(t)(2-F(t))}\right)$
5	$\max\{X_1, X_2, X_3\}$	$\left(\frac{3}{F^2(t)}, \frac{3F(t)-6}{F^2(t)}, \frac{3-3F(t)+F^2(t)}{F^2(t)}\right)$

where the coefficient vector $\boldsymbol{\beta} = (\beta_1, \beta_i, \ldots, \beta_n)$ is called as the maximal signature, the component of which may be negative with $\sum_{i=1}^{n} \beta_i = 1$. In the following, we give the third mixture representation of the inactivity time in terms of expression (2.11). Its proof procedure has the same way as that of Theorem 2.1.22, and hence is omitted.

Theorem 2.1.33 *For any* $t \geq x \geq 0$,

$$P(t - T > x \mid T \leq t) = \sum_{i=1}^{n} \beta_i(t) P(X_{i,i} \leq t - x \mid X_{i,i} \leq t) \tag{2.12}$$

where $\beta_i(t) = \frac{\beta_i F_{i,i}(t)}{F_T(t)}$, $i = 1, 2, \ldots, n$, *and some of these coefficients may be negative. The coefficient vector* $\boldsymbol{\beta}(t) = (\beta_1(t), \beta_2(t), \ldots, \beta_n(t))$ *can be called the conditional domination vector given* $T \leq t$.

Expression (2.12) indicates that the inactivity time $(t - T \mid T \leq t)$ of the system at time t is a mixture of the inactivity time $(t - X_{i,i} \mid X_{i,i} \leq t)$ of the order statistics at time t with coefficients $\beta_i(t)$ for $i = 1, 2, \ldots, n$, that is, $(t - T \mid T \leq t)$ is equivalent in distribution to a mixed system of the inactivity times of parallel systems. For example, considering the system with lifetime $T = \max\{X_1, \min\{X_2, X_3\}\}$, whose maximal signature is $\boldsymbol{\beta} = (0, 2, -1)$, and that we know it is not working at time t, then by some computations, the coefficient vector is given by

$$\boldsymbol{\beta}(t) = \left(0, \frac{2}{2 - F(t)}, \frac{-F(t)}{2 - F(t)}\right).$$

Hence from Theorem 2.1.33, we have

$$\begin{aligned} &P(t - T > x \mid T \leq t) \\ &= \frac{2}{2 - F(t)} P(X_{2,2} \leq t - x \mid X_{2,2} \leq t) - \frac{F(t)}{2 - F(t)} P(X_{3,3} \leq t - x \mid X_{3,3} \leq t). \end{aligned}$$

Table 2.3 Vectors of coefficients in (2.12) with $n = 3$ for coherent systems with 3 IID components.

System	$T = \phi(X_1, X_2, X_3)$	$\boldsymbol{\beta}(t)$
1	$\min\{X_1, X_2, X_3\}$	$\left(\frac{3}{3-3F(t)+F^2(t)}, \frac{-3F(t)}{3-3F(t)+F^2(t)}, \frac{F^2(t)}{3-3F(t)+F^2(t)}\right)$
2	$\min\{X_1, \max\{X_2, X_3\}\}$	$\left(\frac{1}{1+F(t)-F^2(t)}, \frac{F(t)}{1+F(t)-F^2(t)}, \frac{-F^2(t)}{1+F(t)-F^2(t)}\right)$
3	$X_{2,3}$(2-out-of-3)	$\left(0, \frac{3}{3-2F(t)}, \frac{-2F(t)}{3-2F(t)}\right)$
4	$\max\{X_1, \min\{X_2, X_3\}\}$	$\left(0, \frac{2}{2-F(t)}, \frac{-F(t)}{2-F(t)}\right)$
5	$\max\{X_1, X_2, X_3\}$	$(0, 0, 1)$

In Table 2.3, we have presented the vectors of coefficients order 3 in (2.12) for coherent systems of 3 IID components. It is noted that from Table 2.3, $\lim_{t\to\infty} \boldsymbol{\beta}(t) = \boldsymbol{\beta}$, that is, when t closes to infinity, the coefficient vector equals the maximal signature $\boldsymbol{\beta}$.

2.2 On Coherent Systems with Exchangeable Components

This section investigates some ordering properties of the residual lives and inactivity time of general coherent systems with exchangeable components by using signatures, extending the case that components are independent and identically distributed in last section.

2.2.1 Stochastic Comparisons of Residual Lives

Recall the joint survival function is exchangeable if $R(x_1, \ldots, x_n)$ is symmetric in $x_1, \ldots, x_n$. Samaniego (1985) (also see Kochar et al. 1999) showed that a coherent system having IID units is a mixture of k-out-of-n system. Navarro et al. (2005) observed it holds when the components of the system are exchangeable. Khaledi and Shaked (2007) showed that the reliability function of the residual life of the coherent system with IID components is given by

$$P(T - t > x \mid X_{r,n} > t) = \sum_{i=s}^{n} p_i \, P(X_{i,n} - t > x \mid X_{r,n} > t), \qquad r \le s \le n.$$

In fact, from the procedure of the proof of Theorem 2.3 in Khaledi and Shaked (2007), this also holds even when $X_1, \ldots, X_n$ are exchangeable.

In the sequel, suppose $\mathbf{p} = (0, \ldots, 0, p_s, \ldots, p_n), \mathbf{q} = (0, \ldots, 0, q_s, \ldots, q_n)$, for $r \leq s \leq n$. Let $\mathbf{p}$ and $\mathbf{q}$ be the corresponding signatures of two coherent systems $T_1 = \tau_1(X_1, X_2, \ldots, X_n)$, $T_2 = \tau_2(X_1, X_2, \ldots, X_n)$ with common exchangeable components lifetimes $X_1, X_2, \ldots, X_n$, where τ_1 and τ_2 are coherent life functions.

Let $\bar{H}_{r,i,n,t}(x) = P(X_{i,n} - t > x \mid X_{r,n} > t)$, then it can be written as

$$\begin{aligned}\bar{H}_{r,i,n,t}(x) &= \frac{P(X_{i,n} - t > x, X_{r,n} > t)}{P(X_{r,n} > t)} \\ &= \frac{\sum_{j=0}^{r-1} \binom{n}{j} \sum_{k=n-i+1}^{n-j} \binom{n-j}{k} P_{j,k}(t, x)}{P(X_{r,n} > t)}, \end{aligned} \tag{2.13}$$

where $P_{j,k}(t, x) = P(X_1 \leq t, \ldots, X_j \leq t, t < X_{j+1} \leq t + x, \ldots, t < X_{k-1} \leq t + x, X_k > t + x, \ldots, X_n > t + x)$. Therefore, for $x \geq 0$,

$$P(T_1 - t > x \mid X_{r,n} > t) = \sum_{i=s}^{n} p_i \bar{H}_{r,i,n,t}(x) \quad r \leq s \leq n. \tag{2.14}$$

$$P(T_2 - t > x \mid X_{r,n} > t) = \sum_{i=s}^{n} q_i \bar{H}_{r,i,n,t}(x) \quad r \leq s \leq n. \tag{2.15}$$

Theorem 2.2.1 *If $\mathbf{p} \leq_{st} \mathbf{q}$, then, for any $t \geq 0$,*

$$(T_1 - t \mid X_{r,n} > t) \leq_{st} (T_2 - t \mid X_{r,n} > t).$$

Proof By (2.13), for any $x \geq 0$, $\bar{H}_{r,i,n,t}(x) \leq \bar{H}_{r,i+1,n,t}(x)$, hence $\mathbf{p} \leq_{st} \mathbf{q}$ implies that

$$\sum_{i=s}^{n} p_i \bar{H}_{(r,i,n,t)}(x) \leq \sum_{i=s}^{n} q_i \bar{H}_{(r,i,n,t)}(x).$$

This validates the result.

The following two lemmas are useful in proving some desired results.

Lemma 2.2.2 (Shaked and Shanthikumar 2007) *Let Θ_i be a random variable with support $\mathcal{X} \subset \mathbb{R}$ and distribution function F_i, $i = 1, 2$, $X(\theta)$ has the distribution belonging to the family $\{G_\theta, \theta \in \mathcal{X}\}$. Suppose $X(\Theta_i)$ has the distribution function*

$$H_i(y) = \int_{\mathcal{X}} G_\theta(y) \, \mathrm{d}F_i(\theta), \quad y \in \Re, \ i = 1, 2.$$

If $X(\theta) \leq_{hr} X(\theta')$ whenever $\theta \leq \theta'$ and $\Theta_1 \leq_{hr} \Theta_2$, then $X(\Theta_1) \leq_{hr} X(\Theta_2)$.

Lemma 2.2.3 (Karlin 1968) *Let A, B, and C be subsets of the real line. For any σ-finite measures $\mu(z)$, the function*

$$K(x, y) = \int_B L(x, z)M(z, y)d\mu(z)$$

is TP2 for $x \in A$ and $y \in B$ if $L(x, z)$ is TP2 for $x \in A$ and $z \in C$, and $M(z, y)$ is TP2 for $z \in C$ and $y \in B$.

Theorems 2.2.4, 2.2.5 extend corresponding results given by Li and Zhang (2008).

Theorem 2.2.4 *If $\mathbf{p} \leq_{hr} \mathbf{q}$, and $(X_{i,n} - t > x \mid X_{r,n} > t) \leq_{hr} (X_{i+1,n} - t > x \mid X_{r,n} > t)$, then, for any $t \geq 0$,*

$$(T_1 - t \mid X_{r,n} > t) \leq_{hr} (T_2 - t \mid X_{r,n} > t).$$

Proof The condition $(X_{i,n} - t > x \mid X_{r,n} > t) \leq_{hr} (X_{i+1,n} - t > x \mid X_{r,n} > t)$ implies that $(X_{i,n} - t > x \mid X_{r,n} > t) \leq_{hr} (X_{j,n} - t > x \mid X_{r,n} > t)$ for all $s \leq i \leq j \leq n$. By the hazard rate order $\mathbf{p} \leq_{hr} \mathbf{q}$ and Lemma 2.2.2, the desired result follows immediately.

Theorem 2.2.5 *If $\mathbf{p} \leq_{lr} \mathbf{q}$, and $(X_{i,n} - t > x \mid X_{r,n} > t) \leq_{lr} (X_{i+1,n} - t > x \mid X_{r,n} > t)$, then, for any $t \geq 0$,*

$$(T_1 - t \mid X_{r,n} > t) \leq_{lr} (T_2 - t \mid X_{r,n} > t).$$

Proof Denote $h_{r,i,n,t}(x)$ be the density function of $(X_{i,n} - t > x \mid X_{r,n} > t)$, then, for $r \leq s \leq n$, the density functions of $(T_1 - t \mid X_{r,n} > t)$ and $(T_2 - t \mid X_{r,n} > t)$ are respectively

$$h_1(x) = \sum_{i=s}^{n} p_i h_{r,i,n,t}(x)$$

and

$$h_2(x) = \sum_{i=s}^{n} q_i h_{r,i,n,t}(x).$$

For $c > 0$, we consider the function

$$h(x) = h_2(x) - ch_1(x) = \sum_{i=s}^{n} (q_i - cp_i)h_{r,i,n,t}(x).$$

The condition $(X_{i,n} - t > x \mid X_{r,n} > t) \leq_{lr} (X_{i+1,n} - t > x \mid X_{r,n} > t)$ implies that $h_{r,i,n,t}(x)$ is TP2 for $x \in \Re$ and $i = s, \ldots, n$. Since $\mathbf{p} \leq_{lr} \mathbf{q}$, q_i/p_i is increasing in $i = s, \ldots, n$, and hence the sequence $\{q_i - cp_i\}$ has at most one change of sign from negative to positive as i from s to n. According to Karlin (1968) we have $h(x)$ has at most one change of sign from negative to positive as x increases. This validates the result.

In practice, it may be difficult to check the conditions appeared in the theorems, sometimes they can be checked numerically. Example 2.2.6 shows these conditions hold in a specific model.

Example 2.2.6 Consider coherent system with four changeable units, whose joint reliability distribution (Gumble distribution) is given by

$$R(x_1, x_2, x_3, x_4) = e^{-x_1-x_2-x_3-x_4-0.5x_1x_2x_3x_4},$$

for $x_i \geq 0, i = 1, 2, 3, 4$. It can be computed that, for fixed $t \geq 0$,

$$\bar{H}_{2,2,4,t}(x) = \frac{4e^{-3(t+x)} - 3e^{-4(t+x)-0.5(t+x)^4}}{4e^{-3t} - 3e^{-4t-0.5t^4}},$$

$$\bar{H}_{2,3,4,t}(x) = \frac{12e^{-3t-2x} - 8e^{-3(t+x)} + 3e^{-4(t+x)-0.5(t+x)^4} - 6e^{-4t-2x-0.5(t(t+x))^2}}{4e^{-3t} - 3e^{-4t-0.5t^4}},$$

$$\bar{H}_{2,4,4,t}(x) = \frac{12e^{-3t-x} + 4e^{-3(t+x)} - 12e^{-3t-2x} - e^{-4(t+x)-0.5(t+x)^4}}{4e^{-3t} - 3e^{-4t-0.5t^4}} + \frac{6e^{-4t-2x-0.5(t(t+x))^2} - 8e^{-4t-x-0.5t^3(t+x)}}{4e^{-3t} - 3e^{-4t-0.5t^4}}$$

for any $x \geq 0$. Let $t = 1$, it can be verified that $\bar{H}_{2,i,4,t}(x) \leq_{hr} (\leq_{lr})\bar{H}_{2,i+1,4,t}(x)$, $i = 2, 3$.

2.2.2 *Stochastic Comparisons of Inactivity Time*

In this section, assume that **p** and **q** are the signatures of two coherent systems $T_1 = \tau_1(X_1, X_2, \ldots, X_n)$, $T_2 = \tau_2(X_1, X_2, \ldots, X_n)$ with common exchangeable components lifetimes $X_1, X_2, \ldots, X_n$, where τ_1 and τ_2 are coherent life functions. Suppose, for some $s \leq r \leq n$, that the signatures of τ_1 and τ_2 are $\mathbf{p} = (p_1, \ldots, p_s, 0, \ldots, 0)$ and $\mathbf{q} = (q_1, \ldots, q_s, 0, \ldots, 0)$, respectively.

Let $\bar{H}_{i,r,n,(t)}(x) = P(t - X_{i,n} > x \mid X_{r,n} \leq t)$, then it can be written as

$$\bar{H}_{i,r,n,(t)}(x) = \frac{P(t - X_{i,n} > x, X_{r,n} \leq t)}{P(X_{r,n} \leq t)} = \frac{\sum_{j=i}^{n} \binom{n}{j} \sum_{k=0}^{n-j} \binom{n-j}{k} P_{j,k}(t, x)}{P(X_{r,n} \leq t)}, \quad (2.16)$$

where $P_{j,k}(t, x) = P(X_1 \leq t - x, \ldots, X_j \leq t - x, t - x < X_{j+1} \leq t, \ldots, t - x < X_{j+k} \leq t, X_{j+k+1} > t, \ldots, X_n > t)$.

In a similar as in Khaledi and Shaked (2007), it can be shown that, for $s \leq r \leq n$, the reliability functions of the inactivity times of a coherent systems with exchangeable components are

$$P(t - T_1 > x \mid X_{r,n} \leq t) = \sum_{i=s}^{n} p_i \bar{H}_{i,r,n,(t)}(x), \qquad s \leq r \leq n. \tag{2.17}$$

$$P(t - T_2 > x \mid X_{r,n} \leq t) = \sum_{i=s}^{n} q_i \bar{H}_{i,r,n,(t)}(x), \qquad s \leq r \leq n. \tag{2.18}$$

Theorem 2.2.7 *If* $\mathbf{p} \leq_{st} \mathbf{q}$, *then, for* $r \geq s$,

$$(t - T_1 \mid X_{r,n} \leq t) \geq_{st} (t - T_2 \mid X_{r,n} \leq t).$$

Proof By (2.16), for any $x \geq 0$, $\bar{H}_{(i,r,n,t)}(x) \geq \bar{H}_{i+1,r,n,(t)}(x)$, hence $\mathbf{p} \leq_{st} \mathbf{q}$ implies that

$$\sum_{i=s}^{n} p_i \bar{H}_{i,r,n,(t)}(x) \geq \sum_{i=s}^{n} q_i \bar{H}_{i,r,n,(t)}(x).$$

This validates the result.

The following lemma is useful in proving the main conclusions in this section.

Lemma 2.2.8 (Shaked and Shanthikumar 2007) *Let* α *and* β *be real valued functions such that* α *is nonnegative and* β/α *and* α *are non-increasing. If* X_i *has distribution* F_i, $i = 1, 2$ *and* $X_1 \leq_{rh} X_2$, *then*

$$\frac{\int_{-\infty}^{\infty} \beta(x)\, \mathrm{d}F_1(x)}{\int_{-\infty}^{\infty} \alpha(x)\, \mathrm{d}F_1(x)} \geq \frac{\int_{-\infty}^{\infty} \beta(x)\, \mathrm{d}F_2(x)}{\int_{-\infty}^{\infty} \alpha(x)\, \mathrm{d}F_2(x)}.$$

Theorems 2.2.9, 2.2.10 extend corresponding Theorem 10, 11 given by Li and Zhang (2008).

Theorem 2.2.9 *If* $\mathbf{p} \leq_{rh} \mathbf{q}$, *and* $(t - X_{i+1,n} \mid X_{r,n} \leq t) \leq_{hr} (t - X_{i,n} \mid X_{r,n} \leq t)$, *then, for* $r \geq s$,

$$(t - T_1 \mid X_{r,n} \leq t) \geq_{rh} (t - T_2 \mid X_{r,n} \leq t).$$

Proof By (2.17), (2.18), it is enough to prove that

$$\frac{P(t - T_1 > x \mid X_{r,n} \leq t)}{P(t - T_2 > x \mid X_{r,n} \leq t)} = \frac{\sum_{i=1}^{s} p_i \bar{H}_{i,r,n,(t)}(x)}{\sum_{i=1}^{s} q_i \bar{H}_{i,r,n,(t)}(x)}$$

is increasing in x, that is, for all $x_2 \geq x_1 \geq 0$,

$$\frac{\sum_{i=s}^{n} p_i \bar{H}_{i,r,n,(t)}(x_2)}{\sum_{i=s}^{n} p_i \bar{H}_{i,r,n,(t)}(x_1)} \geq \frac{\sum_{i=s}^{n} q_i \bar{H}_{i,r,n,(t)}(x_2)}{\sum_{i=s}^{n} q_i \bar{H}_{i,r,n,(t)}(x_1)}.$$

Let $\alpha(i) = \bar{H}_{i,r,n,(t)}(x_1)$ and $\beta(i) = \bar{H}_{i,r,n,(t)}(x_2)$. It is clear, from (2.16), that $\alpha(i)$ is decreasing in $i = s, \ldots, n$, and by the condition, $\beta(i)/\alpha(i)$ is decreasing in i for all $t \geq 0$. From Lemma 2.2.8, the desired result follows.

Theorem 2.2.10 *If $\mathbf{p} \leq_{lr} \mathbf{q}$, and $(t - X_{i+1,n} \mid X_{r,n} \leq t) \leq_{lr} (t - X_{i,n} \mid X_{r,n} \leq t)$, then, for $r \geq s$,*

$$(t - T_1 \mid X_{r,n} \leq t) \geq_{lr} (t - T_2 \mid X_{r,n} \leq t).$$

Proof Note that, for $s \leq r \leq n$, the density functions of $(t - T_1 \mid X_{r,n} > t)$ and $(t - T_2 \mid X_{r,n} > t)$ are $\sum_{i=1}^{s} p_i h_{i,r,n,(t)}(x)$ and $\sum_{i=1}^{s} q_i h_{i,r,n,(t)}(x)$, respectively, it is enough to prove that

$$\frac{\sum_{i=1}^{s} p_i h_{i,r,n,(t)}(x)}{\sum_{i=1}^{s} q_i h_{i,r,n,(t)}(x)}$$

is increasing in $x \geq 0$. Obviously, this is equivalent to, for all $x_2 \geq x_1 \geq 0$,

$$g(x_1, x_2) = \sum_{i=1}^{s} \sum_{j=1}^{s} p_i q_j \left[h_{i,r,n,(t)}(x_2) h_{j,r,n,(t)}(x_1) - h_{i,r,n,(t)}(x_1) h_{j,r,n,(t)}(x_2)\right] \geq 0.$$

Since $\mathbf{p} \leq_{lr} \mathbf{q}$ and $(t - X_{i+1,n} \mid X_{r,n} \leq t) \leq_{lr} (t - X_{i,n} \mid X_{r,n} \leq t)$, we have

$$\begin{aligned}
g(x_1, x_2) = & \sum_{i=1}^{s} \sum_{j=1}^{s} p_i q_j \left[h_{i,r,n,(t)}(x_2) h_{j,r,n,(t)}(x_1) - h_{i,r,n,(t)}(x_1) h_{j,r,n,(t)}(x_2)\right] \\
= & \sum_{i=1}^{s} \sum_{j=i}^{s} p_i q_j \left[h_{i,r,n,(t)}(x_2) h_{j,r,n,(t)}(x_1) - h_{i,r,n,(t)}(x_1) h_{j,r,n,(t)}(x_2)\right] \\
& + \sum_{i=1}^{s} \sum_{j=1}^{i} p_i q_j \left[h_{i,r,n,(t)}(x_2) h_{j,r,n,(t)}(x_1) - h_{i,r,n,(t)}(x_1) h_{j,r,n,(t)}(x_2)\right] \\
= & \sum_{i=1}^{s} \sum_{j=i}^{s} p_i q_j \left[h_{i,r,n,(t)}(x_2) h_{j,r,n,(t)}(x_1) - h_{i,r,n,(t)}(x_1) h_{j,r,n,(t)}(x_2)\right] \\
& + \sum_{i=1}^{s} \sum_{j=i}^{s} p_j q_i \left[h_{j,r,n,(t)}(x_2) h_{i,r,n,(t)}(x_1) - h_{j,r,n,(t)}(x_1) h_{i,r,n,(t)}(x_2)\right]
\end{aligned}$$

$$= \sum_{i=1}^{s}\sum_{j=i}^{s} p_i p_j \left(\frac{q_j}{p_j} - \frac{q_i}{p_i}\right)\left[h_{i,r,n,(t)}(x_2)h_{j,r,n,(t)}(x_1) - h_{i,r,n,(t)}(x_1)h_{j,r,n,(t)}(x_2)\right]$$

$$\geq 0.$$

This completes the proof of result.

2.3 On Coherent Systems with INID Components

This section presents several useful mixture representations for the reliability function of the residual live (inactivity time) of a coherent system with independent but non-identically distributed (INID) components. Some stochastic comparisons of residual lives (inactivity times) between two systems are conducted based on the stochastic ordering of coefficient vectors (or components) of the two systems.

2.3.1 The Residual Life of General Coherent Systems

In this subsection, several useful mixture representations for the reliability function of the residual live of a coherent system with INID components are presented, which are based on order statistics, signatures and mean reliability functions. We then discuss some stochastic comparisons of residual lives between two systems based on the stochastic ordering of coefficient vectors (or components) of the two systems.

2.3.1.1 Preliminaries

As described before, the reliability function of a coherent system having n IID units can be represented as a mixture of k-out-of-n systems with weights s_k for $k = 1, 2, \ldots, n$; that is, for any $t > 0$,

$$P(T > t) = \sum_{i=1}^{n} s_i P(X_{i,n} > t). \tag{2.19}$$

It also holds even when the components of the system are exchangeable (i.e., when the joint survival function $R(x_1, \ldots, x_n)$ of $X_1, \ldots, X_n$ is symmetric in $x_1, \ldots, x_n$). The result holds for mixed systems as well. A mixed system of order n is a stochastic mixture of coherent systems of order n that selects a system over the class of coherent systems of order n at random according to a fixed probability distribution (see Navarro et al. (2008a), and Samaniego (2007). However, the representation in (2.19) need not necessarily hold when the components are non-identically distributed, as described by Navarro et al. (2008a).

Navarro and Eryilmaz (2007) introduced the concepts of minimal and maximal signatures of coherent systems. With minimal and maximal signatures, the reliability function of a coherent system can be written as general discrete mixtures

$$P(T > t) = \sum_{i=1}^{n} a_i P(X_{1,i} > t), \tag{2.20}$$

$$P(T > t) = \sum_{i=1}^{n} b_i P(X_{i,i} > t), \tag{2.21}$$

where the coefficient vectors $\mathbf{a} = (a_1, \ldots, a_n)$ and $\mathbf{b} = (b_1, \ldots, b_n)$ are, respectively, called the minimal signature (or the domination vector) and the maximal signature, with $\sum_{i=1}^{n} a_i = 1$ and $\sum_{i=1}^{n} b_i = 1$, but some of the coefficients may be negative. Under the IID assumption of the components of systems, the vectors $\mathbf{s}$, $\mathbf{a}$ and $\mathbf{b}$ only depend on the system structure function. Moreover, since $\mathbf{s} = \mathbf{a}A_n$ for a triangular matrix A_n, $\mathbf{s}$ can be obtained from $\mathbf{a}$ and vice versa.

Upon using the distribution of order statistics, the expression in (2.19) can be readily given

$$P(T > t) = \sum_{i=1}^{n} s_i \sum_{j=n-i+1}^{n} \binom{n}{j} [\bar{F}(t)]^j [F(t)]^{n-j}, \tag{2.22}$$

where $F(t)$ and $\bar{F}(t)$ are the distribution and survival functions of the components, respectively. Setting $p = \bar{F}(t)$, the reliability of the system at time t in (2.22) may be represented as

$$h(p) = \sum_{i=1}^{n} s_i \sum_{j=n-i+1}^{n} \binom{n}{j} p^j (1-p)^{n-j}. \tag{2.23}$$

In fact, this polynomial may also be written in the form $h(p) = \sum_{i=1}^{n} d_i p^i$, where $\mathbf{d} = (d_1, \ldots, d_n)$ is generally referred to as the system's domination vector (see Satyanarayana and Prabhakar 1978). The relationship between $\mathbf{s}$ and $\mathbf{d}$ has been described by Samaniego (2007). The coefficients of the polynomial in (2.23) only depend on the system structure, and $h(p)$ strictly increases for $p \in (0, 1)$, such that $h(0) = 0$ and $h(1) = 1$.

Navarro et al. (2008a) showed that if $T = \phi(X_1, \cdots, X_n)$ is the lifetime of a coherent system with IID component lifetimes $X_1, \ldots, X_n$ distributed according to a common continuous distribution F and $\mathbf{s} = (s_1, \ldots, s_n)$ is the system signature, then the distribution of the system residual lifetime $T - t$, given $T > t$, is a mixture of the residual lives of k-out-of-n systems. Specifically, for all $t \geq 0$ and $x \geq 0$,

$$P(T - t > x \mid T > t) = \sum_{i=1}^{n} s_i(t) P(X_{i,n} - t > x \mid X_{i,n} > t),$$

where $s_i(t) = s_i P(X_{i,n} > t)/\bar{F}_T(t)$ for $i = 1, \ldots, n$, given $\bar{F}(t) > 0$. These authors also obtained

$$P(T - t > x \mid T > t) = \sum_{i=1}^{n} a_i(t) P(X_{1,i} - t > x \mid X_{1,i} > t),$$

where $a_i(t) = a_i P(X_{1,i} > t)/\bar{F}_T(t)$ for $i = 1, \ldots, n$, given $\bar{F}(t) > 0$, and $\mathbf{a} = (a_1, \ldots, a_i, \ldots, a_n)$ is the minimal signature vector. The coefficient vector $\mathbf{a}(t) = (a_1(t), \ldots, a_n(t))$, with $\sum_{i=1}^{n} a_i(t) = 1$, has been called the conditional domination vector. Some of the coefficients $a_i(t)$ may be negative, of course.

From the expression in (2.21), we can express the residual lifetime of $T - t$, given $T > t$, as

$$P(T - t > x \mid T > t) = \sum_{i=1}^{n} b_i(t) P(X_{i,i} - t > x \mid X_{i,i} > t),$$

where $b_i(t) = b_i P(X_{i,i} > t)/\bar{F}_T(t)$ for $i = 1, \ldots, n$, given $\bar{F}(t) > 0$, and $\mathbf{b} = (b_1, \ldots, b_i, \ldots, b_n)$ is the maximal signature vector. The coefficient vector $\mathbf{b}(t) = (b_1(t), \ldots, b_i(t), \ldots, b_n(t))$, with $\sum_{i=1}^{n} b_i(t) = 1$, may be called the conditional maximal signature vector. Also, some of these coefficients may be negative.

Based on the expression in (2.20), the following results have been established by Navarro et al. (2008a).

Theorem 2.3.1 *Let $T_1 = \phi_1(X_1, \cdots, X_n)$ and $T_2 = \phi_2(X_1, \cdots, X_n)$ be the lifetimes of two coherent systems, both based on n components with IID lifetimes distributed according to a common continuous distribution F. Let, for all $t \geq 0$, $\mathbf{s}_1(t) = (s_{11}(t), \ldots, s_{1i}(t), \ldots, s_{1n}(t))$ and $\mathbf{s}_2(t) = (s_{21}(t), \ldots, s_{2i}(t), \ldots, s_{2n}(t))$ denote the respective coefficient vectors in* (2.20).

(a) If $\mathbf{s}_1(t) \leq_{st} \mathbf{s}_2(t)$, then $(T_1 - t \mid T_1 > t) \leq_{st} (T_2 - t \mid T_2 > t)$;
(b) If $\mathbf{s}_1(t) \leq_{hr} \mathbf{s}_2(t)$, then $(T_1 - t \mid T_1 > t) \leq_{hr} (T_2 - t \mid T_2 > t)$;
(c) If $\mathbf{s}_1(t) \leq_{lr} \mathbf{s}_2(t)$, then $(T_1 - t \mid T_1 > t) \leq_{lr} (T_2 - t \mid T_2 > t)$.

The lifetimes of coherent systems with independent but non-identically distributed components have been studied subsequently by Navarro et al. (2011). They obtained the signature-based representations for the reliability of these systems, as described in the following theorem.

Theorem 2.3.2 *Let T be the lifetime of a coherent system having independent component lifetimes $X_1, \ldots, X_n$, and $\bar{F}_i$ be the reliability function of X_i, for $i = 1, \ldots, n$. Assume that h and H are the system's reliability polynomial and reliability structure function, respectively, and $\boldsymbol{s} = (s_1, \ldots, s_n)$, $\boldsymbol{a} = (a_1, \ldots, a_n)$ and $\boldsymbol{b} = (b_1, \ldots, b_n)$ are the corresponding signature, minimal signature and maximal signature vectors* *of an n-component system with the same structure as T but with IID component lifetimes $Y_1, \ldots, Y_n$. Then, the reliability function of the system lifetime T can be expressed as*

$$P(T > t) = \sum_{i=1}^{n} s_i P(Y_{i,n} > t) = \sum_{i=1}^{n} s_i \bar{G}_{i,n}(t), \tag{2.24}$$

$$P(T > t) = \sum_{i=1}^{n} a_i P(Y_{1,i} > t) = \sum_{i=1}^{n} a_i \bar{G}_{1,i}(t), \tag{2.25}$$

$$P(T > t) = \sum_{i=1}^{n} b_i P(Y_{i,i} > t) = \sum_{i=1}^{n} b_i \bar{G}_{i,i}(t), \tag{2.26}$$

where $Y_{1,n}, \ldots, Y_{n,n}$ are the order statistics from $Y_1, \ldots, Y_n$ having a common reliability function

$$\bar{G}(t) = h^{-1} H(\bar{F}_1(t), \ldots, \bar{F}_n(t)). \tag{2.27}$$

Clearly, $h(\bar{G}(t)) = H(\bar{F}_1(t), \ldots, \bar{F}_n(t))$.

According to the properties of h and H, the function $\bar{G}$ satisfies the properties of a reliability function and is therefore a proper reliability function. The reliability function $\bar{G}(t)$ can therefore be called the mean reliability function associated with the system and the components' reliability function.

By utilizing the representations in Theorem 2.3.2, some stochastic comparison results for two different systems with two sets of independent components have been obtained in Navarro et al. (2011), which extended the existing results on signature-based results for the IID case to the INID case. By using Theorem 2.2 in Navarro et al. (2011), the following result below can be obtained immediately and its proof is therefore omitted for the sake of brevity.

Theorem 2.3.3 *Let $T_1 = \phi_1(X_{11}, \ldots, X_{1n})$ and $T_2 = \phi_2(X_{21}, \ldots, X_{2n})$ be the lifetimes of two coherent systems with the same signature $\mathbf{s} = (s_1, \ldots, s_i, \ldots, s_n)$ but having different independent components. Let h_1 and h_2 be their reliability polynomials, and H_1 and H_2 be their structure reliability functions, respectively. As described in Theorem 2.3.2, suppose $\bar{G}_1(t) = h_1^{-1}(H_1(\bar{F}_{11}(t), \ldots, \bar{F}_{1n}(t)))$ is the reliability function of a set of IID random variables $Y_{11}, \ldots, Y_{1n}$, and $\bar{G}_2(t) = h_2^{-1}(H_2(\bar{F}_{21}(t), \ldots, \bar{F}_{2n}(t)))$ is the reliability function of another set of IID random variables $Y_{21}, \ldots, Y_{2n}$, for any $t \geq 0$.*

(a) If $\bar{G}_1(t) \leq_{st} \bar{G}_2(t)$, then $T_1 \leq_{st} T_2$;

(b) If $\bar{G}_1(t) \leq_{hr} \bar{G}_2(t)$ and $xh_1'(x)/h_1(x)$ or $xh_2'(x)/h_2(x)$ is decreasing, then $T_1 \leq_{hr} T_2$;

(c) If $\bar{G}_1(t) \leq_{rh} \bar{G}_2(t)$ and $(1-x)xh_1'(x)/(1-h_1(x))$ or $(1-x)h_2'(x)/(1-h_2(x))$ is increasing, then $T_1 \leq_{rh} T_2$.

Remark 2.3.4 Few representation results have been obtained in the literature for the residual life of coherent systems with heterogeneous components. In this section we study the residual life of a coherent system with INID components. We Present mixture representations for the residual lifetime by using order statistics and signature. Some stochastic comparisons of residual lives are then made between two systems with different structures and two different sets of INID components. Some properties are then obtained from these representations. These results form nice extensions of some known results in the literature for the IID case.

2.3.1.2 Stochastic Comparisons of Residual Life

As the first result, we give the mixture representation for the residual lifetime of coherent systems with INID components based on the signature and mean reliability function.

Theorem 2.3.5 *Suppose $\mathbf{s} = (s_1, \ldots, s_n)$ is the signature vector of an n-component system, with INID components lifetimes $X_1, \ldots, X_n$ having corresponding reliability functions $\bar{F}_1, \ldots, \bar{F}_n$. Let h and H be the system's reliability polynomial and reliability structure function, respectively. Then, for any $t > 0$ and $x > 0$,*

$$P(T - t > x \mid T > t) = \sum_{i=1}^{n} s_i(t) P(Y_{i,n} - t > x \mid Y_{i,n} > t), \tag{2.28}$$

where $s_i(t) = s_i P(Y_{i,n} > t)/P(T > t)$ for $i = 1, \ldots, n$, given $P(T > t) > 0$, such that $\sum_{i=1}^{n} s_i(t) = 1$, and $Y_{i,n}$ is the i-th order statistic from a set of IID random variables $Y_1, \ldots, Y_n$ with a common mean reliability function $\bar{G}(t) = h^{-1}H(\bar{F}_1(t), \ldots, \bar{F}_n(t))$.

Proof From Theorem 2.3.3 and (2.27), for any $t > 0$ and $x > 0$, we have

$$\begin{aligned}
&P(T - t > x \mid T > t) \\
&= \frac{\sum_{i=1}^{n} s_i P(Y_{i,n} > t + x)}{P(T > t)} \\
&= \frac{\sum_{i=1}^{n} s_i P(Y_{i,n} > t + x | Y_{i,n} > t) \times P(Y_{i,n} > t)}{P(T > t)} \\
&= \sum_{i=1}^{n} s_i(t) P(Y_{i,n} - t > x \mid Y_{i,n} > t).
\end{aligned}$$

This completes the proof of the theorem.

Theorem 2.3.5 shows that the residual life $(T - t|T > t)$ of a coherent system with INID components at time t may be represented as a mixture of the residual lifetimes $(Y_{i,n} - t|Y_{i,n} > t)$ of the order statistics from a set of IID component lifetimes with coefficients $s_i(t)$, for $i = 1, \ldots, n$. The coefficients $s_i(t)$ depend on H and $\bar{F}_i$ for $i = 1, \ldots, n$; that is, the residual life of the coherent system with INID components, given $T > t$, can be regarded as a discrete mixture of the residual lifetimes of k-out-of-n systems formed by a set of IID random variables.

Now, we shall discuss some stochastic properties of the coefficient vector $\mathbf{s}(t) = (s_1(t), \ldots, s_n(t))$. It can be shown that $\mathbf{s}(t)$ has tail stochastic behavior when the lifetimes of the n components are INID, which is an extension of the corresponding results for the case when the components are IID, presented by Navarro et al. (2008b). That is, if T is a coherent system with signature vector $\mathbf{s} = (s_1, \ldots, s_j, 0, \ldots, 0)$, and $s_j > 0$ for $j \in \{1, 2, \ldots, n\}$, then $\lim_{t\to 0}\mathbf{s}(t) = (\underbrace{0, 0, \ldots, 0}_{j-1 \text{ times}}, 1, \underbrace{0, 0, \ldots, 0}_{n-j \text{ times}})$. Furthermore, we can show that $\mathbf{s}(t_1) \le_{st} \mathbf{s}(t_2)$ for $t_1 \le t_2$ as known in the case when the components are IID. This shows that $(T - t_1|T > t_1) \le_{st} (T - t_2|T > t_2)$ for all $t_1 \le t_2$.

Example 2.3.6 Consider a system with lifetime $T = \max\{X_1, X_2\}$ having INID component lifetimes X_i distributed according to $\bar{F}_i$, for $i = 1, 2$, and the signature vector for this system is clearly $(0, 1)$. For any $t > 0$, the reliability function is given by

$$\begin{aligned} P(T > t) &= P(\max\{X_1, X_2\} > t) \\ &= \bar{F}_1(t) + \bar{F}_2(t) - \bar{F}_1(t)\bar{F}_2(t) \\ &= H(\bar{F}_1(t), \bar{F}_2(t)), \end{aligned}$$

where $H(p_1, p_2) = p_1 + p_2 - p_1 p_2$. If we let $p = p_1 = p_2$, then $h(p) = 2p - p^2$. Its inverse function is $h^{-1}(x) = 1 - \sqrt{1 - x}$, for $0 \le x \le 1$. Hence, for $0 \le t$, we have

$$\bar{G}(t) = 1 - \sqrt{1 - \bar{F}_1(t) - \bar{F}_2(t) + \bar{F}_1(t)\bar{F}_2(t)},$$

$$G_{1:2}(t) = \left(1 - \sqrt{1 - \bar{F}_1(t) - \bar{F}_2(t) + \bar{F}_1(t)\bar{F}_2(t)}\right)^2,$$

$$G_{2:2}(t) = \bar{F}_1(t) + \bar{F}_2(t) - \bar{F}_1(t)\bar{F}_2(t) = H(\bar{F}_1(t), \bar{F}_2(t)).$$

Hence, the conditional reliability function is

$$\begin{aligned} P(T - t > x|T > t) &= \frac{H(\bar{F}_1(t + x), \bar{F}_2(t + x))}{H(\bar{F}_1(t), \bar{F}_2(t))} \\ &= P(Y_{2,2} - t > x|Y_{2,2} > t). \end{aligned}$$

Example 2.3.7 Consider a system with lifetime $T = \max\{X_1, \min\{X_2, X_3\}\}$ having INID component lifetimes X_i distributed according to $\bar{F}_i$, for $i = 1, 2, 3$. The signature vector for this system is $(0, \frac{2}{3}, \frac{1}{3})$. Then, for any $t > 0$, the reliability function is given by

$$\begin{aligned} P(T > t) &= P(\max\{X_1, \min\{X_2, X_3\}\} > t) \\ &= \bar{F}_1(t) + \bar{F}_2(t)\bar{F}_3(t) - \bar{F}_1(t)\bar{F}_2(t)\bar{F}_3(t) \\ &= H(\bar{F}_1(t), \bar{F}_2(t), \bar{F}_3(t)), \end{aligned}$$

and the conditional residual survival function is

$$P(T - t > x | T > t) = \frac{H(\bar{F}_1(t+x), \bar{F}_2(t+x), \bar{F}_3(t+x))}{H(\bar{F}_1(t), \bar{F}_2(t), \bar{F}_3(t))},$$

where $H(p_1, p_2, p_3) = p_1 + p_2 p_3 - p_1 p_2 p_3$. If let $p = p_1 = p_2 = p_3$, then $h(p) = p + p^2 - p^3$. The inverse function of this cubic equation is complicated, and so numerical methods need to be employed to compute the inverse function.

In what follows, we discuss some stochastic comparisons of two coherent systems with a common structure, but with different INID components. In order to obtain these results, we need the following corollary.

Corollary 2.3.8 *Let T_1, T_2 be the lifetimes of two coherent system with two sets of IID component lifetimes X_{1i}, X_{2i} $(i = 1, \ldots, n)$ having respective reliability functions $\bar{F}_1$, $\bar{F}_2$. Assume these two systems have a common signature $s = (s_1, \ldots, s_n)$. Then, if $\bar{F}_1 \leq_{st} \bar{F}_2$, then $\mathbf{s}_1(t) \leq_{st} \mathbf{s}_2(t)$, where $\mathbf{s}_1(t) = (s_{11}(t), \ldots, s_{1n}(t))$, $\mathbf{s}_2(t) = (s_{21}(t), \ldots, s_{2n}(t))$, with $s_{1i}(t) = s_i P(X_{1i,n} > t)/P(T_1 > t)$, $s_{2i}(t) = s_i P(X_{2i,n} > t)/P(T_2 > t)$, $X_{1i,n}$ being the i-th order statistic from X_{1i}, and $X_{2i,n}$ being the i-th order statistic from X_{2i} $(i = 1, \ldots, n)$.*

Proof We only need to show that for any $t \geq 0$ and for all $i = 1, \ldots, n$,

$$\sum_{j=i}^{n} s_{1j}(t) \leq \sum_{j=i}^{n} s_{2j}(t),$$

that is,

$$\sum_{j=i}^{n} s_j P(X_{1j,n} > t)/P(T_1 > t) \leq \sum_{j=i}^{n} s_j P(X_{2j,n} > t)/P(T_2 > t),$$

which is equivalent to

$$\sum_{j=i}^{n} \sum_{k=1}^{n} s_j s_k [P(X_{2j,n} > t)P(X_{1k,n} > t) - P(X_{2k,n} > t)P(X_{1j,n} > t)] \geq 0.$$

Note that

$$\sum_{j=i}^{n}\sum_{k=1}^{n} s_j s_k[\bar{F}_{2j,n}(t)\bar{F}_{1k,n}(t) - \bar{F}_{1j,n}(t)\bar{F}_{2k,n}(t)] = \sum_{j=i}^{n}\sum_{k=1}^{i} s_j s_k[\bar{F}_{2j,n}(t)\bar{F}_{1k,n}(t) - \bar{F}_{1j,n}(t)\bar{F}_{2k,n}(t)],$$

since

$$\sum_{j=i}^{n}\sum_{k=i}^{n} s_j s_k[\bar{F}_{2j,n}(t)\bar{F}_{1k,n}(t) - \bar{F}_{1j,n}(t)\bar{F}_{2k,n}(t)] = 0.$$

Hence, it is enough to show that, for $k \leq j$,

$$\bar{F}_{2j,n}(t)\bar{F}_{1k,n}(t) - \bar{F}_{1j,n}(t)\bar{F}_{2k,n}(t) \geq 0,$$

which is equivalent to

$$\frac{\sum_{m=n-k+1}^{n}\binom{n}{m}\phi_1^m(t)}{\sum_{l=n-j+1}^{n}\binom{n}{l}\phi_1^j(t)} \geq \frac{\sum_{m=n-k+1}^{n}\binom{n}{m}\phi_1^m(t)}{\sum_{l=n-j+1}^{n}\binom{n}{l}\phi_1^j(t)},$$

where $\phi_1(t) = \frac{\bar{F}_1(t)}{F_1(t)}$ and $\phi_2(t) = \frac{\bar{F}_2(t)}{F_2(t)}$. After some computations, it can be simplified to

$$\sum_{m=n-k+1}^{n-j}\sum_{l=n-j+1}^{n}\binom{n}{m}\binom{n}{l}\phi_1^m(t)\phi_2^m(t) \times [\phi_2^{l-m}(t) - \phi_1^{l-m}(t)] \geq 0.$$

From $\bar{F}_1 \leq_{st} \bar{F}_2$, it is easy to see that $\phi_2^{l-m}(t) - \phi_1^{l-m}(t) \geq 0$ for $m < l$, and hence the required result follows.

Theorem 2.3.9 *Let T_1 be the lifetime of a coherent system with independent component lifetimes X_{1i} $(i = 1, \ldots, n)$ having a reliability function $\bar{F}_{1i}$, and T_2 be the lifetime of another coherent system with independent component lifetimes X_{2i} $(i = 1, \ldots, n)$ having a reliability function $\bar{F}_{2i}$. Suppose these two systems have a common structure, and so the same signature. Let $\bar{G}_1(t) = h^{-1}H(\bar{F}_{11}(t), \ldots, \bar{F}_{1n}(t))$ and $\bar{G}_2(t) = h^{-1}H(\bar{F}_{21}(t), \ldots, \bar{F}_{2n}(t))$ be the corresponding mean reliability functions.*

(a) If $\bar{G}_1(t) \leq_{hr} \bar{G}_2(t)$, then

$$(T_1 - t \mid T_1 > t) \leq_{st} (T_2 - t \mid T_2 > t);$$

(b) If $\bar{G}_1(t) \leq_{hr} \bar{G}_2(t)$ and $xh'(x)/h(x)$ is decreasing, then

$$(T_1 - t \mid T_1 > t) \leq_{hr} (T_2 - t \mid T_2 > t);$$

(c) If $\bar{G}_1(t) \le_{rh} \bar{G}_2(t)$, $\bar{G}_1(\bar{G}_1^{-1}(y) - x) \ge \bar{G}_2(\bar{G}_2^{-1}(y) - x)$ *for* $\min\{\bar{G}_1^{-1}(y) - x, \bar{G}_2^{-1}(y) - x\} > 0$, *and* $(1-y)h'(y)/\{h[\bar{G}_1(\bar{G}_1^{-1}(y) - x)] - h(y)\}$ *or* $(1-y)h'(y)/\{h[\bar{G}_2(\bar{G}_2^{-1}(y) - x)] - h(y)\}$ *is increasing in* y *for any* $x > 0$, *then*

$$(T_1 - t \mid T_1 > t) \le_{rh} (T_2 - t \mid T_2 > t),$$

where $\bar{G}_i^{-1}(t) = \sup\{x : \bar{G}_i(x) > t\}$, $i = 1, 2$;

(d) If $\bar{G}_1(t) \le_{lr} \bar{G}_2(t)$ *and* $h'(x)/h'(\bar{G}_1(\bar{G}_2^{-1}(x)))$ *is increasing, then*

$$(T_1 - t \mid T_1 > t) \le_{lr} (T_2 - t \mid T_2 > t).$$

Proof (a) For any $t > 0$ and $x > 0$, we have

$$\begin{aligned}
P(T_1 - t > x \mid T_1 > t) &= \sum_{i=1}^{n} s_{1i}(t) P(Y_{1i,n} - t > x \mid Y_{1i,n} > t) \\
&\le \sum_{i=1}^{n} s_{2i}(t) P(Y_{1i,n} - t > x \mid Y_{1i,n} > t) \\
&\le \sum_{i=1}^{n} s_{2i}(t) P(Y_{2i,n} - t > x \mid Y_{2i,n} > t) \\
&= P(T_2 - t > x \mid T_2 > t),
\end{aligned}$$

where $s_{1i}(t) = s_i P(Y_{1i,n} > t)/P(T_1 > t)$, $s_{2i}(t) = s_i P(Y_{2i,n} > t)/P(T_2 > t)$ for $i = 1, \ldots, n$, given $P(T_j > t) > 0$, $j = 1, 2$, and $Y_{ji,n}$ is the i-th order statistic from a set of IID random variables $Y_{j1}, \ldots, Y_{jn}$ with a common reliability function $\bar{G}_j(t) = h^{-1}H(\bar{F}_{j1}(t), \ldots, \bar{F}_{jn}(t))$, $j = 1, 2$. The first inequality follows from Corollary 2.3.8, and the second one holds since $\bar{G}_1(t) \le_{hr} \bar{G}_2(t)$.

(b) Let $p_i(t) = \bar{G}_i(t)$ for $t > 0, i = 1, 2$. Let h_1 and h_2 denote the hazard rate functions corresponding to $\bar{G}_1$ and $\bar{G}_2$, respectively, and $h_{T_1-t|T_1>t}(x)$ and $h_{T_2-t|T_2>t}(x)$ be the hazard rate functions of $(T_1 - t \mid T_1 > t)$ and $(T_2 - t \mid T_2 > t)$, respectively. Then, $P(T_i - t > x \mid T_i > t) = \frac{h(p_i(t+x))}{h(p_i(t))}$, $i = 1, 2$. Note that

$$\begin{aligned}
h_{T_1-t|T_1>t}(x) &= h_1(t + x) \times \frac{y_1 h'(y_1)}{h(y_1)} \\
&\ge h_2(t + x) \times \frac{y_1 h'(y_1)}{h(y_1)} \quad (\bar{G}_1(t) \le_{hr} \bar{G}_2(t)) \\
&\ge h_2(t + x) \times \frac{y_2 h'(y_2)}{h(y_2)} \quad (xh'(x)/h(x) \downarrow \text{ in } x) \\
&= h_{T_2-t|T_2>t}(x),
\end{aligned}$$

where $y_i = \bar{G}_i(t + x)$, $i = 1, 2$. Thus the result holds.

(c) If we set $y_i = \bar{G}_i(t + x)$ for $t > 0$, $i = 1, 2$, then $p_i(t) = \bar{G}_i(\bar{G}_i^{-1}(y_i) - x)$. Let r_1 and r_2 denote the reversed hazard rate functions corresponding to $\bar{G}_1$ and $\bar{G}_2$, respectively, and $r_{T_1-t|T_1>t}(x)$ and $r_{T_2-t|T_2>t}(x)$ be the reversed hazard rate functions of $(T_1 - t \mid T_1 > t)$ and $(T_2 - t \mid T_2 > t)$, respectively. Then, we have

$$\begin{aligned} r_{T_1-t|T_1>t}(x) &= r_1(t+x) \times \frac{(1-y_1)h'(y_1)}{h(p_1(t)) - h(y_1)} \\ &= r_1(t+x) \times \frac{(1-y_1)h'(y_1)}{h[\bar{G}_1(\bar{G}_1^{-1}(y_1)-x)] - h(y_1)} \\ &\leq r_2(t+x) \times \frac{(1-y_1)h'(y_1)}{h[\bar{G}_1(\bar{G}_1^{-1}(y_1)-x)] - h(y_1)} \\ &\leq r_2(t+x) \times \frac{(1-y_2)h'(y_2)}{h[\bar{G}_1(\bar{G}_1^{-1}(y_2)-x)] - h(y_2)} \\ &\leq r_2(t+x) \times \frac{(1-y_2)h'(y_2)}{h[\bar{G}_2(\bar{G}_2^{-1}(y_2)-x)] - h(y_2)} \\ &= r_{T_2-t|T_2>t}(x), \end{aligned}$$

where the first inequality follows from $\bar{G}_1(t) \leq_{rh} \bar{G}_2(t)$, the second one from the fact that $(1-y)h'(y)/\{h[\bar{G}_1(\bar{G}_1^{-1}(y)-x)] - h(y)\}$ is increasing in y, and the last one holds from the fact $\bar{G}_1(\bar{G}_1^{-1}(y)-x) \geq \bar{G}_2(\bar{G}_2^{-1}(y)-x)$. Thus the result holds.

The proof of (d) is quite easy and therefore is omitted.

The following example shows that the condition $\bar{G}_1(\bar{G}_1^{-1}(y)-x) \geq \bar{G}_2(\bar{G}_2^{-1}(y)-x)$ of (c) can not be deleted.

Example 2.3.10 Let Y_1 and Y_2 be distributed as two parameter Weibull with respective positive scale parameters λ_1, λ_2, and a common shape parameter $k = 2$, that is, $\bar{G}_1(t) = e^{-(\lambda_1 t)^2}$ and $\bar{G}_2(t) = e^{-(\lambda_2 t)^2}$, for $t > 0$. Then, for $0 < t < 1$ and $x > 0$, we have $\bar{G}_i^{-1}(t) - x = \frac{1}{\lambda_i}\sqrt{-\ln t} - x$ for $i = 1, 2$, so that, $\bar{G}_i(\bar{G}_i^{-1}(t) - x) = te^{-(\lambda_i x)^2} e^{2\lambda_i x\sqrt{-\ln t}}$. We then obtain

$$\begin{aligned} &\bar{G}_2(\bar{G}_2^{-1}(t)-x) - \bar{G}_1(\bar{G}_1^{-1}(t)-x) \\ &= te^{-(\lambda_1 x)^2} e^{2\lambda_1 x\sqrt{-\ln t}} \\ &\times [e^{-(\lambda_2 x)^2 + (\lambda_1 x)^2} e^{2\lambda_2 x\sqrt{-\ln t} - 2\lambda_1 x\sqrt{-\ln t}} - 1] \\ &= te^{-(\lambda_1 x)^2} e^{2\lambda_1 x\sqrt{-\ln t}} \\ &\times [e^{(\lambda_1-\lambda_2)[(\lambda_1+\lambda_2)x^2 - 2x\sqrt{-\ln t}]} - 1]. \end{aligned}$$

Clearly, $\bar{G}_1(t) \leq_{rh} \bar{G}_2(t)$ does not guarantee that the last term above in the above expression is always nopositive or nonnegative for $0 < t < 1$ and $x > 0$.

The following example shows that the condition $(1-y)h'(y)/\{h[\bar{G}_1(\bar{G}_1^{-1}(y) - x)] - h(y)\}$ or $(1-y)h'(y)/\{h[\bar{G}_2(\bar{G}_2^{-1}(y) - x)] - h(y)\}$ is increasing in y for any $x > 0$ in (c) also can not be deleted.

Example 2.3.11 Consider two systems with respective lifetimes $T_1 = \min\{X_{11}, X_{12}, X_{13}\}$ and $T_2 = \min\{X_{21}, X_{22}, X_{23}\}$, where X_{ij} is distributed as Exponential with parameter λ_{ij} (> 0) for $i = 1, 2$ and $j = 1, 2, 3$. Suppose $\bar{G}_i(t) = h^{-1}H(\bar{F}_{i1}(t), \bar{F}_{i2}(t), \bar{F}_{i3}(t))$ is the reliability function of IID component lifetimes Y_{i1}, Y_{i2}, Y_{i3}, for $i = 1, 2$. It can be computed from (12), $\bar{G}_i(t) = \sqrt[3]{e^{-(\lambda_{i1}+\lambda_{i2}+\lambda_{i3})t}}$, for $i = 1, 2$.

Clearly, $\lambda_{11} + \lambda_{12} + \lambda_{13} \geq \lambda_{21} + \lambda_{22} + \lambda_{23}$ means $Y_{11} \leq_{rh} Y_{21}$. However, it is easily seen that

$$\frac{(1-y)h'(y)}{h[\bar{G}_1(\bar{G}_1^{-1}(y) - x)] - h(y)} = \frac{3(1-y)}{y(e^{(\lambda_{11}+\lambda_{12}+\lambda_{13})x} - 1)}$$

is decreasing in $y \geq 0$ for any $x \geq 0$.

In the following theorem, we present a stochastic comparison of two coherent systems with different structures and different sets of INID components.

Theorem 2.3.12 *Let $T_1 = \phi_1(X_{11}, \cdots, X_{1n})$ be the lifetime of a coherent system with independent component lifetimes X_{1i} $(i = 1, \ldots, n)$ having a reliability distribution $\bar{F}_{1i}$, with a coefficient vector $\mathbf{s}_1(t) = (s_{11}(t), \ldots, s_{1n}(t))$. Also, let $T_2 = \phi_2(X_{21}, \cdots, X_{2n})$ be the lifetime of another coherent system with independent component lifetimes X_{2i} $(i = 1, \ldots, n)$ having a reliability distribution $\bar{F}_{2i}$, with a coefficient vector $\mathbf{s}_2(t) = (s_{21}(t), \ldots, s_{2n}(t))$. Let h_1 and h_2 be the reliability polynomials, and H_1 and H_2 be the structure reliability functions, respectively. Let $\bar{G}_1(t) = h_1^{-1}H_1(\bar{F}_{11}(t), \ldots, \bar{F}_{1n}(t))$, and $\bar{G}_2(t) = h_2^{-1}H_2(\bar{F}_{21}(t), \ldots, \bar{F}_{2n}(t))$ for all $t \geq 0$. Then if $\mathbf{s}_1(t) \leq_{st} \mathbf{s}_2(t)$ and $\bar{G}_1(t) \leq_{hr} \bar{G}_2(t)$, we have $(T_1 - t \mid T_1 > t) \leq_{st} (T_2 - t \mid T_2 > t)$.*

Proof Let $Y_{11,n}, \ldots, Y_{1n,n}$ be the order statistics obtained from the IID random variables $Y_{11}, \ldots, Y_{1n}$ having reliability function $\bar{G}_1(t)$, and $Y_{21,n}, \ldots, Y_{2n,n}$ be the order statistics obtained from $Y_{21}, \ldots, Y_{2n}$ having reliability function $\bar{G}_2(t)$. Then, for any $t > 0$ and $x > 0$, the reliability functions of the random variables $(T_1 - t \mid T_1 > t)$ and $(T_2 - t \mid T_2 > t)$ can be expressed as

$$\begin{aligned} P(T_1 - t > x \mid T_1 > t) &= \sum_{i=1}^{n} s_{1i}(t) P(Y_{1i,n} - t > x \mid Y_{1i,n} > t) \\ &= \sum_{i=1}^{n} s_{1i}(t) \bar{G}_{1i,n}(x|t), \end{aligned}$$

and

$$\begin{aligned} P(T_2 - t > x \mid T_2 > t) &= \sum_{i=1}^{n} s_{2i}(t) P(Y_{2i,n} - t > x \mid Y_{2i,n} > t) \\ &= \sum_{i=1}^{n} s_{2i}(t) \bar{G}_{2i,n}(x|t), \end{aligned}$$

where $\bar{G}_{1i,n}(x|t) = P(Y_{1i,n} - t > x \mid Y_{1i,n} > t)$ and $\bar{G}_{2i,n}(x|t) = P(Y_{2i,n} - t > x \mid Y_{2i,n} > t)$.

From Theorem 1.A.6 of Shaked and Shanthikumar (2007), $\mathbf{s}_1(t) \leq_{st} \mathbf{s}_2(t)$ implies that

$$\sum_{i=1}^{n} s_{1i}(t)\bar{G}_{1i,n}(x|t) \leq \sum_{i=1}^{n} s_{2i}(t)\bar{G}_{1i,n}(x|t).$$

Also by Theorem 1.B.34 of Shaked and Shanthikumar (2007), the condition $\bar{G}_1(t) \leq_{hr} \bar{G}_2(t)$ implies that $\bar{G}_{1i,n}(t) \leq_{hr} \bar{G}_{2i,n}(t)$ for $i = 1, \ldots, n$, and so $\bar{G}_{1i,n}(x|t) \leq_{hr} \bar{G}_{2i,n}(x|t)$. This means that $\bar{G}_{1i,n}(x|t) \leq_{st} \bar{G}_{2i,n}(x|t)$, which in turn implies

$$\sum_{i=1}^{n} s_{2i}(t)\bar{G}_{1i,n}(x|t) \leq \sum_{i=1}^{n} s_{2i}(t)\bar{G}_{2i,n}(x|t),$$

as required.

Example 2.3.13 Consider two coherent systems with lifetimes $T_1 = \min\{X_{11}, \max\{X_{12}, X_{13}\}\}$ and $T_2 = \max\{X_{21}, \min\{X_{22}, X_{23}\}\}$. Let X_{i1}, X_{i2}, X_{i3} be distributed according to reliability functions $\bar{F}_{i1}, \bar{F}_{i2}, \bar{F}_{i3}$, respectively, for $i = 1, 2$. Assume $\bar{G}_i(t)$ is the corresponding mean reliability function of IID component lifetimes Y_{i1}, Y_{i2}, Y_{i3}, for $i = 1, 2$. After some computations, it can be shown that the coefficient vectors of the systems T_1 and T_2 are

$$\mathbf{s}_1(t) = \left(\frac{\bar{G}_1(t)}{6 - 3\bar{G}_1(t)}, \frac{6 - 4\bar{G}_1(t)}{6 - 3\bar{G}_1(t)}, 0\right),$$

$$\mathbf{s}_2(t) = \left(0, \frac{6\bar{G}_2(t) - 4\bar{G}_2^2(t)}{3 + 3\bar{G}_2(t) - 3\bar{G}_2^2(t)}, \frac{3 - 3\bar{G}_2(t) + \bar{G}_2^2(t)}{3 + 3\bar{G}_2(t) - 3\bar{G}_2^2(t)}\right),$$

respectively.

Clearly, $\mathbf{s}_1(t) \leq_{st} \mathbf{s}_2(t)$ for any $t \geq 0$. If let Y_{11} and Y_{21} be distributed as Exponential with parameters 2 and 1, respectively, then, $Y_{11} \leq_{hr} Y_{21}$. From Theorem 6, for any $t \geq 0$ and $x > 0$, we have

$$\begin{aligned} P(T_1 - t > x \mid T_1 > t) &= \sum_{i=1}^{n} s_{1i}(t)P(Y_{1i,n} - t > x \mid Y_{1i,n} > t) \\ &= \frac{2e^{-6x-8t} - 3e^{-6t-6x} + 6e^{-4t-4x} - 4e^{-6t-4x}}{(2 - e^{-2t})(3e^{-4t} - 2e^{-6t})}, \\ P(T_2 - t > x \mid T_2 > t) \end{aligned}$$

$$= \sum_{i=1}^{n} s_{2i}(t) P(Y_{2i,n} - t > x \mid Y_{2i,n} > t)$$

$$= \frac{(6e^{-t} - 4e^{-2t})(3e^{-2t-2x} - 2e^{-3t-3x})}{(3 + 3e^{-t} - 3e^{-2t})(3e^{-2t} - 2e^{-3t})}$$

$$+ \frac{(3 - e^{-t} + e^{-2t})(3e^{-t-x} - 3e^{-2t-2x} + e^{-3t-3x})}{(3 + 3e^{-t} - 3e^{-2t})(3e^{-t} - 3e^{-2t} + e^{-3t})}.$$

Using Mathematica software, it can be checked that $P(T_2 - t > x \mid T_2 > t) \geq P(T_1 - t > x \mid T_1 > t)$, for all $x > 0$, for each fixed $t \geq 0$; that is, $(T_2 - t > x \mid T_2 > t) \geq_{st} (T_1 - t > x \mid T_1 > t)$.

Next, we show that the expressions in (2.25) and (2.26) can be extended to coherent systems with INID components.

Theorem 2.3.14 *Let $\mathbf{a} = (a_1, \ldots, a_n)$ and $\mathbf{b} = (b_1, \ldots, b_n)$ be the minimal and maximal signatures of an n-component system with independent component lifetimes $X_1, \ldots, X_n$. Then, the reliability function of the residual lifetime $T - t$ at time t can be expressed as*

$$P(T - t > x | T > t) = \sum_{i=1}^{n} a_i(t) P(Y_{1,i} - t > x | Y_{1,i} > t)$$

$$= \sum_{i=1}^{n} a_i(t) \bar{G}_{1,i}(x|t)$$

$$= \sum_{i=1}^{n} b_i(t) P(Y_{i,i} - t > x | Y_{i,i} > t)$$

$$= \sum_{i=1}^{n} b_i(t) \bar{G}_{i,i}(x|t),$$

where $Y_{1,n}, \ldots, Y_{n,n}$ are the order statistics obtained from IID random variables $Y_1, \ldots, Y_n$ having a common reliability function $\bar{G}(t) = h^{-1}H(\bar{F}_1(t), \ldots, \bar{F}_n(t))$, and $a_i(t) = \frac{a_i P(Y_{1,i}>t)}{P(T>t)}$, $b_i(t) = \frac{b_i P(Y_{i,i}>t)}{P(T>t)}$ with $\sum_{i=1}^{n} a_i(t) = 1$ and $\sum_{i=1}^{n} b_i(t) = 1$. Some of these coefficients may be negative.

Theorem 2.3.14 shows that the residual life $(T - t|T > t)$ of the coherent system with INID components at time t may be represented as a mixture of the residual lifetimes $(Y_{1,i} - t|Y_{1,i} > t)$ of the smallest order statistics from a set of IID component lifetimes with coefficients $a_i(t)$ for $i = 1, \ldots, n$, or a mixture of the residual lifetimes $(Y_{i,i} - t|Y_{i,i} > t)$ of the largest order statistics from a set of IID component lifetimes with coefficients $b_i(t)$ for $i = 1, \ldots, n$. Also, the coefficients $a_i(t)$ and $b_i(t)$ depend on H and $\bar{F}_i$, for $i = 1, \ldots, n$.

2.3.1.3 Conclusions

This section presents several useful mixture representations for the reliability function of the residual life of a coherent system with independent but non-identically distributed components. These new results are used to compare the residual lives of coherent systems under different conditions based on the stochastic ordering of coefficient vectors (or components). The results form extensions of some known results when component lifetimes are independent and identically distributed. Furthermore, the utility of the results is illustrated in several examples in which the systems' residual reliabilities are computed and compared. For the case of systems with dependent and non-identically distributed components, it is a very difficult problem. This will be worth further study and discussion in the future.

2.3.2 The Inactivity Time of General Coherent Systems

In this part, several useful mixture representations are presented for the reliability function of the inactivity time of systems with heterogeneous components based on order statistics, (pseudo) signatures and mean reliability functions. Some stochastic comparisons of inactivity times between two systems are discussed. These results form nice extensions of some existing results for the case when the components are independent and identically distributed.

2.3.2.1 Introduction

Zhang (2010b) proved that the distribution of the system's inactivity time $t - T$, given $T \leq t$, is a mixture of the distribution of the inactivity times of $(n - i + 1)$-out-of-n systems, that is,

$$P(t - T > x \mid T \leq t) = \sum_{i=1}^{n} s_i(t) P(X_{i,n} \leq t - x \mid X_{i,n} \leq t), \tag{2.29}$$

where $s_i(t) = s_i P(X_{i,n} \leq t) / F_T(t)$, given $F_T(t) > 0$. Zhang (2010b) also proved

$$P(t - T > x \mid T \leq t) = \sum_{i=1}^{n} a_i(t) P(X_{1,i} \leq t - x \mid X_{1,i} \leq t), \tag{2.30}$$

$$P(t - T > x \mid T \leq t) = \sum_{i=1}^{n} b_i(t) P(X_{i,i} \leq t - x \mid X_{i,i} \leq t), \tag{2.31}$$

where $a_i(t) = a_i P(X_{1,i} \leq t)/F_T(t)$, $b_i(t) = b_i P(X_{i,i} \leq t)/F_T(t)$, given $F_T(t) > 0$. The coefficient vectors $\mathbf{a}(t) = (a_1(t), \ldots, a_n(t))$ and $\mathbf{b}(t) = (b_1(t), \ldots, b_n(t))$, with $\sum_{i=1}^n a_i(t) = 1$ and $\sum_{i=1}^n b_i(t) = 1$, may be called the conditional minimal, and conditional maximal signature vectors, respectively.

Based on the expression (2.29), Zhang (2010b) has established the following results.

Theorem 2.3.15 *Let T_i be the lifetime of a coherent system with n IID component lifetimes distributed as F, and $\mathbf{s}_i(t) = (s_{i1}(t), \ldots, s_{in}(t))$ as the coefficient vector described in* (2.29), $i = 1, 2$.

(a) If $\mathbf{s}_1(t) \leq_{st} \mathbf{s}_2(t)$, then $(t - T_1 \mid T_1 \leq t) \geq_{st} (t - T_2 \mid T_2 \leq t)$;
(b) If $\mathbf{s}_1(t) \leq_{rh} \mathbf{s}_2(t)$, then $(t - T_1 \mid T_1 \leq t) \geq_{hr} (t - T_2 \mid T_2 \leq t)$;
(c) If $\mathbf{s}_1(t) \leq_{lr} \mathbf{s}_2(t)$, then $(t - T_1 \mid T_1 \leq t) \geq_{lr} (t - T_2 \mid T_2 \leq t)$.

Recently, the lifetimes of coherent systems with heterogenous (independent but not identically distributed components have been considered by Navarro et al. (2011). They presented representations for the reliability of these systems as given in the following theorem.

Theorem 2.3.16 *Let T be the lifetime of a coherent system having heterogeneous component lifetimes $X_1, \ldots, X_n$ distributed according to reliability functions $\bar{F}_1, \ldots, \bar{F}_n$. Assume that h and H are the system's reliability polynomial and reliability structure function, respectively, and $\mathbf{s} = (s_1, \ldots, s_n)$, $\boldsymbol{a} = (a_1, \ldots, a_n)$ and $\boldsymbol{b} = (b_1, \ldots, b_n)$ are the corresponding signature, minimal signature and maximal signature vectors. Then,*

$$P(T \leq t) = \sum_{i=1}^{n} s_i P(Y_{i,n} \leq t) = \sum_{i=1}^{n} s_i G_{i,n}(t), \tag{2.32}$$

$$P(T \leq t) = \sum_{i=1}^{n} a_i P(Y_{1,i} \leq t) = \sum_{i=1}^{n} a_i G_{1,i}(t), \tag{2.33}$$

$$P(T \leq t) = \sum_{i=1}^{n} b_i P(Y_{i,i} \leq t) = \sum_{i=1}^{n} b_i G_{i,i}(t), \tag{2.34}$$

where $Y_{1,n}, \ldots, Y_{n,n}$ are the order statistics of IID component lifetimes $Y_1, \ldots, Y_n$ with a reliability

$$\bar{G}(t) = h^{-1}(H(\bar{F}_1(t), \ldots, \bar{F}_n(t))). \tag{2.35}$$

According to the properties of h and H, the function $\bar{G}$ satisfies the properties of a reliability function and is therefore a proper reliability function; see Navarro et al. (2011). The function $\bar{G}(t)$ can be called the mean reliability function associated with the system structure and the components' reliability function; see Navarro and Spizzichino (2010) for more details.

Some stochastic comparisons of systems with heterogeneous components were obtained by Navarro et al. (2011), which extended the existing results for the case when the components are IID. From Theorem 2.2 in Navarro et al. (2011), the result below can be obtained immediately.

Theorem 2.3.17 *Let T_i be the lifetime of a coherent system having heterogeneous component lifetimes $X_{i1}, \ldots, X_{in}$ distributed as $\bar{F}_{i1}, \ldots, \bar{F}_{in}$, and h_i and H_i be the system's reliability polynomial and structure function, respectively, $i = 1, 2$. Let $\bar{G}_i(t) = h_i^{-1} H_i(\bar{F}_{i1}(t), \ldots, \bar{F}_{in}(t))$ be the reliability function of IID variables $Y_{i1}, \ldots, Y_{in}$, $i = 1, 2$. Then, we have:*

(a) If $Y_{11} \leq_{st} Y_{21}$, then $T_1 \leq_{st} T_2$;

(b) If $Y_{11} \leq_{hr} Y_{21}$ and $xh_1'(x)/h_1(x)$ or $xh_2'(x)/h_2(x)$ is decreasing, then $T_1 \leq_{hr} T_2$;

(c) If $Y_{11} \leq_{rh} Y_{21}$ and $(1-x)xh_1'(x)/(1-h_1(x))$ or $(1-x)h_2'(x)/(1-h_2(x))$ is increasing, then $T_1 \leq_{rh} T_2$.

Following this line of work, we study here the inactivity time of a coherent system with heterogeneous components. Specifically, we present several mixture representations for the reliability function of the inactivity time. Some stochastic properties for the coefficient vector are discussed as well. Also, some stochastic comparisons based on the inactivity times of two coherent systems are conducted. These results form nice extensions of some existing results for the case when the components are IID. Finally, some examples with practical applications of the results are presented.

2.3.2.2 Mixture Representations of Inactivity Time

In this section, we give some mixture representations of the inactivity time of coherent systems with heterogeneous components based on order statistics, signature and mean reliability function.

Theorem 2.3.18 *Let T be the lifetime of a coherent system having heterogeneous component lifetimes $X_1, \ldots, X_n$ distributed according to reliability functions $\bar{F}_1, \ldots, \bar{F}_n$, respectively, and h and H be the system's reliability polynomial and reliability structure function, respectively. Further, let $\bar{G}(t) = h^{-1} H(\bar{F}_1(t), \ldots, \bar{F}_n(t))$ be the reliability function of IID component lifetimes $Y_1, \ldots, Y_n$. Assuming that $s = (s_1, \ldots, s_n)$ is the signature vector of T, then, for any $t \geq x > 0$,*

$$P(t - T > x \mid T \leq t) = \sum_{i=1}^{n} s_i(t) P(t - Y_{i,n} > x \mid Y_{i,n} \leq t), \tag{2.36}$$

where coefficients $s_i(t) = s_i P(Y_{i,n} \leq t)/P(T \leq t)$ for $i = 1, \ldots, n$, given $P(T \leq t) > 0$, such that $\sum_{i=1}^{n} s_i(t) = 1$, and $Y_{i,n}$ is the i-th order statistic among $Y_1, \ldots, Y_n$.

Proof From Theorem 2.3.16, for any $t > x > 0$,

$$\begin{aligned}P(t - T > x \mid T \le t) &= \frac{\sum_{i=1}^{n} s_i P(Y_{i,n} \le t - x)}{P(T \le t)} \\ &= \frac{\sum_{i=1}^{n} s_i P(Y_{i,n} \le t - x | Y_{i,n} \le t) \times P(Y_{i,n} \le t)}{P(T \le t)} \\ &= \sum_{i=1}^{n} s_i(t) P(t - Y_{i,n} > x \mid Y_{i,n} \le t).\end{aligned}$$

Hence, the result holds.

The result shows that the inactivity time $(t - T|T \le t)$ of the coherent system with heterogeneous components at time t can be represented as a mixture of the inactivity times $(t - Y_{i,n}|Y_{i,n} \le t)$ of $(n - i + 1)$-out-of-n systems having a set of IID component lifetimes with coefficients $s_i(t)$ for $i = 1, \ldots, n$. The coefficients $s_i(t)$ depend on both H and F_i for $i = 1, \ldots, n$.

Example 2.3.19 Consider coherent systems with three heterogeneous components having lifetimes X_i distributed according to $\bar{F}_i$, $i = 1, 2, 3$. First, we investigate the system with lifetime $T_1 = \min\{X_1, X_2, X_3\}$, whose signature vector is $(1, 0, 0)$. Then, for any $t > 0$, the reliability function is given by

$$\begin{aligned}P(T_1 > t) &= P(\min\{X_1, X_2, X_3\} > t) \\ &= \bar{F}_1(t)\bar{F}_2(t)\bar{F}_3(t) \\ &= H_1(\bar{F}_1(t), \bar{F}_2(t), \bar{F}_3(t)),\end{aligned}$$

where the reliability structure function $H_1(p_1, p_2, p_3) = p_1 p_2 p_3$. If we let $p = p_1 = p_2 = p_3$, then the reliability polynomial $h_1(p) = p^3$. The inverse function of h_1 is $h_1^{-1}(x) = \sqrt[3]{x}$, for $0 \le x \le 1$. Hence, for $0 \le t \le 1$, the corresponding mean reliability function is

$$\bar{G}_1(t) = \sqrt[3]{\bar{F}_1(t)\bar{F}_2(t)\bar{F}_3(t)}.$$

Then, the reliability function of the inactivity time of the system is

$$P(t - T_1 > x|T_1 \le t) = s_1(t)P(t - Y_{1,3} > x|Y_{1,3} \le t) = \frac{1 - H_1(\bar{F}_1(t - x), \bar{F}_2(t - x), \bar{F}_3(t - x))}{1 - H_1(\bar{F}_1(t), \bar{F}_2(t), \bar{F}_3(t))}.$$

For the system with lifetime $T_2 = \max\{X_1, X_2, X_3\}$, whose signature vector is $(0, 0, 1)$, for any $t > 0$, it can be shown that the reliability function is

$$\begin{aligned}P(T_2 > t) &= \Sigma_{i=1}^{3} \bar{F}_i(t) - \Sigma_{1 \le i < j \le 3} \bar{F}_i(t)\bar{F}_j(t) + \bar{F}_1(t)\bar{F}_2(t)\bar{F}_3(t) \\ &= H_2(\bar{F}_1(t), \bar{F}_2(t), \bar{F}_3(t)).\end{aligned}$$

This system's reliability polynomial is $h_2(p) = p^3 - 3p^2 + 3p$, and $h_2^{-1}(x) = 1 + \sqrt[3]{x-1}$, for $0 \le x \le 1$. Hence, the corresponding mean reliability function is

$$\bar{G}_2(t) = 1 + \sqrt[3]{H_2(\bar{F}_1(t), \bar{F}_2(t), \bar{F}_3(t)) - 1}.$$

For the systems $T_3 = X_{2,3}$, $T_4 = \max\{X_1, \min\{X_2, X_3\}\}$, and $T_5 = \min\{X_1, \max\{X_2, X_3\}\}$, the corresponding reliability polynomials are $h_3(p) = 3p^2 - 2p^3$, $h_4(p) = p + p^2 - p^3$, and $h_5(p) = 2p^2 - p^3$ for $0 \le p \le 1$, but their inverse functions are complicated.

In the following, we give two other mixture representations for the inactivity time based on conditional minimal and maximum signatures, showing that the expressions in (2.33) and (2.34) can be extended to coherent systems with heterogeneous components.

Theorem 2.3.20 *Let T be the lifetime of a coherent system having heterogeneous component lifetimes $X_1, \ldots, X_n$ distributed according to reliability functions $\bar{F}_1, \ldots, \bar{F}_n$, respectively, and h and H be the system's reliability polynomial and reliability structure function, respectively. Let $\bar{G}(t) = h^{-1}H(\bar{F}_1(t), \ldots, \bar{F}_n(t))$ be the reliability function of IID component lifetimes $Y_1, \ldots, Y_n$. Assuming that $\boldsymbol{a} = (a_1, \ldots, a_n)$ and $\boldsymbol{b} = (b_1, \ldots, b_n)$ are the minimal and maximal signature vectors of T, the reliability function of the inactivity time $t - T$, given $T \le t$, can be expressed as*

$$\begin{aligned} P(t - T > x | T \le t) &= \sum_{i=1}^{n} a_i(t) P(t - Y_{1,i} > x | Y_{1,i} \le t) \\ &= \sum_{i=1}^{n} b_i(t) P(t - Y_{i,i} > x | Y_{i,i} \le t), \end{aligned}$$

where $Y_{1,n}, \ldots, Y_{n,n}$ are the order statistics among $Y_1, \ldots, Y_n$, and $a_i(t) = a_i P(Y_{1,i} \le t)/P(T \le t)$, $b_i(t) = b_i P(Y_{i,i} \le t)/P(T \le t)$, given $P(T \le t) > 0$, such that $\sum_{i=1}^{n} a_i(t) = 1$ and $\sum_{i=1}^{n} b_i(t) = 1$. Clearly, some of these coefficients may be negative.

These results show that the reliability function of the inactivity time $(t - T | T \le t)$ of the coherent system with heterogeneous components at time t may be represented as a mixture of the reliability function of the inactivity time $(t - Y_{1,i} | Y_{1,i} \le t)$ of series systems from a set of IID component lifetimes with coefficients $a_i(t)$ for $i = 1, \ldots, n$, or a mixture of the reliability function of the inactivity time $(t - Y_{i,i} | Y_{i,i} \le t)$ of parallel systems from a set of IID component lifetimes with coefficients $b_i(t)$ for $i = 1, \ldots, n$. Also, the coefficients $a_i(t)$ and $b_i(t)$ depend on both H and F_i for $i = 1, \ldots, n$.

Example 2.3.21 Consider the system with lifetime $T = \max\{X_1, X_2, X_3\}$ having heterogeneous component lifetimes X_1, X_2, X_3 distributed according to reliability functions $\bar{F}_1, \bar{F}_2, \bar{F}_3$, respectively. From Navarro and Eryilmaz (2007), the minimal signature and maximum signature vectors of this system are $(3, -3, 1)$ and $(0, 0, 1)$, respectively. According to Example 2.2, the corresponding mean reliability function is

$$\bar{G}(t) = 1 + \sqrt[3]{H(\bar{F}_1(t), \bar{F}_2(t), \bar{F}_3(t)) - 1}$$
$$= 1 + \sqrt[3]{\Sigma_{i=1}^{3}\bar{F}_i(t) - \Sigma_{1\le i<j\le 3}\bar{F}_i(t)\bar{F}_j(t) + \bar{F}_1(t)\bar{F}_2(t)\bar{F}_3(t) - 1}.$$

Zhang (2010b) obtained the conditional minimal signature and maximum signature vectors in this case as $\mathbf{a}(t) = \left(\frac{3}{G^2(t)}, \frac{3G(t)-6}{G^2(t)}, \frac{3-3G(t)+G^2(t)}{G^2(t)}\right)$ and $\mathbf{b}(t) = (0, 0, 1)$, respectively. Thus, from Theorem 2.3,

$$P(t - T > x|T \le t) = \frac{3}{G^2(t)}P(t - Y_{1,1} > x|Y_{1,1} \le t) + \frac{3G(t) - 6}{G^2(t)}P(t - Y_{1,2} > x|Y_{1,2} \le t)$$
$$+ \frac{3 - 3G(t) + G^2(t)}{G^2(t)}P(t - Y_{1,3} > x|Y_{1,3} \le t)$$
$$= P(t - Y_{3,3} > x|Y_{3,3} \le t),$$

where $Y_{1,3}, ..., Y_{3,3}$ are the order statistics among Y_1, Y_2, Y_3 with distribution function $G(t) = 1 - \bar{G}(t)$.

We now discuss some stochastic properties of the coefficient vector $\mathbf{s}(t) = (s_1(t), \ldots, s_n(t))$. It can be shown that $\mathbf{s}(t)$ has tail stochastic behavior when the lifetimes of the n components are heterogeneous, which is an extension of the case when the components are IID, as described by Zhang (2010b). Specifically, if T is the lifetime of a coherent system with a signature vector $\mathbf{s} = (0, \ldots, 0, s_j, \ldots, s_n)$, such that $s_j > 0$ for $j \in \{1, 2, \ldots, n\}$, then $\lim_{t\to 0}\mathbf{s}(t) = (\underbrace{0, 0, \ldots, 0}_{j-1 \text{ times}}, 1, \underbrace{0, 0, \ldots, 0}_{n-j \text{ times}})$. Moreover, it can be shown that $\mathbf{s}(t_1) \ge_{st} \mathbf{s}(t_2)$ for $t_1 \le t_2$ as known in the case when the components are IID, which in turn means that $(t_1 - T|T \le t_1) \ge_{st} (t_2 - T|T \le t_2)$ for $t_1 \le t_2$.

2.3.2.3 Stochastic Comparisons of Inactivity Times

At first, we give the stochastic order of two coefficient vectors based on the usual stochastic order of the components of two systems, which becomes useful in later theorems.

Theorem 2.3.22 *Let T_i be the lifetime of a coherent system with IID component lifetimes $X_{i1}, \ldots, X_{in}$ having respective reliability function $\bar{F}_i$ for $i = 1, 2$. Suppose these two systems have the same signature $s = (s_1, \ldots, s_n)$. Then, we have:*

$$\textit{If } X_{11} \le_{st} X_{21}, \textit{ then } \mathbf{s}_1(t) \ge_{st} \mathbf{s}_2(t),$$

where $\mathbf{s}_1(t) = (s_{11}(t), \ldots, s_{1n}(t))$, $\mathbf{s}_2(t) = (s_{21}(t), \ldots, s_{2n}(t))$, *with* $s_{1i}(t) = s_i P(X_{1i,n} \le t)/P(T_1 \le t)$ *and* $s_2(t) = s_i P(X_{2i,n} \le t)/P(T_2 \le t)$, *given* $P(T_i \le t) > 0$ $(i = 1, 2)$, *and* $X_{1i,n}$ *is the* i*-th order statistic from* $X_{11}, \ldots, X_{1n}$ *and* $X_{2i,n}$ *is the* i*-th order statistic from* $X_{21}, \ldots, X_{2n}$.

Proof The result is equivalent to that, for any $t \ge 0$ and for all $i = 1, \ldots, n$,

$$\sum_{j=i}^{n} s_{1j}(t) \ge \sum_{j=i}^{n} s_{2j}(t);$$

that is,

$$\sum_{j=i}^{n} s_j P(X_{1j,n} \le t)/P(T_1 \le t) \ge \sum_{j=i}^{n} s_j P(X_{2j,n} \le t)/P(T_2 \le t),$$

which in turn is equivalent to

$$\sum_{j=i}^{n}\sum_{k=1}^{n} s_j s_k [P(X_{1j,n} \le t)P(X_{2k,n} \le t) - P(X_{1k,n} \le t)P(X_{2j,n} \le t)] \ge 0.$$

Note that

$$\sum_{j=i}^{n}\sum_{k=1}^{n} s_j s_k [F_{1j,n}(t)F_{2k,n}(t) - F_{2j,n}(t)F_{1k,n}(t)] = \sum_{j=i}^{n}\sum_{k=1}^{i} s_j s_k [F_{1j,n}(t)F_{2k,n}(t) - F_{2j,n}(t)F_{1k,n}(t)],$$

where $F_{1j,n}(t)$ and $F_{2j,n}(t)$ are the distribution functions of $X_{1j,n}$ and $X_{2j,n}$, respectively. Hence, it is enough to show that, for $k \le j$,

$$F_{1j,n}(t)F_{2k,n}(t) - F_{2j,n}(t)F_{1k,n}(t) \ge 0,$$

which is equivalent to

$$\frac{\sum_{i=j}^{n} \binom{n}{i} \phi_1^i(t)}{\sum_{m=k}^{n} \binom{n}{m} \phi_1^m(t)} \ge \frac{\sum_{i=j}^{n} \binom{n}{i} \phi_2^i(t)}{\sum_{m=k}^{n} \binom{n}{m} \phi_2^m(t)}, \tag{2.37}$$

where $\phi_1(t) = F_1(t)/\bar{F}_1(t)$ and $\phi_2(t) = F_2(t)/\bar{F}_2(t)$. The inequality in Eq. (14) can be simplified to

$$\sum_{i=j}^{n}\sum_{m=k}^{j-1} \binom{n}{m}\binom{n}{i} \phi_1^m(t)\phi_2^m(t)[\phi_1^{i-m}(t) - \phi_2^{i-m}(t)] \ge 0.$$

Now, $\bar{F}_1 \leq_{st} \bar{F}_2$ implies that $\phi_1^{i-m}(t) - \phi_2^{i-m}(t) \geq 0$ for $m < i$. Hence, the required result follows.

In the following, we shall discuss some stochastic comparisons based on the inactivity times of two coherent systems with two sets of different heterogeneous components under some specified conditions. They will be useful in designing more reliable systems.

Theorem 2.3.23 *Let T_i be the lifetime of a coherent system having heterogeneous component lifetimes $X_{i1}, \ldots, X_{in}$ distributed according to reliability functions $\bar{F}_{i1}, \ldots, \bar{F}_{in}$, for $i = 1, 2$. Assume that T_1 and T_2 have the same structure and hence the same signature vector $s = (s_1, \ldots, s_n)$. Let h and H be the system's reliability polynomial and reliability structure function, respectively, and $\bar{G}_i(t) = h^{-1}H(\bar{F}_{i1}(t), \ldots, \bar{F}_{in}(t))$ be the reliability function of IID component lifetimes $Y_{i1}, \ldots, Y_{in}$, for $i = 1, 2$. Then, we have the following:*

(a) If $Y_{11} \leq_{rh} Y_{21}$, $\bar{G}_1(\bar{G}_1^{-1}(y) + x) \leq \bar{G}_2(\bar{G}_2^{-1}(y) + x)$ for $\min\{\bar{G}_1^{-1}(y) + x, \bar{G}_2^{-1}(y) + x\} > 0$, and $(1 - y)h'(y)/\{h(y) - h[\bar{G}_1(\bar{G}_1^{-1}(y) + x)]\}$ or $(1 - y)h'(y)/\{h(y) - h[\bar{G}_2(\bar{G}_2^{-1}(y) + x)]\}$ is increasing in y for any $x > 0$, then $(t - T_1 \mid T_1 \leq t) \leq_{rh} (t - T_2 \mid T_2 \leq t)$;

(b) If $Y_{11} \leq_{hr} Y_{21}$ and $yh'(y)/h(y)$ is decreasing, then $(t - T_1 \mid T_1 \leq t) \leq_{hr} (t - T_2 \mid T_2 \leq t)$;

(c) If $Y_{11} \leq_{lr} Y_{21}$ and $yh'(y)/h(y)$ is decreasing, then $(t - T_1 \mid T_1 \leq t) \leq_{lr} (t - T_2 \mid T_2 \leq t)$.

Proof (a) Let $p_i(t) = \bar{G}_i(t)$ for $t > 0, i = 1, 2$. Let $r_i(t) = \frac{d}{dt}(log(1 - G_i(t))$ be the reversed hazard rate functions of Y_{ij}, for $i = 1, 2$ and $j = 1, \ldots, n$, respectively, and $r_{t-T_1|T_1\leq t}(x)$ and $r_{t-T_2|T_2\leq t}(x)$ be the reversed hazard rate functions of $(t - T_1 \mid T_1 \leq t)$ and $(t - T_2 \mid T_2 \leq t)$, respectively. Then, $P(t - T_i > x \mid T_i \leq t) = \frac{1-h(p_i(t-x))}{1-h(p_i(t))}$ for $i = 1, 2$. For $t > x > 0$, we have

$$
\begin{aligned}
r_{t-T_1|T_1\leq t}(x) &= r_1(t-x) \times \frac{(1 - p_1(t-x))h'(p_1(t-x))}{h(p_1(t-x)) - h(p_1(t))} \\
&= r_1(t-x) \times \frac{(1 - y_1)h'(y_1)}{h(y_1) - h[\bar{G}_1(\bar{G}_1^{-1}(y_1) + x)]} \\
&\leq r_2(t-x) \times \frac{(1 - y_1)h'(y_1)}{h(y_1) - h[\bar{G}_1(\bar{G}_1^{-1}(y_1) + x)]} \\
&\leq r_2(t-x) \times \frac{(1 - y_2)h'(y_2)}{h(y_2) - h[\bar{G}_1(\bar{G}_1^{-1}(y_2) + x)]} \\
&\leq r_2(t-x) \times \frac{(1 - y_2)h'(y_2)}{h(y_2) - h[\bar{G}_2(\bar{G}_2^{-1}(y_2) + x)]} \\
&= r_{t-T_2|T_2\leq t}(x),
\end{aligned}
$$

where $y_i = p_i(t - x)$, $i = 1, 2$, and the first inequality holds since $\bar{G}_1(t) \leq_{rh} \bar{G}_2(t)$, the second inequality follows from the fact that $(1 - y)h'(y)/\{h(y) -$

$h[\bar{G}_1(\bar{G}_1^{-1}(y)+x)]\}$ is increasing in y, and the last one follows from the fact that $\bar{G}_1(\bar{G}_1^{-1}(y)+x) \le \bar{G}_2(\bar{G}_2^{-1}(y)+x)$. Thus, the result follows.

The proof of the results in (b) and (c) are similar to that in (a), and so we refrain from presenting them.

The following example shows that the condition $\bar{G}_1(\bar{G}_1^{-1}(y)+x) \le \bar{G}_2(\bar{G}_2^{-1}(y)+x)$ in (a) can not be deleted.

Example 2.3.24 Assume Y_1 and Y_2 are distributed as Uniform with support intervals $[a, b]$ $(a < b)$ and $[c, d]$ $(c < d)$, respectively, that is, $\bar{G}_1(t) = (b-t)/(b-a)$ for $t \in (a, b)$, and $\bar{G}_2(t) = (d-t)/(d-c)$ for $t \in (c, d)$. It is clear that $a < c$ implies $Y_1 \le_{rh} Y_2$. Then, for $0 < t < 1$, we have $\bar{G}_1^{-1}(t) = b - t(b-a)$ for $a < b$ and $\bar{G}_2^{-1}(t) = d - t(d-c)$ for $c < d$. Then we have, for $a < c$ and any $x > 0$,

$$\bar{G}_2(\bar{G}_2^{-1}(t)+x) - \bar{G}_1(\bar{G}_1^{-1}(t)+x) = x\left[\frac{(a-c)+(d-b)}{(b-a)(d-c)}\right].$$

Clearly, for any $d > c$, $b > a$ and $a < c$, $\bar{G}_1(t) \le_{rh} \bar{G}_2(t)$ does not guarantee that the above term is always non-positive or non-negative for $0 < t < 1$ and $x > 0$.

The following example shows that the condition $(1-y)h'(y)/\{h(y) - h[\bar{G}_1(\bar{G}_1^{-1}(y)+x)]\}$ or $(1-y)h'(y)/\{h(y) - h[\bar{G}_2(\bar{G}_2^{-1}(y)+x)]\}$ is increasing in y for any $x > 0$ in (a) also can not be deleted.

Example 2.3.25 Consider the system with lifetime $T_i = \min\{X_{i1}, X_{i2}, X_{i3}\}$, where X_{ij} is distributed as Exponential with parameter λ_{ij} (> 0) for $i = 1, 2$ and $j = 1, 2, 3$. Suppose $\bar{G}_i(t) = h^{-1}H(\bar{F}_{i1}(t), \bar{F}_{i2}(t), \bar{F}_{i3}(t))$ be the reliability function of IID component lifetimes Y_{i1}, Y_{i2}, Y_{i3}, for $i = 1, 2$. Therefore, we have $\bar{G}_i(t) = \sqrt[3]{e^{-(\lambda_{i1}+\lambda_{i2}+\lambda_{i3})t}}$, $i = 1, 2$.

It is clear that $\lambda_{11} + \lambda_{12} + \lambda_{13} \ge \lambda_{21} + \lambda_{22} + \lambda_{23}$ means $Y_{11} \le_{rh} Y_{21}$. However, it is easily seen that

$$\frac{(1-y)h'(y)}{h(y) - h[\bar{G}_1(\bar{G}_1^{-1}(y)+x)]} = \frac{3(1-y)}{y(1-e^{-(\lambda_{11}+\lambda_{12}+\lambda_{13})x})}$$

is decreasing in $y \ge 0$ for any $x \ge 0$.

Example 2.3.26 Consider a system with lifetime $T_i = \min\{X_{i1}, X_{i2}\}$, where X_{i1}, X_{i2} are heterogeneous component lifetimes distributed according to $\bar{F}_{i1}$, $\bar{F}_{i2}$, for $i = 1, 2$. The signature vector of T_1 and T_2 is $(1, 0)$. Then, for any $t > 0$, the reliability function is given by

$$P(T_i > t) = H(\bar{F}_{i1}(t), \bar{F}_{i2}(t)),$$

where $H(p_1, p_2) = p_1 p_2$. Clearly, $h(p) = p^2$ for $0 \le p \le 1$. The inverse function of h is $h^{-1}(x) = \sqrt[2]{x}$, for $0 \le x \le 1$. Assume Y_{11} and Y_{21} are distributed as Uniform with support intervals $[a, b]$ $(a < b)$ and $[c, d]$ $(c < d)$, respectively, and

$\bar{G}_i(t) = h^{-1}H(\bar{F}_{i1}(t), \bar{F}_{i2}(t))$ is the reliability function of IID component lifetimes $Y_{i1}, Y_{i2}, i = 1, 2$. From Example 2.3.25, we have, for $a < c$ and $a - c + d - b > 0$, $\bar{G}_2(\bar{G}_2^{-1}(t) + x) \geq \bar{G}_1(\bar{G}_1^{-1}(t) + x)$ for any $x \geq 0$. Furthermore, it can be checked that

$$(1 - y)h'(y)/\{h(y) - h[\bar{G}_1(\bar{G}_1^{-1}(y) + x)]\} = 2y(1 - y)/(2y + 2xy - 2x - x^2 - 1)$$

is increasing in $y \geq 0$, for each $x \geq 0$. It follows that all the conditions of (a) are satisfied, and so, from Theorem 2.3.24, $(t - T_1 \mid T_1 \leq t) \leq_{rh} (t - T_2 \mid T_2 \leq t)$.

Moreover, it is clear that $b < d$ implies $Y_{11} \leq_{hr} (\leq_{lr}) Y_{21}$. It can also be easily checked that $yh'(y)/h(y)$ is decreasing in $y \geq 0$. Then, from Theorem 2.3.23, $(t - T_1 \mid T_1 \leq t) \leq_{hr} (\leq_{lr})(t - T_2 \mid T_2 \leq t)$.

Theorem 2.3.27 *Let T_i be the lifetime of a coherent system with a coefficient vector $\mathbf{s}_i(t) = (s_{i1}(t), \ldots, s_{in}(t))$ and having heterogeneous component lifetimes $X_{i1}, \ldots, X_{in}$ distributed according to reliability functions $\bar{F}_{i1}, \ldots, \bar{F}_{in}$, respectively. Let h_i and H_i be the system's reliability polynomial and reliability structure function, respectively, and let $\bar{G}_i(t) = h_i^{-1}H_i(\bar{F}_{i1}(t), \ldots, \bar{F}_{in}(t))$ be the reliability function of IID component lifetimes $Y_{i1}, \ldots, Y_{in}$, for $i = 1, 2$. Then, we have:*

If $\mathbf{s}_1(t) \leq_{st} \mathbf{s}_2(t)$ and $Y_{11} \leq_{rh} Y_{21}$, then $(t - T_1 \mid T_1 \leq t) \geq_{st} (t - T_2 \mid T_2 \leq t)$.

Proof Let $Y_{11,n}, \ldots, Y_{1n,n}$ be the order statistics obtained from IID random variables $Y_{11}, \ldots, Y_{1n}$ having reliability function $\bar{G}_1(t)$, and $Y_{21,n}, \ldots, Y_{2n,n}$ be the order statistics obtained from another set of IID random variables $Y_{21}, \ldots, Y_{2n}$ having reliability function $\bar{G}_2(t)$. From equation (2.33), for any $t > x > 0$, we have

$$\begin{aligned}
P(t - T_1 > x \mid T_1 \leq t) &= \sum_{i=1}^{n} s_{1i}(t) P(t - Y_{1i,n} > x \mid Y_{1i,n} \leq t) \\
&\geq \sum_{i=1}^{n} s_{2i}(t) P(t - Y_{1i,n} > x \mid Y_{1i,n} \leq t) \\
&\geq \sum_{i=1}^{n} s_{2i}(t) P(t - Y_{2i,n} > x \mid Y_{2i,n} \leq t) \\
&= P(t - T_2 > x \mid T_2 \leq t),
\end{aligned}$$

where the first inequality follows from the facts that $\mathbf{s}_1(t) \leq_{st} \mathbf{s}_2(t)$ and $P(t - Y_{1i,n} > x \mid Y_{1i,n} \leq t)$ is decreasing in i. The second inequality holds from the fact that, by Theorem 1.B.60 of Shaked and Shanthikumar (2007), the condition $\bar{G}_1(t) \leq_{rh} \bar{G}_2(t)$ implies that $(t - Y_{1i,n} \mid Y_{1i,n} \leq t) \geq_{rh} (t - Y_{2i,n} \mid Y_{2i,n} \leq t)$ for $i = 1, \ldots, n$, and so $(t - Y_{1i,n} \mid Y_{1i,n} \leq t) \geq_{st} (t - Y_{2i,n} \mid Y_{2i,n} \leq t)$. Hence, the proof is complete.

Example 2.3.28 Consider two coherent systems with lifetimes $T_1 = \min\{X_{11}, \max\{X_{12}, X_{13}\}\}$ and $T_2 = \max\{X_{21}, \min\{X_{22}, X_{23}\}\}$. Let X_{i1}, X_{i2}, X_{i3} be distributed according to reliability functions $\bar{F}_{i1}, \bar{F}_{i2}, \bar{F}_{i3}$, respectively, for $i = 1, 2$. Assume $\bar{G}_i(t)$ is the corresponding mean reliability function of IID component lifetimes Y_{i1}, Y_{i2}, Y_{i3}, for $i = 1, 2$. After some computations, it can be shown that the coefficient vectors of the systems T_1 and T_2 are

$$\mathbf{s}_1(t) = \left(\frac{3 - 3G_1(t) + G_1^2(t)}{3(1 + G_1(t) - G_1^2(t))}, \frac{2G_1(t)(3 - 2G_1(t))}{3(1 + G_1(t) - G_1^2(t))}, 0\right), \mathbf{s}_2(t) = \left(0, \frac{6 - 4G_2(t)}{6 - 3G_2(t)}, \frac{G_2(t)}{6 - 3G_2(t)}\right),$$

respectively, where $G_i(t) = 1 - \bar{G}_i(t)$, for $i = 1, 2$.

If let Y_{11} and Y_{21} be distributed as Exponential with parameters 2 and 1, respectively, then, $Y_{11} \leq_{rh} Y_{21}$, and $\mathbf{s}_1(t) \leq_{st} \mathbf{s}_2(t)$ for any $t \geq 0$. Therefore, from Theorem 2.1, for any $t \geq 0$ and $x > 0$, we find

$$\begin{aligned} P(t - T_1 > x \mid T_1 \leq t) &= \sum_{i=1}^{n} s_{1i}(t) P(t - Y_{1i,n} > x \mid Y_{1i,n} \leq t) \\ &= \frac{1}{3(1 - e^{-2t} - e^{-4t})} \left[\frac{(1 + e^{-2t} + e^{-4t})(1 - e^{6x-6t})}{1 - e^{-6t}} + \frac{2(1 + e^{-2t} - 2e^{-4t})(1 - 3e^{4x-4t} + 2e^{6x-6t})}{1 - 3e^{-4t} + 2e^{-6t}} \right], \\ P(t - T_2 > x \mid T_2 \leq t) &= \sum_{i=1}^{n} s_{2i}(t) P(t - Y_{2i,n} > x \mid Y_{2i,n} \leq t) \\ &= \frac{1}{3(1 + e^{-t})} \left[\frac{(2 + 4e^{-t})(1 - 3e^{2x-2t} + 2e^{3x-3t})}{1 - 3e^{-2t} + 2e^{-3t}} + \frac{(1 - e^{x-t})^3}{(1 - e^{-t})^2} \right]. \end{aligned}$$

Now, by using Mathematica software, it can be checked that $P(t - T_2 > x \mid T_2 \leq t) \leq P(t - T_1 > x \mid T_1 \leq t)$, for all $x > 0$, for each fixed $t \geq 0$; that is, $(t - T_1 \mid T_1 \leq t) \geq_{st} (t - T_2 \mid T_2 \leq t)$.

2.4 Ordered Properties of Residual Lives and Inactivity Time of Coherent Systems

In this part, we mainly investigate the aging properties and stochastic comparisons of general coherent systems with different sizes of component.

2.4.1 *Introduction*

Recently, Khaledi and Shaked (2007) further pointed out that the signature of a coherent system, which functions with probability 1 as long as at least $n - s + 1$ $(1 \leq s \leq n)$ components function, must be of the form $(0, \ldots, 0, p_s, \ldots, p_n)$. It was also showed there that the reliability function of the residual life of a coherent system with IID components can be written as

$$P(T_X - t > x \mid X_{r,n} > t) = \sum_{i=s}^{n} p_i \, P(X_{i,n} - t > x \mid X_{r,n} > t), \qquad r \le s \le n. \tag{2.38}$$

In a similar as in Khaledi and Shaked (2007), it can be shown that, the reliability function of the inactivity time of a coherent system with IID components is

$$P(t - T_X > x \mid X_{r,n} \le t) = \sum_{i=1}^{s} p_i P(t - X_{i,n} > x \mid X_{r,n} \le t) \qquad s \le r \le n. \tag{2.39}$$

In fact, (2.38) and (2.39) hold even in some more general contexts, for example, when $X_1, \ldots, X_n$ are exchangeable.

2.4.2 Ordered Properties of Residual Lives

In the sequel, denote $h_{r,i,n,t}(x)$, $\bar{H}_{r,i,n,t}(x)$ the density and reliability functions of random variable $\mathrm{GRL}_{r,i,n}(t) = [X_{i:n} - t | X_{r:n} > t]$ $(r \le i \le n)$, respectively. In order to get the main results, we firstly present several lemmas in the following.

Lemma 2.4.1 (Hu et al. 2007a) *For $r \le i \le n$ and any $t \ge 0$,*

$$GRL_{r,i+1,n}(t) \ge_{lr} GRL_{r,i,n}(t) \ge_{lr} GRL_{r,i-1,n-1}(t).$$

Lemma 2.4.2 (Kochar and Kirmani 1995) *Let $\psi_1(x, u)$ and $\psi_2(x, u)$ be two positive real-valued functions such that:*

(i) for $u_1 \le u_2$, $\frac{\psi_1(x,u_2)}{\psi_1(x,u_1)}$ and $\frac{\psi_2(x,u_2)}{\psi_1(x,u_1)}$ are both increasing in x;

(ii) for each fixed x, $\frac{\psi_2(x,u)}{\psi_1(x,u)}$ is increasing in u.

Then $X \le_{lr} Y$ implies that the ratio

$$\frac{E\psi_2(x, Y)}{E\psi_1(x, X)}$$

is increasing in x provided that the expectations exist.

Lemma 2.4.3 (Misra and Meulen (2003)) *Let Θ be a subset of the real line $\Re$, and let X be a nonnegative random variable having a distribution belonging to the family $\mathcal{P} = \{H(\cdot|\theta),\ \theta \in \Theta\}$ which satisfies that*

$$H(\cdot|\theta_2) \ge_{st} [\le_{st}] H(\cdot|\theta_1) \quad whenever\ \theta_1,\ \theta_2 \in \Theta,\ and\ \theta_2 > \theta_1 > 0.$$

Let $\Psi(x,\ \theta)$ be a real valued function defined on $\Re \times \Theta$, and be measurable in x for each θ such that $\mathbb{E}_\theta[\Psi(x,\ \theta)]$ exists. Then

(1) *$\mathbb{E}_\theta[\Psi(X,\ \theta)]$ is increasing in θ, if $\Psi(x,\ \theta)$ is increasing in θ and increasing [decreasing] in x;*

(2) $\mathbb{E}_\theta[\Psi(X, \theta)]$ *is decreasing in* θ*, if* $\Psi(x, \theta)$ *is decreasing in* θ *and decreasing [increasing] in* x.

Theorem 2.4.4 *Let* $X_1, X_2, \ldots, X_n$ *be the independent and identically distributed components lifetimes of the components of a coherent system with life function* τ*. Denote* $T_X^n = \tau_1(X_1, X_2, \ldots, X_n)$*,* $T_X^{n-1} = \tau_2(X_1, X_2, \ldots, X_{n-1})$*. For some* $r \le s \le n$*, suppose that the signatures of* τ_1 *and* τ_2 *are of the forms* $\mathbf{p} = (0, \ldots, 0, p_s, \ldots, p_n)$*,* $\boldsymbol{q} = (0, \ldots, 0, q_s, \ldots, q_{n-1})$*, respectively. If* $\mathbf{p} \ge_{lr} \boldsymbol{q}$ *then, for any* $t \ge 0$ *and* $r \le s$*,*

$$[T_X^n - t \mid X_{r,n} > t] \ge_{lr} [T_X^{n-1} - t \mid X_{r,n-1} > t].$$

Proof Denote $h_{r,n,t}(x)$, $h_{r,n-1,t}(x)$ the density functions of $[T_X^n - t \mid X_{r,n} > t]$ and $[T_X^{n-1} - t \mid X_{r,n-1} > t]$, respectively. Then, by Lemma 2.2 in Li and Zhao (2006) and Eq. (2.38), for $r \le s \le n$, it can be obtained

$$h_{r,n,t}(x) = \frac{\sum_{i=s}^{n} p_i \sum_{k=0}^{r-1} \binom{n}{k} \varphi^k(t) h_{1,i-k,n-k,t}(x)}{\sum_{k=0}^{r-1} \binom{n}{k} \varphi^k(t)},$$

and

$$h_{r,n-1,t}(x) = \frac{\sum_{i=s}^{n-1} q_i \sum_{k=0}^{r-1} \binom{n-1}{k} \varphi^k(t) h_{1,i-k,n-1-k,t}(x)}{\sum_{k=0}^{r-1} \binom{n-1}{k} \varphi^k(t)}.$$

Hence, their ratio can be represented as

$$\frac{h_{r,n,t}(x)}{h_{r,n-1,t}(x)} = \frac{\sum_{i=s}^{n} p_i \sum_{k=0}^{r-1} \binom{n}{k} \varphi^k(t) h_{1,i-k,n-k,t}(x)}{\sum_{i=s}^{n-1} q_i \sum_{k=0}^{r-1} \binom{n-1}{k} \varphi^k(t) h_{1,i-k,n-1-k,t}(x)} \times \frac{\sum_{k=0}^{r-1} \binom{n-1}{k} \varphi^k(t)}{\sum_{k=0}^{r-1} \binom{n}{k} \varphi^k(t)}$$

$$\propto \frac{\sum_{i=s}^{n} p_i \sum_{k=0}^{r-1} \binom{n}{k} \varphi^k(t) h_{1,i-k,n-k,t}(x)}{\sum_{i=s}^{n-1} q_i \sum_{k=0}^{r-1} \binom{n-1}{k} \varphi^k(t) h_{1,i-k,n-1-k,t}(x)}$$

$$= \frac{E_x\left[\sum_{k=0}^{r-1}\binom{n}{k}\varphi^k(t)h_{1,I-k,n-k,t}(x)\right]}{E_x\left[\sum_{k=0}^{r-1}\binom{n-1}{k}\varphi^k(t)h_{1,J-k,n-1-k,t}(x)\right]}$$

$$= \frac{E\psi_2(x, I)}{E\psi_1(x, J)},$$

where random variables I, J, have respect discrete distributions **p**, **q**, and

$$\psi_2(x,i) = \sum_{k=0}^{r-1}\binom{n}{k}\varphi^k(t)h_{1,i-k,n-k,t}(x),$$

$$\psi_1(x,i) = \sum_{k=0}^{r-1}\binom{n-1}{k}\varphi^k(t)h_{1,i-k,n-1-k,t}(x).$$

By Lemma 2.4.1, we can investigate that $\frac{\psi_1(x,i)}{\psi_1(x,i-1)}$ and $\frac{\psi_2(x,i)}{\psi_1(x,i-1)}$ are both increasing in x. And for each fixed $x > 0$,

$$\frac{\psi_2(x,i)}{\psi_1(x,i)} - \frac{\psi_2(x,i-1)}{\psi_1(x,i-1)}$$

$$= \frac{\sum_{k=0}^{r-1}\binom{n}{k}\varphi^k(t)h_{1,i-k,n-k,t}(x)}{\sum_{k=0}^{r-1}\binom{n-1}{k}\varphi^k(t)h_{1,i-k,n-1-k,t}(x)} - \frac{\sum_{k=0}^{r-1}\binom{n}{k}\varphi^k(t)h_{1,i-k-1,n-k,t}(x)}{\sum_{k=0}^{r-1}\binom{n-1}{k}\varphi^k(t)h_{1,i-k-1,n-1-k,t}(x)}$$

$$\stackrel{sgn}{=} \sum_{j=0}^{r-1}\binom{n}{j}\varphi^j(t)h_{1,i-j,n-j,t}(x)\sum_{k=0}^{r-1}\binom{n-1}{k}\varphi^k(t)h_{1,i-k-1,n-1-k,t}(x)$$

$$-\sum_{j=0}^{r-1}\binom{n}{j}\varphi^j(t)h_{1,i-j-1,n-j,t}(x)\sum_{k=0}^{r-1}\binom{n-1}{k}\varphi^k(t)h_{1,i-k,n-1-k,t}(x)$$

$$= \sum_{j=0}^{r-1}\sum_{k=0}^{r-1}\binom{n}{j}\binom{n-1}{k}\psi^{j+k}(t)[1-\bar{F}_t(x)]^{2i-j-k-3}[\bar{F}_t(x)]^{2(n-i)}f_t^2(x)$$

$$\times\left\{(i-j)\binom{n-j}{i-j}(i-1-k)\binom{n-1-k}{i-1-k} - (i-j-1)\binom{n-j}{i-j-1}(i-k)\binom{n-1-k}{i-k}\right\}$$

$$= \sum_{j=0}^{r-1}\sum_{k=0}^{r-1}a_{j,k}(t,x)\times\{i-k-1+(n-i)(j-k)\}$$

$$= \sum_{j=0}^{r-1}\sum_{k=j}^{r-1}a_{j,k}(t,x)\times\{i-k-1+(n-i)(j-k)\}$$

$$+\sum_{j=0}^{r-1}\sum_{k=0}^{j} a_{j,k}(t,x)\times\{i-k-1+(n-i)(j-k)\}-\sum_{j=0}^{r-1} a_{j,j}(t,x)(i-j-1)$$
$$=\sum_{j=0}^{r-1}\sum_{k=j}^{r-1} a_{j,k}(t,x)\times\{i-k-1+(n-i)(j-k)\}$$
$$+\sum_{j=0}^{r-1}\sum_{k=j}^{r-1} a_{k,j}(t,x)\times\{i-j-1+(n-i)(k-j)\}-\sum_{j=0}^{r-1} a_{j,j}(t,x)(i-j-1)$$
$$=\sum_{j=0}^{r-1}\sum_{k=j}^{r-1} a_{j,k}(t,x)\times\{2(i-1)-j-k\}-\sum_{j=0}^{r-1} a_{j,j}(t,x)(i-j-1)$$
$$=\sum_{j=0}^{r-1}\sum_{k=j+1}^{r-1} a_{j,k}(t,x)\times\{2(i-1)-j-k\}+\sum_{j=0}^{r-1} a_{j,j}(t,x)(i-j-1)\geq 0,$$

where $a_{j,k}(t,x)$ $=\binom{n}{j}\binom{n}{k}\frac{(n-k)!(n-j)!}{n(n-i+1)!(i-k-1)!(i-j-1)!(n-i)!}\psi^{j+k}(t)[1-\bar{F}_t(x)]^{2i-j-k-3}[\bar{F}_t(x)]^{2(n-i)}f_t^2(x)$, that is, $\frac{\psi_2(x,i)}{\psi_1(x,i)}$ is increasing in i. Thus all the conditions in Lemma 2.4.2 are satisfied, then the desired result is obtained.

Example 2.4.5 Consider the two systems of order 4 depicted in Fig. 2.5, it can be evaluated that the signature of the left one is $\mathbf{q}=\left(0,0,\frac{1}{2},\frac{1}{2}\right)$ and that of the right one is $\mathbf{p}=\left(0,\frac{1}{6},\frac{7}{12},\frac{1}{4}\right)$. It can be verified that $\mathbf{p}\leq_{lr}\mathbf{q}$ and hence $\mathbf{p}\leq_{hr}\mathbf{q}$. By the two theorems above the system in the left side is better in the sense that it has stochastically longer general residual life.

Theorem 2.4.6 *Let $X_1, X_2, \ldots, X_n$ be the independent and identically distributed components lifetimes of the components of a coherent system with life function τ. For $r\leq s\leq n$, suppose that the signature of τ is of the form $p=(0,\ldots,0,p_s,\ldots,p_n)$.*

(a) If X is IFR, then $[T_X-t\mid X_{r,n}>t]$ is stochastically decreasing in $t\geq 0$.

(b) If $[T_X-t\mid X_{r,n}>t]$ is stochastically increasing in $t\geq 0$, then X is DFR.

Proof (a) By (2.38), for any $t\geq 0$,

$$P(T_X-t>x\mid X_{r,n}>t)=\sum_{i=s}^{n} p_i\bar{H}_{r,i,n,t}(x).$$

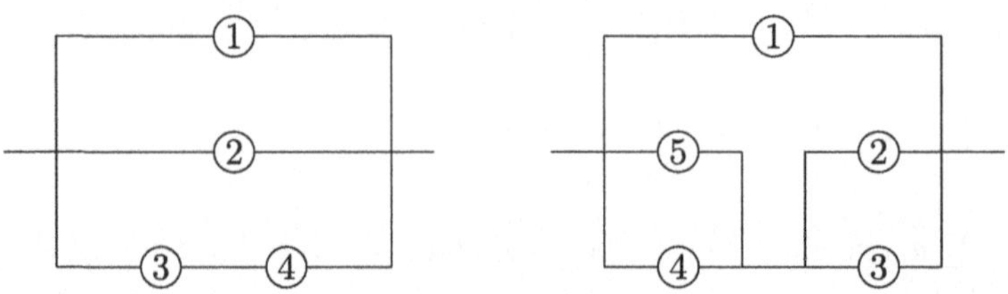

Fig. 2.5 Two coherent systems of different order with likely ratio ordered signatures

According to Theorem 2.2 in Li and Zhao (2006), X is IFR implies that $\bar{H}_{r,i,n,t}(x)$ is stochastically decreasing in $t \geq 0$, hence, $[T_X - t \mid X_{r,n} > t]$ is stochastically decreasing in $t \geq 0$.

(b) $[T_X - t \mid X_{r,n} > t]$ is stochastically increasing in $t \geq 0$ implies that

$$\frac{\partial P(T_X - t > x \mid X_{r,n} > t)}{\partial t} = \sum_{i=s}^{n} p_i \frac{\partial \bar{H}_{r,i,n,t}(x)}{\partial t} \geq 0.$$

Since p_i is nonnegative for all $i = s, \ldots, n$, there exists at least an j $(s \leq j \leq n)$ satisfying with $\frac{\partial \bar{H}_{r,j,n,t}(x)}{\partial t} \geq 0$. Again from Theorem 2.2 in Li and Zhao (2006), it follows that X is DFR.

Theorem 2.4.7 *Let $X_1, X_2, \ldots, X_n$ be the independent and identically distributed components lifetimes of the components of a coherent system with life function τ. Denote $T_X = \tau(X_1, X_2, \ldots, X_n)$. Suppose that the signature of τ is of the form $p = (0, \ldots, 0, p_s, \ldots, p_n)$. For any $t \geq 0$,*

(a) If X is NBU, then $[T_X - t \mid X_{r,n} > t] \leq_{st} T_X$.

(b) If $[T_X - t \mid X_{r,n} > t] \geq_{st} T_X$, then X is NWU.

Proof (a) For any $t \geq 0$,

$$P(T_X - t > x \mid X_{r,n} > t) = \sum_{i=s}^{n} p_i \bar{H}_{r,i,n,t}(x),$$

and

$$P(T_X > x) = \sum_{i=s}^{n} p_i P(X_{i,n} > x).$$

According to Theorem 2.3 in Li and Zhao (2006), X is NBU implies that $\bar{H}_{r,i,n,t}(x) \leq P(X_{i,n} > x)$, for $s \leq i \leq n$. It follows that $[T_X - t \mid X_{r,n} > t] \leq_{st} T_X$.

(b) $[T_X - t \mid X_{r,n} > t] \geq_{st} T_X$ implies that $\sum_{i=s}^{n} p_i\{\bar{H}_{r,i,n,t}(x) - P(X_{i,n} > x)\} \geq 0$. Since p_i is nonnegative for all $i = s, \ldots, n$, there exists at least an j $(s \leq j \leq n)$ such that $\bar{H}_{r,i,n,t}(x) \geq P(X_{i,n} > x)$. Again from Theorem 2.3 in Li and Zhao (2006), It holds that X is NWU.

The following theorem extends Theorem in Li and Zhao (2008), which needs a strong condition. The proof is clear and hence is omitted.

Theorem 2.4.8 *Let $X_1, X_2, \ldots, X_n$ be the independent and identically distributed components lifetimes of the components of a coherent system with life function τ. Denote $T_X = \tau(X_1, X_2, \ldots, X_n)$, $T_Y = \tau(Y_1, Y_2, \ldots, Y_n)$. Suppose that the signature of τ is of the form $p = (0, \ldots, 0, p_s, \ldots, p_n)$. If $X_{i:n} \leq_{hr} Y_{j:n}$ for all $i, j = 1, 2, \ldots, n$, then, for any $t \geq 0$,*

$$[T_X - t \mid X_{r,n} > t] \leq_{hr} [T_Y - t \mid Y_{r,n} > t].$$

2.4.3 Ordered Properties of Inactivity Time

In the sequel, denote $h_{i,r,n,t}(x)$, $\bar{H}_{i,r,n,t}(x)$ the density and reliability functions of random variable $[t - X_{i:n}|X_{r:n} \leq t]$ $(i \leq r \leq n)$, respectively. In order to get the main results, we firstly present several lemmas in the following The following lemma is useful in proving the main conclusions in this section.

Theorem 2.4.9 *If X is DRHR, then, for $s \leq r \leq n$, $(t - T_X \mid X_{r,n} \leq t)$ is stochastically increasing in $t \geq 0$.*

Proof By (2.39), for $s \leq r \leq n$,

$$P(t - T_X > x \mid X_{r,n} \leq t) = \sum_{i=1}^{s} p_i \bar{H}_{i,r,n,t}(x)$$

According to Theorem 2.2 in Hu et al. (2007b), X is DRHR implies that $\bar{H}_{i,r,n,t}(x)$ is stochastically increasing in $t \geq 0$, hence, $[t - T_X \mid X_{r,n} \leq t]$ is stochastically increasing in $t \geq 0$.

Theorem 2.4.10 *Let $X_1, X_2, \ldots, X_n$ be the independent and identically distributed components lifetimes of the components of a coherent system with life function τ. Denote $T_X = \tau(X_1, X_2, \ldots, X_n)$, $T_Y = \tau(Y_1, Y_2, \ldots, Y_n)$. Suppose that the signature of τ is of the form $p = (0, \ldots, 0, p_s, \ldots, p_n)$. If $X_{i:n} \leq_{rh} Y_{j:n}$ for all $i, j = 1, 2, \ldots, n$, then, for any $t \geq 0$,*

$$[T_X - t \mid X_{r,n} \leq t] \geq_{rh} [T_Y - t \mid Y_{r,n} \leq t].$$

Chapter 3
Residual Lifetime and Inactivity Time of Coherent Systems Under Double Monitoring

This chapter focuses on the residual lifetime and the inactivity time of the k-out-of-n system and general coherent one consisting of independent and identically distributed components or independent but not identically distributed components, under double monitoring. Several mixture representations for the residual lifetime and inactivity time by using order statistics and signature are presented. Then, some stochastic comparisons of residual lives or inactivity times are obtained between two systems with different structures and two different sets of components.

3.1 On Coherent System with IID Components

This section considers the residual lifetime and the inactivity time of the k-out-of-n system and general coherent one consisting of independent and identically distributed components under double monitoring. Some ordered properties of residual lives or inactivity times are conducted.

3.1.1 On the k-Out-of-n System

This part is for the residual lifetime and the inactivity time of the $(n-k+1)$-out-of-n system consisting of independent and identically distributed components, under the condition that the total number of failures of the components at time t_1 is $r(<k)$, and at time $t_2(>t_1)$, the system is still working with probability one or it has failed. Under these conditions, some ordered properties of one system or between two systems with two sets of independent components are conducted. The results are stronger than those in Poursaeed (2010).

Z. Zhang, *Conditional Order Statistics and Conditional Spacings*,
https://doi.org/10.1007/978-981-96-7151-9_3

3.1.1.1 Introduction

To obtain more information or be safe of a system, it is always continuously inspected. But because of some reasons such as cost, it is enough to inspect it discretely. That is, the system is inspected under different times. The simpler one is the system is under inspection at two different time. Based on this, Poursaeed (2010) researched some ordered properties of the expected value of the residual lifetime

$$(X_{k:n} - t_2 | X_{r:n} < t_1 < X_{r+1:n}, X_{k:n} > t_2) \tag{3.1}$$

of $(n-k+1)$-out-of-n systems with n independent components having common distribution F, given that before time t_1, exactly r $(r < k)$ components have failed and at t_2 $(t_2 > t_1)$, the system is still working. Also, they investigate the mean past time

$$(t_2 - X_{k:n} | X_{r:n} < t_1 < X_{r+1:n}, X_{k:n} < t_2), \tag{3.2}$$

of $(n-k+1)$-out-of-n systems with n independent components having common distribution F, given that before time t_1, exactly r $(r < k)$ components have failed and at t_2 $(t_2 > t_1)$, the system has failed.

In the following of this section, we investigate some order properties of the expressions (3.1) and (3.2). Also, we investigate the stochastic comparisons of coherent systems under the condition that the total number of failures of the components at time t_1 is r $(r < k)$, and at time t_2 $(t_2 > t_1)$ the system is still working or it has failed. Under these conditions, some ordered properties of one system or between two systems with two sets of independent components are obtained both for the hazard rate order and the likelihood ratio order.

3.1.1.2 The Stochastic Results of the Residual Life

Assume that for $t_2 > t_1 > 0$, $X_{r,k,n}^{t_1,t_2} = (X_{k:n} - t_2 | X_{r:n} < t_1 < X_{r+1:n}, X_{k:n} > t_2)$. It is easy to get, for any $x > 0$,

$$\begin{aligned} P(X_{r,k,n}^{t_1,t_2} > x) &= P(X_{k:n} - t_2 > x \mid X_{r:n} < t_1 < X_{r+1:n}, X_{k:n} > t_2) \\ &= \frac{\sum_{i=n-k+1}^{n-r} \binom{n-r}{i} \left[\frac{\bar{F}(t_2+x)}{\bar{F}(t_1)}\right]^i \left[1 - \frac{\bar{F}(t_2+x)}{\bar{F}(t_1)}\right]^{n-r-i}}{\sum_{i=n-k+1}^{n-r} \binom{n-r}{i} \left[\frac{\bar{F}(t_2)}{\bar{F}(t_1)}\right]^i \left[1 - \frac{\bar{F}(t_2)}{\bar{F}(t_1)}\right]^{n-r-i}}. \end{aligned} \tag{3.3}$$

Theorem 3.1.1 *Let $X_1, X_2, \ldots, X_{max\{n_1,n_2\}}$ be IID random variables. Then for $t_2 > t_1 > 0$ and for $r \in \{1, 2, \ldots, \min(n_1, n_2)\}$,*

$$X_{r:k_1:n_1}^{t_1,t_2} \le_{lr} X_{r:k_2:n_2}^{t_1,t_2}, \quad \textit{whenever } k_1 \le k_2 \textit{ and } n_1 - k_1 \ge n_2 - k_2.$$

Proof Denote by $f_{r:k_1:n_1}^{t_1,t_2}(x)$ and $f_{r:k_2:n_2}^{t_1,t_2}(x)$ the respective density functions of $X_{r,k_1,n_1}^{t_1,t_2}$ and $X_{r,k_2,n_2}^{t_1,t_2}$. Then from (3.3), for any $x > 0$,

$$f^{t_1,t_2}_{r:k_1:n_1}(x)=\frac{f_{X^{t_1,t_2}_{k_1-r:n_1-r}}(x)}{\sum_{i=n_1-k_1+1}^{n_1-r}\binom{n_1-r}{i}\left[\frac{\bar F(t_2)}{\bar F(t_1)}\right]^i\left[1-\frac{\bar F(t_2)}{\bar F(t_1)}\right]^{n_1-r-i}},$$

$$f^{t_1,t_2}_{r:k_2:n_2}(x)=\frac{f_{X^{t_1,t_2}_{k_2-r:n_2-r}}(x)}{\sum_{i=n_2-k_2+1}^{n_2-r}\binom{n_2-r}{i}\left[\frac{\bar F(t_2)}{\bar F(t_1)}\right]^i\left[1-\frac{\bar F(t_2)}{\bar F(t_1)}\right]^{n_2-r-i}},$$

where $X^{t_1,t_2}_{k_1-r:n_1-r}$ ($X^{t_1,t_2}_{k_2-r:n_2-r}$) denotes the (k_1-r)th ((k_2-r)th) order statistic among n_1-r (n_2-r) IID random variables with the same distribution as $(X-t_2\mid X>t_1)$, and $f_{X^{t_1,t_2}_{k_1-r:n_1-r}}(x)$ ($f_{X^{t_1,t_2}_{k_2-r:n_2-r}}(x)$) is the corresponding density function.

According to Theorem 1.C.37 of Shaked and Shanthikumar (2007), for fixed r, $X^{t_1,t_2}_{k_1-r:n_1-r}\le_{lr}X^{t_1,t_2}_{k_2-r:n_2-r}$, whenever $k_1\le k_2$, and $n_1-k_1\ge n_2-k_2$, hence $f_{X^{t_1,t_2}_{k_2-r:n_2-r}}(x)/f_{X^{t_1,t_2}_{k_1-r:n_1-r}}(x)$ is increasing in $x>0$. It follows that

$$\frac{f^{t_1,t_2}_{r:k_2:n_2}(x)}{f^{t_1,t_2}_{r:k_1:n_1}(x)}\propto\frac{f_{X^{t_1,t_2}_{k_2-r:n_2-r}}(x)}{f_{X^{t_1,t_2}_{k_1-r:n_1-r}}(x)}$$

is increasing in $x>0$, which validates the result.

According to the theorem above, the following two corollaries can be obtained which show that $X^{t_1,t_2}_{r:k:n}$ is decreasing in n but increasing in k in the likelihood ratio order, respectively.

Corollary 3.1.2 *For fixed r and k, and $t_2>t_1>0$, $X^{t_1,t_2}_{r:k:n}\ge_{lr}X^{t_1,t_2}_{r:k:n+1}$.*

Corollary 3.1.3 *For fixed r and n, and $t_2>t_1>0$, $X^{t_1,t_2}_{r:k:n}\ge_{lr}X^{t_1,t_2}_{r:k-1:n}$.*

The following theorem shows that $X^{t_1,t_2}_{r:k:n}$ is decreasing in r in the likelihood ratio order.

Theorem 3.1.4 *For fixed n and k, and $t_2>t_1>0$, $X^{t_1,t_2}_{r:k:n}\le_{lr}X^{t_1,t_2}_{r-1:k:n}$.*

Proof Denote by $f^{t_1,t_2}_{r:k:n}(x)$ and $f^{t_1,t_2}_{r-1:k:n}(x)$ the respective density functions of $X^{t_1,t_2}_{r,k,n}$ and $X^{t_1,t_2}_{r-1,k,n}$. Then, from (3.3), for any $x>0$,

$$f^{t_1,t_2}_{r:k:n}(x)=\frac{f_{X^{t_1,t_2}_{k-r:n-r}}(x)}{\sum_{i=n-k+1}^{n-r}\binom{n-r}{i}\left[\frac{\bar F(t_2)}{\bar F(t_1)}\right]^i\left[1-\frac{\bar F(t_2)}{\bar F(t_1)}\right]^{n-r-i}},$$

$$f^{t_1,t_2}_{r-1:k:n}(x)=\frac{f_{X^{t_1,t_2}_{k-r+1:n-r+1}}(x)}{\sum_{i=n-k+1}^{n-r+1}\binom{n-r+1}{i}\left[\frac{\bar F(t_2)}{\bar F(t_1)}\right]^i\left[1-\frac{\bar F(t_2)}{\bar F(t_1)}\right]^{n-r+1-i}}.$$

From Corollary 1.C.38 of Shaked and Shanthikumar (2007), $X^{t_1,t_2}_{k-r:n-r}(x)\le_{lr}X^{t_1,t_2}_{k-r+1:n-r+1}(x)$, hence

$$\frac{f_{r:k:n}^{t_1,t_2}(x)}{f_{r-1:k:n}^{t_1,t_2}(x)} \propto \frac{f_{X_{k-r:n-r}^{t_1,t_2}}(x)}{f_{X_{k-r+1:n-r+1}^{t_1,t_2}}(x)}$$

is decreasing in $x > 0$, which establishes the desired result.

For a $(n-k+1)$-out-of-n system with independent random units $X_1, X_2, \ldots, X_n$, which have identical distribution F as X. For another $(n-k+1)$-out-of-n system with independent random units $Y_1, Y_2, \ldots, Y_n$, which have common distribution G as Y. For $t_2 > t_1 > 0$, let $\bar{F}_{r:k:n}^{t_1,t_2}(x)$, $\bar{G}_{r,k,n}^{t_1,t_2}(x)$ be the reliability functions of the random variables $X_{r,k,n}^{t_1,t_2} = (X_{k:n} - t_2 | X_{r:n} < t_1 < X_{r+1:n}, X_{k:n} > t_2)$, $Y_{r:k:n}^{t_1,t_2} = (Y_{k:n} - t_2 | Y_{r:n} < t_1 < Y_{r+1:n}, Y_{k:n} > t_2)$, respectively.

The following two theorems conduct stochastic comparisons of two $(n-k+1)$-out-of-n systems based on their hazard rate ordering and likelihood ratio order.

Theorem 3.1.5 *If* $X \le_{hr} Y$, *then* $X_{r:k:n}^{t_1,t_2} \le_{hr} Y_{r:k:n}^{t_1,t_2}$.

Proof By the expression in (3.3), we have

$$\bar{F}_{r:k:n}^{t_1,t_2}(x) = \frac{P(X_{k-r:n-r}^{t_1,t_2} > x)}{\sum_{i=n-k+1}^{n-r} \binom{n-r}{i} \left[\frac{\bar{F}(t_2)}{\bar{F}(t_1)}\right]^i \left[1 - \frac{\bar{F}(t_2)}{\bar{F}(t_1)}\right]^{n-r-i}},$$

$$\bar{G}_{r:k:n}^{t_1,t_2}(x) = \frac{P(Y_{k-r:n-r}^{t_1,t_2} > x)}{\sum_{i=n-k+1}^{n-r} \binom{n-r}{i} \left[\frac{\bar{G}(t_2)}{\bar{G}(t_1)}\right]^i \left[1 - \frac{\bar{G}(t_2)}{\bar{G}(t_1)}\right]^{n-r-i}},$$

where $X_{k-r:n-r}^{t_1,t_2}$, $Y_{k-r:n-r}^{t_1,t_2}$ denote the $(k-r)$th order statistic among the $n-r$ IID random variables with the same distribution as $(X - t_2 \mid X > t_1)$ and $(Y - t_2 \mid Y > t_1)$, respectively.

The condition $X \le_{hr} Y$ implies that $(X - t_2 \mid X > t_1) \le_{hr} (Y - t_2 \mid Y > t_1)$, hence $X_{k-r:n-r}^{t_1,t_2} \le_{hr} Y_{k-r:n-r}^{t_1,t_2}$, which implies that the ratio

$$\frac{\bar{G}_{r:k:n}^{t_1,t_2}(x)}{\bar{F}_{r:k:n}^{t_1,t_2}(x)} \propto \frac{P(Y_{k-r:n-r}^{t_1,t_2} > x)}{P(X_{k-r:n-r}^{t_1,t_2} > x)}$$

is increasing in $x > 0$, therefore the desired result follows immediately.

Theorem 3.1.6 *If* $X \le_{lr} Y$, *then* $X_{r:k:n}^{t_1,t_2} \le_{lr} Y_{r:k:n}^{t_1,t_2}$.

Proof Denote by $f_{r:k:n}^{t_1,t_2}(x)$ and $g_{r:k:n}^{t_1,t_2}(x)$ the respective density functions of $X_{r,k,n}^{t_1,t_2}$ and $Y_{r,k,n}^{t_1,t_2}$. Then from (3.3), for any $x > 0$,

$$f_{r:k:n}^{t_1,t_2}(x) = \frac{f_{X_{k-r:n-r}^{t_1,t_2}}(x)}{\sum_{i=n-k+1}^{n-r} \binom{n-r}{i} \left[\frac{\bar{F}(t_2)}{\bar{F}(t_1)}\right]^i \left[1 - \frac{\bar{F}(t_2)}{\bar{F}(t_1)}\right]^{n-r-i}},$$

$$g_{r:k:n}^{t_1,t_2}(x) = \frac{g_{Y_{k-r:n-r}^{t_1,t_2}}(x)}{\sum_{i=n-k+1}^{n-r}\binom{n-r}{i}\left[\frac{\bar{G}(t_2)}{\bar{G}(t_1)}\right]^i\left[1-\frac{\bar{G}(t_2)}{\bar{G}(t_1)}\right]^{n-r-i}},$$

where $f_{X_{k-r:n-r}^{t_1,t_2}}(x)$, $g_{X_{k-r:n-r}^{t_1,t_2}}(x)$ are density distributions of order statistics $X_{k-r:n-r}^{t_1,t_2}$ and $Y_{k-r:n-r}^{t_1,t_2}$, respectively.

The condition $X \leq_{lr} Y$ implies that $(X - t_2 \mid X > t_1) \leq_{lr} (Y - t_2 \mid Y > t_1)$, hence $X_{k-r:n-r}^{t_1,t_2} \leq_{lr} Y_{k-r:n-r}^{t_1,t_2}$, which implies that the ratio

$$\frac{g_{r:k:n}^{t_1,t_2}(x)}{f_{r:k:n}^{t_1,t_2}(x)} \propto \frac{g_{Y_{k-r:n-r}^{t_1,t_2}}(x)}{f_{X_{k-r:n-r}^{t_1,t_2}}(x)}$$

is increasing in $x > 0$, hence the proof of desired result is complete.

In practice, it may be difficult to check the results appeared in above theorems. The next example shows that these results hold when the components have exponential distributions.

Example 3.1.7 Consider one 3-out-of-4 system with independent components having common distribution function F as X, and another 3-out-of-4 system with independent components having common distribution G as Y. Let X and Y are exponential distributions with parameters $\mu = 2$, $\lambda = 1$, respectively. It is clear that $X \leq_{lr} Y$.

Set $r = 1$, $t_1 = 1$, $t_2 = 2$. Then, by some simple computations, it is obtained that, for any $x \geq 0$,

$$\frac{g_{1:3:4}^{1,2}(x)}{f_{1:3:4}^{1,2}(x)} \propto \frac{g_{Y_{2:3}^{1,2}}(x)}{f_{X_{2:3}^{1,2}}(x)} = \frac{e^{4x+4} - e^{3x+3}}{2(e^{2x+2} - 1)}.$$

Let $f(x) = \frac{e^{4x+4} - e^{3x+3}}{e^{2x+2} - 1}$, then $f'(x) \overset{sgn}{=} e^{3x+3}(2e^{3x+3} - e^{2x+2} - 4e^{x+1} + 3) = e^{3x+3}g(x)$, where $g(x) = 2e^{3x+3} - e^{2x+2} - 4e^{x+1} + 3$. Note that

$$g(0) = 2e^3 - e^2 - 4e + 3 > 0, \text{ and } g'(x) = e^{x+1}(6e^{2x+2} - 2e^{x+1} - 4) = e^{x+1}h(x),$$

where $h(x) = 6e^{2x+2} - 2e^{x+1} - 4$. Note that

$$h(0) = 6e^2 - 2e - 4 > 0, \text{ and } h'(x) = 2e^{x+1}(6e^{x+1} - 2) > 0 \text{ for } x \geq 0,$$

thus, for $x \geq 0$, $h(x) \geq 0$, and hence $g'(x) \geq 0$, and then, $f'(x) \geq 0$. It follows that $f(x)$ is increasing in $x \geq 0$, which shows that $X_{1:3:4}^{1,2} \leq_{lr} Y_{1:3:4}^{1,2}$, and hence $X_{1:3:4}^{1,2} \leq_{hr} Y_{1:3:4}^{1,2}$.

In expression (3.3), when $r = k - 1$, it is obtained

$$P(X_{k:n} - t_2 > x \mid X_{k-1:n} < t_1 < X_{k:n}, X_{k:n} > t_2) = \frac{\bar{F}^{n-k+1}(t_2 + x)}{\bar{F}^{n-k+1}(t_2)}.$$

It is clear that this distribution above does not depend on t_1, hence the condition $t_1 < X_{k:n}$ may be deleted. Furthermore, this survival distribution can be regarded as that of $Z_{1:n-k+1} = \min(Z_1, \ldots, Z_{n-k+1})$, where $Z_1, \ldots, Z_{n-k+1}$ are IID random variables with the same law as $(X - t_2|X > t_2)$.

Therefore, if X has memoryless property, then

$$P(X_{k:n} - t_2 > x \mid X_{k-1:n} < t_1, X_{k:n} > t_2) = \bar{F}^{n-k+1}(x),$$

which is just the distribution of $\min\{X_1, X_1, \ldots, X_{n-k+1}\}$.

Moreover, it is well known that the exponential distribution is the only continuous distribution which has memoryless property. Hence in the following we give a characterization of exponential distribution.

Theorem 3.1.8 *The random variable X_1 has an exponential distribution if and only if for a fixed $k \in \{1, 2, \ldots, n\}$, $\min\{X_1, X_2, \ldots, X_{n-k+1}\}$ and $(X_{k:n} - t_2 \mid X_{k-1:n} < t_1, X_{k:n} > t_2)$ have the same distribution for all $t_2 > t_1 > 0$.*

3.1.1.3 The Stochastic Results of the Inactivity Time

Denote by $X_{r,k,n}^{t_1,(t_2)}$ the random variable $(t_2 - X_{k:n}|X_{r:n} < t_1 < X_{r+1:n}, X_{k:n} < t_2)$. It can be obtained, for $t_2 - x > t_1 > 0$,

$$\begin{aligned} P(X_{r,k,n}^{t_1,(t_2)} > x) &= P(t_2 - X_{k:n} > x \mid X_{r:n} < t_1 < X_{r+1:n}, X_{k:n} < t_2) \\ &= \frac{\sum_{i=k-r}^{n-r} \binom{n-r}{i} \left[1 - \frac{\bar{F}(t_2-x)}{\bar{F}(t_1)}\right]^i \left[\frac{\bar{F}(t_2-x)}{\bar{F}(t_1)}\right]^{n-r-i}}{\sum_{i=k-r}^{n-r} \binom{n-r}{i} \left[1 - \frac{\bar{F}(t_2)}{\bar{F}(t_1)}\right]^i \left[\frac{\bar{F}(t_2)}{\bar{F}(t_1)}\right]^{n-r-i}} \\ &= \frac{P(X_{n-k+1:n-r}^{t_1,(t_2)} > x)}{\sum_{i=k-r}^{n-r} \binom{n-r}{i} \left[1 - \frac{\bar{F}(t_2)}{\bar{F}(t_1)}\right]^i \left[\frac{\bar{F}(t_2)}{\bar{F}(t_1)}\right]^{n-r-i}}, \end{aligned} \quad (3.4)$$

where $X_{n-k+1:n-r}^{t_1,(t_2)}$ denotes the $(n-k+1)$th order statistic among $n-r$ IID variables with the same distribution as $(t_2 - X \mid X > t_1)$.

According to (3.4), Corollaries 3.1.2, 3.1.3 and Theorem 3.1.4 in former section, it is easy to obtain the following theorem.

Theorem 3.1.9 *For* $t_2 > t_1 > 0$,

(1) *and fixed* n, k, $X_{r:k:n}^{t_1,(t_2)} \geq_{lr} X_{r-1:k:n}^{t_1,(t_2)}$;
(2) *and fixed* r, k, $X_{r:k:n}^{t_1,(t_2)} \geq_{lr} X_{r:k:n-1}^{t_1,(t_2)}$;
(3) *and fixed* n, r, $X_{r:k:n}^{t_1,(t_2)} \leq_{lr} X_{r:k-1:n}^{t_1,(t_2)}$.

For a $(n-k+1)$-out-of-n system with independent random units $X_1, X_2, \ldots, X_n$, which have identical distribution F as X. For another $(n-k+1)$-out-of-n system with independent random units $Y_1, Y_2, \ldots, Y_n$, which have common distribution G as Y. For $t_2 - x > t_1 > 0$, let $\bar{F}_{r:k:n}^{t_1,(t_2)}(x)$, $\bar{G}_{r,k,n}^{t_1,(t_2)}(x)$ be the reliability functions of the random variables $X_{r,k,n}^{t_1,(t_2)} = (t_2 - X_{k:n} | X_{r:n} < t_1 < X_{r+1:n}, X_{k:n} < t_2)$, $Y_{r:k:n}^{t_1,(t_2)} = (t_2 - Y_{k:n} | Y_{r:n} < t_1 < Y_{r+1:n}, Y_{k:n} < t_2)$, respectively.

The following two theorems conduct stochastic comparisons of two $(n-k+1)$-out-of-n systems based on their hazard rate ordering and likelihood ratio order.

Theorem 3.1.10 *If* $X \leq_{hr} Y$, *then* $X_{r:k:n}^{t_1,(t_2)} \geq_{hr} Y_{r:k:n}^{t_1,(t_2)}$.

Proof By the expression in (3.4), for $t_2 - x > t_1 > 0$, we have

$$\bar{F}_{r:k:n}^{t_1,(t_2)}(x) = \frac{P(X_{n-k+1:n-r}^{t_1,(t_2)} > x)}{\sum_{i=k-r}^{n-r} \binom{n-r}{i} \left[1 - \frac{\bar{F}(t_2)}{\bar{F}(t_1)}\right]^i \left[\frac{\bar{F}(t_2)}{\bar{F}(t_1)}\right]^{n-r-i}},$$

$$\bar{G}_{r:k:n}^{t_1,(t_2)}(x) = \frac{P(Y_{n-k+1:n-r}^{t_1,(t_2)} > x)}{\sum_{i=k-r}^{n-r} \binom{n-r}{i} \left[1 - \frac{\bar{G}(t_2)}{\bar{G}(t_1)}\right]^i \left[\frac{\bar{G}(t_2)}{\bar{G}(t_1)}\right]^{n-r-i}},$$

where $Y_{n-k+1:n-r}^{t_1,(t_2)}$ denotes the $(n-k+1)$th order statistic among the independent and identically variables with the same distribution as $(t_2 - Y \mid Y > t_1)$.

The condition $X \leq_{hr} Y$ implies that $(t_2 - X \mid X > t_1) \geq_{hr} (t_2 - Y \mid Y > t_1)$, hence $X_{n-k+1:n-r}^{t_1,(t_2)} \geq_{hr} Y_{n-k+1:n-r}^{t_1,(t_2)}$, which implies that the ratio

$$\frac{\bar{G}_{r:k:n}^{t_1,(t_2)}(x)}{\bar{F}_{r:k:n}^{t_1,(t_2)}(x)} \propto \frac{P(Y_{n-k+1:n-r}^{t_1,(t_2)} > x)}{P(X_{n-k+1:n-r}^{t_1,(t_2)} > x)}$$

is decreasing in $x > 0$, which validates the desired result.

Theorem 3.1.11 *If* $X \leq_{lr} Y$, *then* $X_{r:k:n}^{t_1,(t_2)} \geq_{lr} Y_{r:k:n}^{t_1,(t_2)}$.

Proof Denote $f_{r:k:n}^{t_1,(t_2)}(x)$ and $g_{r:k:n}^{t_1,(t_2)}(x)$ the respective density functions of $X_{r,k,n}^{t_1,(t_2)}$ and $Y_{r,k,n}^{t_1,(t_2)}$. Then from (3.4), for $t_2 - x > t_1 > 0$,

$$f_{r:k:n}^{t_1,(t_2)}(x) = \frac{f_{X_{n-k+1:n-r}^{t_1,(t_2)}}(x)}{\sum_{i=n-k+1}^{n-r} \binom{n-r}{i} \left[1 - \frac{\bar{F}(t_2)}{\bar{F}(t_1)}\right]^i \left[\frac{\bar{F}(t_2)}{\bar{F}(t_1)}\right]^{n-r-i}},$$

$$g_{r:k:n}^{t_1,(t_2)}(x) = \frac{g_{Y_{n-k+1:n-r}^{t_1,(t_2)}}(x)}{\sum_{i=n-k+1}^{n-r}\binom{n-r}{i}\left[1-\frac{\bar{G}(t_2)}{\bar{G}(t_1)}\right]^i\left[\frac{\bar{G}(t_2)}{\bar{G}(t_1)}\right]^{n-r-i}},$$

where $f_{X_{n-k+1:n-r}^{t_1,(t_2)}}(x)$, $g_{X_{n-k+1:n-r}^{t_1,(t_2)}}(x)$ are density functions of order statistics $X_{n-k+1:n-r}^{t_1,(t_2)}$ and $Y_{n-k+1:n-r}^{t_1,(t_2)}$, respectively.

The condition $X \leq_{lr} Y$ implies that $(t_2 - X \mid X > t_1) \geq_{lr} (t_2 - Y \mid Y > t_1)$, hence $X_{n-k+1:n-r}^{t_1,(t_2)} \geq_{lr} Y_{n-k+1:n-r}^{t_1,(t_2)}$, which implies that the ratio

$$\frac{g_{r:k:n}^{t_1,(t_2)}(x)}{f_{r:k:n}^{t_1,(t_2)}(x)} \propto \frac{g_{Y_{n-k+1:n-r}^{t_1,(t_2)}}(x)}{f_{X_{n-k+1:n-r}^{t_1,(t_2)}}(x)}$$

is decreasing in $x > 0$, which validates the desired result.

Example 3.1.12 Consider two 3-out-of-4 systems as Example 3.1.7. Let X and Y be exponential distributions with parameters $\mu = 2$, $\lambda = 1$, respectively. It is clear that $X \leq_{lr} Y$.

Set $r = 1$, $t_1 = 1$, $t_2 = 2$. Then by some computations, it is get that, for any $x \geq 0$,

$$\frac{g_{1:3:4}^{1,(2)}(x)}{f_{1:3:4}^{1,(2)}(x)} \propto \frac{g_{Y_{2:3}^{1,(2)}}(x)}{f_{X_{2:3}^{1,(2)}}(x)} = \frac{e^{4-4x} - e^{3-3x}}{2(e^{2-2x} - 1)}.$$

By Example 3.1.7, $X_{1:3:4}^{1,(2)} \geq_{lr} Y_{1:3:4}^{1,(2)}$, and hence $X_{1:3:4}^{1,(2)} \geq_{hr} Y_{1:3:4}^{1,(2)}$.

3.1.2 *On the Coherent System*

Based on the stochastically ordered *signatures* between systems, this section investigates some stochastic comparisons of the *residual lives* and the *inactivity times* of general coherent systems (or mixed systems) with n independent and identically distributed (IID) components, given that before time t_1 ($t_1 > 0$), exactly r ($r < n$) components have failed and at time t_2 ($t_2 > t_1$) (under double monitoring), the system are still working or they have failed.

3.1.2.1 Introduction

The signature of coherent systems is broadly useful in reliability analysis. It is a very important tool for investigating the performance of a coherent structure and comparing structures with each other. It has been proven to be quite a useful proxy for a system's design, as it is a distribution-free measure that efficiently captures the precise features of a system s design which influence its performance. One wants to consider a level playing field, as it is clear that a poor system design with good components typically outperforms a good system design with poor components. If, however, two different systems have common components lifetimes, then any difference between them must be attributable to the system design alone.

Kochar and Ma (1999) established several preservation theorems for signatures which demonstrate that signatures are quite useful in the comparison of competing systems. Li and Zhang (2008), and Zhang (2010a) subsequently investigated the residual lives and inactivity times of coherent systems with IID or exchangeable components and obtained a number of preservation theorems for signatures. Samaniego et al. (2009) defined the dynamic signature of a used system, and by means of it some known representation and preservation theorems for system signatures are generalized to dynamic versions.

Mixture representations have proven to be useful in the comparison on the performance of competing systems. Navarro et al. (2008a) showed that if $T = \tau(X_1, \dots, X_n)$ is the lifetime of a coherent system with IID component lifetimes $X_1, \dots, X_n$ distributed according to a common continuous distribution F, then the distribution of the system lifetime T given $T > t$ is a mixture of the residual lifetimes of k-out-of-n systems. That is, for all $t \geq 0$ and $x \geq 0$,

$$P(T - t > x \mid T > t) = \sum_{i=1}^{n} p_i(t) P(X_{i:n} - t > x \mid X_{i:n} > t), \qquad (3.5)$$

where coefficients $p_i(t) = p_i P(X_{i:n} > t)/\bar{F}(t)$ for $i = 1, \dots, n$, and $\bar{F}(t) > 0$.

By using the mixture representation (3.5), the following preservation theorem was obtained for system signatures.

Theorem 3.1.13 *Let $T_1 = \tau_1(X_1, \dots, X_n)$ and $T_2 = \tau_2(X_1, \dots, X_n)$ be the lifetimes of two mixed systems, both based on n components with IID lifetimes distributed according to the common continuous distribution F. And, for all $t \geq 0$, let $\mathbf{p}(t) = (p_1(t), \dots, p_i(t), \dots, p_n(t))$ and $\mathbf{q}(t) = (q_1(t), \dots, q_i(t), \dots, q_n(t))$ be their corresponding coefficient vectors.*

(a) If $\mathbf{p}(t) \leq_{st} \mathbf{q}(t)$, then, $(T_1 - t \mid T_1 > t) \leq_{st} (T_2 - t \mid T_2 > t)$;
(b) If $\mathbf{p}(t) \leq_{hr} \mathbf{q}(t)$, then, $(T_1 - t \mid T_1 > t) \leq_{hr} (T_2 - t \mid T_2 > t)$;
(c) If $\mathbf{p}(t) \leq_{lr} \mathbf{q}(t)$, then, $(T_1 - t \mid T_1 > t) \leq_{lr} (T_2 - t \mid T_2 > t)$.

Samaniego et al. (2009) defined the dynamic signature of a used system with lifetime T. Specifically, let $\mathbf{p}$ be the signature of a coherent system based on n IID components. Suppose the system is operating at time t, and the event $\{T > t\} \bigcap \{X_{i:n} < t < X_{i+1:n}\}$ is noted. Then at time t, there are $n - i$ working components. The dynamic signature of the system at time t is the $n - i$ dimensional vector $\mathbf{p}$ whose k-th element is $p_k = P(T = \{X_{k:n} | T > t\} \bigcap \{X_{i:n} < t < X_{i+1:n}\})$ for $k = i + 1, \dots, n$. By means of it some known representations and preservation theorems for system signatures are generalized to dynamic versions.

The purpose of this section is to extend the results above to the coherent systems (or mixed systems) given that before time t_1 ($t_1 \geq 0$), exactly r components have failed and at t_2 ($t_2 \geq t_1$) (under double monitoring), the system is still working or it has failed. Specifically, a mixture representation of the reliability function of the residual life of a system under double monitoring is obtained, and some stochastic properties and preservation theorems for coefficient vectors are obtained. Moreover, a mixture representation of the reliability function of the inactivity time of a system under double monitoring is obtained, and some preservation theorems for signatures are established.

3.1.2.2 Stochastic Comparisons of Residual Life

This subsection investigates the residual lifetime of a coherent system, under the condition that before time t_1 ($t_1 \geq 0$), exactly r ($r < n$) components have failed and at time t_2 ($t_2 \geq t_1$), the system is still working.

Let $T_1 = \tau_1(X_1, X_2, \ldots, X_n)$, $T_2 = \tau_2(X_1, X_2, \ldots, X_n)$ be the lifetimes of two coherent systems of size n, both based on components with IID lifetimes $X_1, X_2, \ldots, X_n$ having a common distribution function F, where τ_1 and τ_2 are coherent life functions. Assume that $\mathbf{p}$ and $\mathbf{q}$ are the signatures of τ_1 and τ_2, respectively. Under double monitoring, if both systems are still working at time t_2 with probability one, then the corresponding signatures must be of the forms $\mathbf{p} = (0, \ldots, 0, p_s, \ldots, p_n)$, and $\mathbf{q} = (0, \ldots, 0, q_s, \ldots, q_n)$, $r < s \leq n$.

The following theorem gives the mixture representation of the reliability function of the residual life of a coherent system under double monitoring.

Theorem 3.1.14 *For $t_2 \geq t_1 \geq 0$, $r < s \leq n$ and any $x > 0$,*

$$P(T_1 - t_2 > x \mid A_r(t_1), T_1 > t_2) = \sum_{i=s}^{n} p_i(r, t_1, t_2) P(X_{i:n} - t_2 > x, A_r(t_1) | X_{i:n} > t_2, A_r(t_1)),$$

where $A_r(t_1)$ indicates the event $(X_{r:n} < t_1 < X_{r+1:n})$, and

$$p_i(r, t_1, t_2) = \frac{p_i P(X_{i:n} > t_2, A_r(t_1))}{\sum_{i=s}^{n} p_i P(X_{i:n} > t_2, A_r(t_1))} \tag{3.6}$$

such that $\sum_{i=s}^{n} p_i(r, t_1, t_2) = 1$.

Proof Note that, for $t_2 \geq t_1 \geq 0$, and $r < i \leq n$,

$$P(X_{i:n} - t_2 > x, A_r(t_1)) = P(X_{i:n} - t_2 > x, A_r(t_1) | X_{i:n} > t_2, A_r(t_1)) P(X_{i:n} > t_2, A_r(t_1)),$$

hence, for $r < s \leq n$ and any $x > 0$,

$$P(T_1 - t_2 > x \mid A_r(t_1), T_1 > t_2)$$

$$
\begin{aligned}
&= \frac{P(T_1 - t_2 > x, A_r(t_1))}{P(A_r(t_1), T_1 > t_2)} \\
&= \frac{\sum_{i=s}^{n} P(T_1 = X_{i:n}, T_1 - t_2 > x, A_r(t_1))}{\sum_{i=s}^{n} P(A_r(t_1), T_1 > t_2, T_1 = X_{i:n})} \\
&= \frac{\sum_{i=s}^{n} P(T_1 = X_{i:n}) P(T_1 - t_2 > x, A_r(t_1) \mid T_1 = X_{i:n})}{\sum_{i=s}^{n} P(T_1 = X_{i:n}) P(A_r(t_1), T_1 > t_2 \mid T_1 = X_{i:n})} \\
&= \frac{\sum_{i=s}^{n} p_i P(T_1 - t_2 > x, A_r(t_1) \mid T_1 = X_{i:n})}{\sum_{i=s}^{n} p_i P(A_r(t_1), T_1 > t_2 \mid T_1 = X_{i:n})} \\
&= \frac{\sum_{i=s}^{n} p_i P(X_{i:n} - t_2 > x, A_r(t_1))}{\sum_{i=s}^{n} p_i P(X_{i:n} > t_2, A_r(t_1))} \\
&= \sum_{i=s}^{n} p_i(r, t_1, t_2) P(X_{i:n} - t_2 > x, A_r(t_1) | X_{i:n} > t_2, A_r(t_1)).
\end{aligned}
$$

This completes the proof.

Remark The function $p_i(r, t_1, t_2)$ may be identified as the probability of $P(T_1 = X_{i:n} | T_1 > t_2, A_r(t_1))$ as follows:

$$
\begin{aligned}
P(T_1 = X_{i:n} | T_1 > t_2, A_r(t_1)) &= \frac{P(T_1 = X_{i:n}, T_1 > t_2, A_r(t_1))}{P(T_1 > t_2, A_r(t_1))} \\
&= \frac{P(T_1 = X_{i:n}) P(T_1 > t_2, A_r(t_1) | T_1 = X_{i:n})}{P(T_1 > t_2, A_r(t_1))} \\
&= \frac{P(T_1 = X_{i:n}) P(X_{i:n} > t_2, A_r(t_1) | T_1 = X_{i:n})}{P(T_1 > t_2, A_r(t_1))} \\
&= \frac{p_i P(X_{i:n} > t_2, A_r(t_1))}{P(T_1 > t_2, A_r(t_1))} \\
&= p_i(r, t_1, t_2),
\end{aligned}
$$

where the forth equality follows from the fact that the events $\{T_1 = X_{i:n}\}$ and $\{X_{i:n} > t_2, A_r(t_1)\}$ are independent when the component lifetimes are IID

Remark It follows from the result of Theorem 3.1.14, that the residual lifetime $(T_1 - t_2 \mid A_r(t_1), T_1 > t_2)$ of the system under double monitoring is a mixture of the residual lifetime $(X_{i:n} - t_2, A_r(t_1) | X_{i:n} > t_2, A_r(t_1))$ of an $(n - i + 1)$-out-of-n system under double monitoring with coefficients $p_i(r, t_1, t_2)$ for $i = s, s + 1, \ldots, n$. The coefficients vector

$$
\mathbf{p}(r, t_1, t_2) = (0, \ldots, 0, p_s(r, t_1, t_2), \ldots, p_n(r, t_1, t_2)), \quad r < s \leq n
$$

is the conditional distribution of the ordered component lifetimes that would cause the system to fail given $(A_r(t_1), T_1 > t_2)$. The coefficients depend both on the system structure function τ_1 and F. It should be noted that $\mathbf{p}(r, t, t) = (0, \ldots, 0, p_s, \ldots, p_n) \equiv \mathbf{p}, \; r < s \leq n$.

The following result shows that any coherent system with signature $\mathbf{p} = (0, \ldots, 0, p_s, \ldots, p_n)$ has a tail stochastic behavior similar to that of parallel system.

Theorem 3.1.15 *Assume that a coherent system with lifetime T_1 and signature $\mathbf{p} = (0, \ldots, 0, p_s, \ldots, p_n)$, then $\lim_{t_2 \to \infty} \mathbf{p}(r, t_1, t_2) = (\underbrace{0, 0, \ldots, 0}_{n-1 \text{ times}}, 1)$ for fixed $t_1 > 0$.*

Proof Note that, for $t_2 \geq t_1 \geq 0$, and $r < i \leq n$,

$$P(X_{i:n} > t_2, A_r(t_1)) = \binom{n}{r} F^r(t_1) \sum_{j=n-i+1}^{n-r} \binom{n-r}{j} \bar{F}^j(t_2)(\bar{F}(t_1) - \bar{F}(t_2))^{n-r-j}$$

$$= \binom{n}{r} F^r(t_1)\bar{F}^{n-r}(t_1) \sum_{j=n-i+1}^{n-r} \binom{n-r}{j} \frac{\bar{F}^j(t_2)}{\bar{F}^j(t_1)} \left(1 - \frac{\bar{F}(t_2)}{\bar{F}(t_1)}\right)^{n-r-j}$$

$$= \binom{n}{r} F^r(t_1)\bar{F}^{n-r}(t_1) \sum_{j=n-i+1}^{n-r} (-1)^{j-n+i-1} \binom{j-1}{n-i} \binom{n-r}{j} \bar{F}_{t_1}^j(\Delta t),$$

where the last equality follows from David and Nagaraja (2003), and $\Delta t = t_2 - t_1$. Thus it can be easily obtained that

$$\lim_{t_2 \to \infty} \frac{P(X_{i:n} > t_2, A_r(t_1))}{P(X_{k:n} > t_2, A_r(t_1))} = \begin{cases} +\infty, & \text{if } i > k, \\ 1, & \text{if } i = k, \\ 0, & \text{if } i < k. \end{cases}$$

It follows that

$$\lim_{t_2 \to \infty} p_k(r, t_1, t_2) = \lim_{t_2 \to \infty} \frac{p_k P(X_{k:n} > t_2, A_r(t_1))}{\sum_{i=s}^{n} p_i P(X_{i:n} > t_2, A_r(t_1))} = \begin{cases} 1, & \text{if } k = n, \\ 0, & \text{if } k < n, \end{cases}$$

Hence the result holds.

Theorem 3.1.16 *Let T_1, T_2 be the lifetimes of two coherent systems of size n, both based on components with IID lifetimes $X_1, X_2, \ldots, X_n$ distributed according to a common distribution function F, where $\mathbf{p} = (0, \ldots, 0, p_s, \ldots, p_n)$ and $\mathbf{q} = (0, \ldots, 0, q_s, \ldots, q_n)$ are the corresponding signatures. For any $t_2 \geq t_1 \geq 0$, suppose $\mathbf{p}(r, t_1, t_2) = (0, \ldots, 0, p_s(r, t_1, t_2), \ldots, p_n(r, t_1, t_2))$, $\mathbf{q}(r, t_1, t_2) = (0, \ldots, 0, q_s(r, t_1, t_2), \ldots, q_n(r, t_1, t_2))$ are the corresponding coefficient vectors, for $r < s \leq n$. Then, the following results holds.*

(a) If $\mathbf{p}(r, t_1, t_2) \leq_{st} \mathbf{q}(r, t_1, t_2)$, then, for any $t_2 \geq t_1 \geq 0$,

$$(T_1 - t_2 \mid A_r(t_1), T_1 > t_2) \leq_{st} (T_2 - t_2 \mid A_r(t_1), T_2 > t_2);$$

(b) If $\mathbf{p}(r, t_1, t_2) \leq_{hr} \mathbf{q}(r, t_1, t_2)$, then, for any $t_2 \geq t_1 \geq 0$,

$$(T_1 - t_2 \mid A_r(t_1), T_1 > t_2) \leq_{hr} (T_2 - t_2 \mid A_r(t_1), T_2 > t_2);$$

(c) If $\mathbf{p}(r, t_1, t_2) \leq_{lr} \mathbf{q}(r, t_1, t_2)$, *then, for any* $t_2 \geq t_1 \geq 0$,

$$(T_1 - t_2 \mid A_r(t_1), T_1 > t_2) \leq_{lr} (T_2 - t_2 \mid A_r(t_1), T_2 > t_2).$$

Proof It is well-known from Shaked and Shanthikumar (2007) that $X_{i:n} \leq_{lr} X_{i+1:n}$ for $i = 1, \ldots, n-1$, which implies that

$$(X_{i:n} - t_2, A_r(t_1)|X_{i:n} > t_2, A_r(t_1)) \leq_{lr} (X_{i+1:n} - t_2, A_r(t_1)|X_{i+1:n} > t_2, A_r(t_1))$$

(see Theorem 1.C.6 of Shaked and Shanthikumar (2007)), and hence

$$(X_{i:n} - t_2, A_r(t_1)|X_{i:n} > t_2, A_r(t_1)) \leq_{st} (\leq_{hr})(X_{i+1:n} - t_2, A_r(t_1)|X_{i+1:n} > t_2, A_r(t_1)).$$

Using the conditions, and Theorems 1.A.6, 1.B.14 and 1.C.17 of Shaked and Shanthikumar (2007), respectively, it is easy to show (a), (b) and (c). This completes the proof.

Example 3.1.17 Consider the two coherent systems of order 5 depicted in Fig. 3.1. The signature of the system on the left with lifetime $\max\{X_1, X_2, \min\{X_3, X_4, X_5\}\}$ is $\mathbf{q} = \left(0, 0, \frac{3}{10}, \frac{3}{10}, \frac{2}{5}\right)$. With some computations, one can show that, for any $t_2 \geq t_1 \geq 0$ and $r = 1$ (or 2), the corresponding coefficient vector is

$$\mathbf{q}(r, t_1, t_2) = \left(0, 0, \frac{3\bar{F}^2(t_2)}{A(t_1, t_2)}, \frac{9\bar{F}(t_1)\bar{F}(t_2) - 6\bar{F}^2(t_2)}{A(t_1, t_2)}, \frac{4\bar{F}^2(t_2) - 12\bar{F}(t_1)\bar{F}(t_2) + 12\bar{F}^2(t_1)}{A(t_1, t_2)}\right),$$

where $A(t_1, t_2) = \bar{F}^2(t_2) + 12\bar{F}^2(t_1) - 3\bar{F}(t_1)\bar{F}(t_2)$. The signature of the system on the right with lifetime $\max\{\min\{X_1, X_2\}, \min\{X_3, X_4\}, X_5\}$ is $\mathbf{p} = \left(0, 0, \frac{2}{5}, \frac{2}{5}, \frac{1}{5}\right)$, and for any $t_2 \geq t_1 \geq 0$ and $r = 1$ (or 2), the corresponding coefficient vector is

$$\mathbf{p}(r, t_1, t_2) = \left(0, 0, \frac{2\bar{F}^2(t_2)}{B(t_1, t_2)}, \frac{6\bar{F}(t_1)\bar{F}(t_2) - 4\bar{F}^2(t_2)}{B(t_1, t_2)}, \frac{\bar{F}^2(t_2) - 3\bar{F}(t_1)\bar{F}(t_2) + 3\bar{F}^2(t_1)}{B(t_1, t_2)}\right),$$

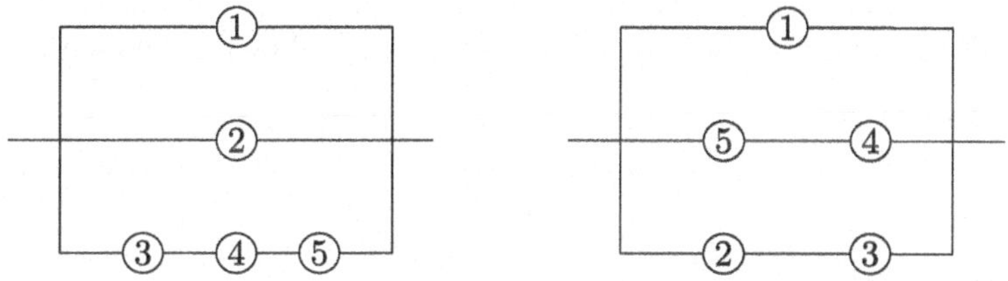

Fig. 3.1 Two coherent systems with likelihood ratio ordered coefficients vectors

where $B(t_1, t_2) = -\bar{F}^2(t_2) + 3\bar{F}^2(t_1) + 3\bar{F}(t_1)\bar{F}(t_2)$. It can be verified that $\mathbf{p}(r, t_1, t_2) \leq_{lr} \mathbf{q}(r, t_1, t_2)$ and hence $\mathbf{p}(r, t_1, t_2) \leq_{hr} (\leq_{st})\mathbf{q}(r, t_1, t_2)$. By Theorem 3.1.16, the system on the left is better in the sense that it has stochastically longer general residual life under the condition that at time t_1 ($t_1 \geq 0$), given that exactly 1 (or 2) component(s) has (have) failed and at time t_2 ($t_2 \geq t_1$), the systems are still working.

Samaniego et al. (2009) have shown that if an operating system of order n is inspected at time t ($t \geq 0$) and it is noted that precisely r failures have occurred, then the vector $\mathbf{s} \in [0, 1]^{n-r}$ whose jth element is the probability that the $(r + j)$th component failure is fatal to the system for $j = 1, 2, \ldots, n - r$, is a distribution-free measure of the design of the residual life of the system. Based on this fact they defined the dynamic signature of a working but used system having age t ($t \geq 0$). If $t_2 = t_1 = t$ in Theorem 3.1.14, then, by Samaniego et al. (2009), the residual reliability function of working but used system can be represented as a mixture of the distribution of order statistic from a random sample of size $n - r$ from the same distribution as $(X_1 - t|X_1 > t)$ with the dynamic signatures being weights, and hence some preservation theorems for the dynamic signatures were established.

3.1.2.3 Stochastic Comparisons of the Inactivity Time

In this part, we investigate the inactivity times of coherent systems, under the condition that before time t_1 ($t_1 \geq 0$), exactly r ($r < n$) components have failed and at time t_2 ($t_2 \geq t_1$), both systems have failed.

Let $T_1 = \tau_1(X_1, X_2, \ldots, X_n)$, $T_2 = \tau_2(X_1, X_2, \ldots, X_n)$ be the lifetimes of two coherent systems of size n, both based on components with IID lifetimes $X_1, X_2, \ldots, X_n$ having common distribution function F, where τ_1 and τ_2 are coherent life functions. Assume that $\mathbf{p}$ and $\mathbf{q}$ are the signatures of τ_1 and τ_2, respectively. Given that at time t_1 ($t_1 \geq 0$), exactly r ($r < n$) components have failed and at time t_2 ($t_2 \geq t_1$), both systems have failed, the corresponding signatures must be of the forms $\mathbf{p} = (0, \ldots, 0, p_l, \ldots, p_m, 0, \ldots, 0)$, $\mathbf{q} = (0, \ldots, 0, q_l, \ldots, q_m, 0, \ldots, 0)$, for $r < l \leq m \leq n$.

Similar to Theorem 3.2.2 in last subsection, it can be shown that if T_1 is the lifetime of a coherent system with signature $\mathbf{p} = (0, \ldots, 0, p_l, \ldots, p_m, 0, \ldots, 0)$ for $r < l \leq m \leq n$, then the reliability function of its inactivity time of the coherent system is a mixture of the reliability functions of the inactivity times of i-out-of-n systems under double monitoring, as described in the following theorem.

Theorem 3.1.18 *For $r < l \leq m \leq n$, $0 \leq t_1 \leq t_2$ and any $x > 0$,*

$$P(t_2 - T_1 > x \mid A_r(t_1), T_1 < t_2) = \sum_{i=l}^{m} p_i(r, t_1, t_2) P(t_2 - X_{i:n} > x, A_r(t_1)|X_{i:n} < t_2, A_r(t_1)),$$

where

$$p_i(r, t_1, t_2) = \frac{p_i P(X_{i:n} < t_2, A_r(t_1))}{\sum_{i=l}^{m} p_i P(X_{i:n} < t_2, A_r(t_1))} \tag{3.7}$$

for $i = l, \ldots, m$, *such that* $\sum_{i=l}^{m} p_i(r, t_1, t_2) = 1$.

Remark The function $p_i(r, t_1, t_2)$ may be identified as the probability of $P(T_1 = X_{i:n}|T_1 < t_2, A_r(t_1))$ as follows:

$$\begin{aligned} P(T_1 = X_{i:n}|T_1 < t_2, A_r(t_1)) &= \frac{P(T_1 = X_{i:n}, T_1 < t_2, A_r(t_1))}{P(T_1 < t_2, A_r(t_1))} \\ &= \frac{P(T_1 = X_{i:n})P(T_1 < t_2, A_r(t_1)|T_1 = X_{i:n})}{P(T_1 < t_2, A_r(t_1))} \\ &= \frac{P(T_1 = X_{i:n})P(X_{i:n} < t_2, A_r(t_1)|T_1 = X_{i:n})}{P(T_1 < t_2, A_r(t_1))} \\ &= \frac{p_i P(X_{i:n} < t_2, A_r(t_1))}{P(T_1 < t_2, A_r(t_1))} \\ &= p_i(r, t_1, t_2), \end{aligned}$$

where the forth equality follows from the fact that the events $\{T_1 = X_{i:n}\}$ and $\{X_{i:n} < t_2, A_r(t_1)\}$ are independent when the components lifetimes are IID

Remark It follows from the result of Theorem 3.1.18, that the inactivity time $(t_2 - T_1 \mid A_r(t_1), T_1 < t_2)$ of the system under double monitoring is a mixture of the inactivity times $(t_2 - X_{i:n}, A_r(t_1)|X_{i:n} < t_2, A_r(t_1))$ of i-out-of-n systems under double monitoring with coefficients $p_i(r, t_1, t_2)$ for $i = l, l+1, \ldots, m$. And the vector of coefficients

$$\mathbf{p}(r, t_1, t_2) = (0, \ldots, 0, p_l(r, t_1, t_2), \ldots, p_m(r, t_1, t_2), 0, \ldots, 0), \quad n \geq r > s$$

is the conditional distribution of the ordered component lifetime that is fatal to the system given $(A_r(t_1), T_1 < t_2)$. Also the coefficients depend both on the coherent function τ_1 and F.

The result below shows that any coherent system with signature $\mathbf{p} = (0, \ldots, 0, p_l, \ldots, p_m, 0, \ldots, 0)$ has its stochastic behavior similar to that of k-out-of-n system.

Theorem 3.1.19 *Assume that a coherent system with signature* $\boldsymbol{p} = (0, \ldots, 0, p_l, \ldots, p_m, 0, \ldots, 0)$ *and lifetime* T_1, *then* $\lim_{t_2 \to \infty} \boldsymbol{p}(r, t_1, t_2) = (\underbrace{0, \ldots, 0}_{l-1 \text{ times}}, 1, \underbrace{0, \ldots, 0}_{n-l \text{ times}})$ *for fixed* $t_1 > 0$.

The following lemma is useful in proving Theorem 3.1.21.

Lemma 3.1.20 (Shaked and Shanthikumar (2007)) *Let α and β be real valued functions such that α is nonnegative and β/α and α are non-increasing. If X_i has distribution F_i, $i = 1, 2$ and $X_1 \leq_{rh} X_2$, then*

$$\frac{\int_{-\infty}^{\infty} \beta(x)\,\mathrm{d}F_1(x)}{\int_{-\infty}^{\infty} \alpha(x)\,\mathrm{d}F_1(x)} \geq \frac{\int_{-\infty}^{\infty} \beta(x)\,\mathrm{d}F_2(x)}{\int_{-\infty}^{\infty} \alpha(x)\,\mathrm{d}F_2(x)}.$$

Theorem 3.1.21 *Suppose $\boldsymbol{p} = (0, \ldots, 0, p_l, \ldots, p_m, 0, \ldots, 0)$, $\boldsymbol{q} = (0, \ldots, 0, q_l, \ldots, q_m, 0, \ldots, 0)$ are the respective signatures of two coherent systems with lifetimes T_1, T_2 having common IID component lifetimes $X_1, X_2, \ldots, X_n$, for $r < l \leq m \leq n$. For any $0 \leq t_1 \leq t_2$,*

$$\boldsymbol{p}(r, t_1, t_2) = (0, \ldots, 0, p_l(r, t_1, t_2), \ldots, p_m(r, t_1, t_2), 0, \ldots, 0),$$

$$\boldsymbol{q}(r, t_1, t_2) = (0, \ldots, 0, q_l(r, t_1, t_2), \ldots, q_m(r, t_1, t_2), 0, \ldots, 0)$$

are the corresponding vectors of coefficients, for $r < s \leq n$. The following are true:

(a) If $\boldsymbol{p}(r, t_1, t_2) \leq_{st} \boldsymbol{q}(r, t_1, t_2)$, then, for $t_2 \geq t_1 \geq 0$,

$$(t_2 - T_1 \mid A_r(t_1), T_1 < t_2) \geq_{st} (t_2 - T_2 \mid A_r(t_1), T_2 < t_2);$$

(b) If $\boldsymbol{p}(r, t_1, t_2) \leq_{rh} \boldsymbol{q}(r, t_1, t_2)$, then, for $t_2 \geq t_1 \geq 0$,

$$(t_2 - T_1 \mid A_r(t_1), T_1 < t_2) \geq_{hr} (t_2 - T_2 \mid A_r(t_1), T_2 < t_2);$$

(c) If $\boldsymbol{p}(r, t_1, t_2) \leq_{lr} \boldsymbol{q}(r, t_1, t_2)$, then, for $t_2 \geq t_1 \geq 0$,

$$(t_2 - T_1 \mid A_r(t_1), T_1 < t_2) \geq_{lr} (t_2 - T_2 \mid A_r(t_1), T_2 < t_2).$$

Proof (a) Note that $(t_2 - X_{i:n}, A_r(t_1)|X_{i:n} < t_2, A_r(t_1))$ is decreasing in i in the likelihood ratio order, which implies that $P(t_2 - X_{i:n} > x, A_r(t_1)|X_{i:n} < t_2, A_r(t_1))$ is decreasing in $i = l, \ldots, m$. Hence

$$\begin{aligned} P(t_2 - T_1 > x \mid A_r(t_1), T_1 < t_2) &= \sum_{i=l}^{m} p_i(r, t_1, t_2) P(t_2 - X_{i:n} > x, A_r(t_1)|X_{i:n} < t_2, A_r(t_1)) \\ &\geq \sum_{i=l}^{m} q_i(r, t_1, t_2) P(t_2 - X_{i:n} > x, A_r(t_1)|X_{i:n} < t_2, A_r(t_1)) \\ &= P(t_2 - T_2 > x \mid A_r(t_1), T_2 < t_2). \end{aligned}$$

The inequality follows from Shaked and Shanthikumar (2007). This proves the result of (a).

(b) It is enough to prove that

$$\frac{P(t_2 - T_1 > x \mid A_r(t_1), T_1 < t_2)}{P(t_2 - T_2 > x \mid A_r(t_1), T_2 < t_2)} = \frac{\sum_{i=l}^{m} p_i(r, t_1, t_2) P(t_2 - X_{i:n} > x, A_r(t_1) | X_{i:n} < t_2, A_r(t_1))}{\sum_{i=l}^{m} q_i(r, t_1, t_2) P(t_2 - X_{i:n} > x, A_r(t_1) | X_{i:n} < t_2, A_r(t_1))}$$

is increasing in $x \geq 0$, that is, for all $x_2 \geq x_1 \geq 0$,

$$\frac{\sum_{i=l}^{m} p_i(r, t_1, t_2) P(t_2 - X_{i:n} > x_2, A_r(t_1) | X_{i:n} < t_2, A_r(t_1))}{\sum_{i=l}^{m} p_i(r, t_1, t_2) P(t_2 - X_{i:n} > x_1, A_r(t_1) | X_{i:n} < t_2, A_r(t_1))}$$
$$\geq \frac{\sum_{i=l}^{m} q_i(r, t_1, t_2) P(t_2 - X_{i:n} > x_2, A_r(t_1) | X_{i:n} < t_2, A_r(t_1))}{\sum_{i=l}^{m} q_i(r, t_1, t_2) P(t_2 - X_{i:n} > x_1, A_r(t_1) | X_{i:n} < t_2, A_r(t_1))}.$$

Let $\alpha(i) = P(t_2 - X_{i:n} > x_1, A_r(t_1) | X_{i:n} < t_2, A_r(t_1))$ and $\beta(i) = P(t_2 - X_{i:n} > x_2, A_r(t_1) | X_{i:n} < t_2, A_r(t_1))$. Note that $\alpha(i)$ is decreasing in $i = l, \ldots, m$, and it is easy to show that $P(t_2 - X_{i:n} > x, A_r(t_1) | X_{i:n} < t_2, A_r(t_1))$ is RR2 in $i = l, \ldots, m$ and $x \geq 0$, which implies that $\beta(i)/\alpha(i)$ is decreasing in i. Hence from Lemma 3.1.20, the desired result of (b) follows.

(c) Let $f_{r:n}^{t_1,(t_2)}(x)$, $g_{r:n}^{t_1,(t_2)}(x)$ denote the density functions of $(t_2 - T_1 \mid A_r(t_1), T_1 < t_2)$ and $(t_2 - T_2 \mid A_r(t_1), T_2 < t_2)$, respectively. By Theorem 6, we have

$$f_{r:n}^{t_1,(t_2)}(x) = \sum_{i=l}^{m} p_i(r, t_1, t_2) f_{X_{i:r:n}^{t_1,(t_2)}}(x),$$

$$g_{r:n}^{t_1,(t_2)}(x) = \sum_{i=l}^{m} q_i(r, t_1, t_2) f_{X_{i:r:n}^{t_1,(t_2)}}(x),$$

where $f_{X_{i:r:n}^{t_1,(t_2)}}(x)$ is the density function of the random variable $(t_2 - X_{i:n}, A_r(t_1) | X_{i:n} < t_2, A_r(t_1))$.

For $c > 0$, let us consider the function

$$h(x) = f_{r:n}^{t_1,(t_2)}(x) - c g_{r:n}^{t_1,(t_2)}(x) = \sum_{i=l}^{m} [p_i(r, t_1, t_2) - c q_i(r, t_1, t_2)] f_{X_{i:r:n}^{t_1,(t_2)}}(x).$$

From Theorem 1.C.37 of Shaked and Shanthikumar (2007), again, for fixed r and n, $f_{X_{i:r:n}^{t_1,(t_2)}}(x)$ is RR2 in $x \in \Re^+$ and $i = l, \ldots, m$. Because $\mathbf{p}(r, t_1, t_2) \leq_{lr} \mathbf{q}(r, t_1, t_2)$, $p_i(r, t_1, t_2)/q_i(r, t_1, t_2)$ is decreasing in $i = l, \ldots, m$, the sequence $\{p_i(r, t_1, t_2) - cq_i(r, t_1, t_2)\}$ has at most one change of sign from positive to negative as i ranges from l to m. According to Karlin (1968), we have $h(x)$ has at most one change of sign from positive to negative as x increases. This proves the result of (c).

Example 3.1.22 Consider the two systems of order 5 depicted in Fig. 3.2, the signature of system on the left with lifetime $\max\{\min\{X_1, X_2\}, \min\{X_3, X_4, X_5\}\}$ is $\mathbf{p} = \left(0, \frac{3}{5}, \frac{3}{10}, \frac{1}{10}, 0\right)$, and for any $t_2 \geq t_1 \geq 0$ and $r = 1$, the corresponding coefficient vector is

$$\mathbf{p}(1, t_1, t_2) = \left(0, \frac{3P(X_{2:5} < t_2, E_1(t_1))}{5B(t_1, t_2)}, \frac{3P(X_{3:5} < t_2, E_1(t_1))}{10B(t_1, t_2)}, \frac{P(X_{4:5} < t_2, E_1(t_1))}{10B(t_1, t_2)}, 0\right),$$

where $B(t_1, t_2) = \sum_{i=2}^{4} p_i P(X_{i:5} < t_2, E_1(t_1))$. The signature of the system on the right with lifetime $\max\{\min\{X_1, X_2\}, \min\{X_3, \max\{X_4, X_5\}\}\}$ is $\mathbf{q} = \left(0, \frac{1}{5}, \frac{3}{5}, \frac{1}{5}, 0\right)$, and for any $t_2 \geq t_1 \geq 0$, the corresponding coefficient vector is

$$\mathbf{q}(1, t_1, t_2) = \left(0, \frac{P(X_{2:5} < t_2, A_1(t_1))}{5C(t_1, t_2)}, \frac{3P(X_{3:5} < t_2, A_1(t_1))}{5C(t_1, t_2)}, \frac{P(X_{4:5} < t_2, A_1(t_1))}{5C(t_1, t_2)}, 0\right),$$

where $C(t_1, t_2) = \sum_{i=2}^{4} q_i P(X_{i:5} < t_2, E_1(t_1))$. It can be verified that $\mathbf{p}(1, t_1, t_2) \leq_{lr} \mathbf{q}(1, t_1, t_2)$, and hence $\mathbf{p}(1, t_1, t_2) \leq_{rh} (\leq_{st})\mathbf{q}(1, t_1, t_2)$. By the theorem above the system in the left side of Fig. 3.2 is better than the system on the right in the sense of inactivity time under the condition that at time t_1 ($t_1 \geq 0$), exactly 1 component has failed and at time t_2 ($t_2 \geq t_1$), both systems have failed.

3.1.2.4 Conclusions

In this part, the mixture representations of the reliability functions of the residual life and inactivity time of a coherent system with n IID components are obtained, given some particular information on the state of the coherent system upon inspection at

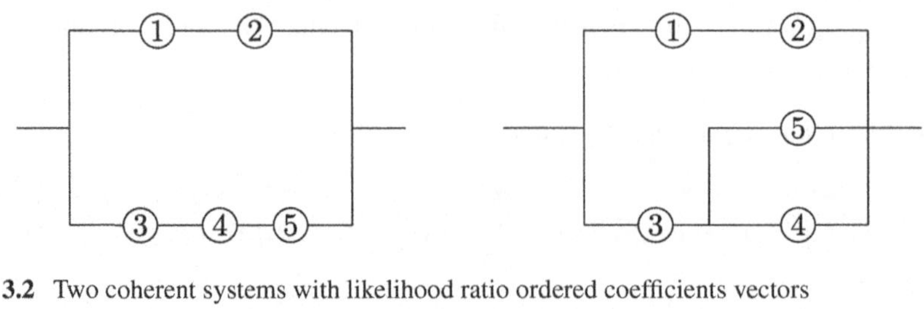

Fig. 3.2 Two coherent systems with likelihood ratio ordered coefficients vectors

time t_1 and t_2 ($t_2 \geq t_1$). Some preservation results of the *residual live* and the *inactivity time* of the system are obtained. The application of these results is illustrated in a few examples in which the system's reliabilities are computed and compared. The results extend previous ones in the literature and are useful for comparing similar systems that have different structure functions.

3.2 The Residual Life and Inactivity Time of k-Out-of-n System with INID Components

This section investigates some properties of the mean residual life (MRL) function, and the mean past lifetime (MPL) function of the parallel system with independent but not identically distributed (INID) components under the condition that, at time $t_1 (t_2 > 0)$ the total number of failures of the components is r, and at time t_2 ($t_2 > t_1$) the system is still functioning, or it has been failed. A few representations and corresponding properties of the MRL and MPL functions under several conditions are obtained. Also some useful examples related to several lifetime distributions and the corresponding MRL (MPL) functions are given.

3.2.1 *Preliminaries*

Let $X_1, X_2, \ldots, X_n$ be absolutely continuous random variables with a respective distribution functions F_i, $i = 1, 2, \ldots, n$, representing the lifetimes of n independent components of a parallel system, and $X_{i:n}$ be the ith smallest order statistic among $X_1, X_2, \ldots, X_n$. Denote the residual life at age $t (\geq 0)$ of a random life X_i by $(X_i - t | X_i > t)$, with reliability function $\bar{F}_i(x|t) = \bar{F}_i(t + x)/\bar{F}_i(t)$, density function $f_i(x|t) = f_i(t + x)/\bar{F}_i(t), x \geq 0$, given that $\bar{F}_i(t) > 0$. For $i = 1, 2, \ldots, n$, the MRL function (denoted by $m_i(t)$) of each component is defined by

$$m_i(t) = \mathrm{E}(X_i - t | X_i > t) = \int_0^\infty \bar{F}_i(x|t)dx = \frac{1}{\bar{F}_i(t)} \int_t^\infty \bar{F}_i(x)dx.$$

Denote the past lifetime at time $t (\geq 0)$ of a random life X_i by $(t - X_i | X_i \leq t)$, with reliability function $\bar{F}_i(x|(t)) = F_i(t - x)/F_i(t)$, density function $f_i(x|(t)) = f_i(t - x)/F_i(t), t - x \geq 0$, given $F_i(t) > 0$. For $i = 1, 2, \ldots, n$, the MPL function (denoted by $h_i(t)$) of each component is defined as follows

$$h_i(t) = \mathrm{E}(t - X_i | X_i \leq t) = \int_0^\infty F_i(x|(t))dx = \frac{1}{F_i(t)} \int_0^t F_i(x)dx.$$

The MRL and MPL functions are very important in reliability and survival theories because they describe the aging process. They have been studied by reliabilists, statisticians, survival analysts and others, and many useful results have been derived concerning them. A review of its theory and applications was given by Kotz and Shanbhag (1980), and Guess and Proschan (1988). Over the last two decades, many results have been get on this subject.

Asadi and Bairamov (2005) introduced a new definition for the MRL function of a parallel system with n independent and identically distributed (IID) units under the condition that, at time t at least $(n - k + 1)$ components of the system are still working, that is, they presented the following measure

$$\mathrm{E}(X_{n:n} - t | X_{k:n} \geq t), k = 1, 2, \ldots, n,$$

and obtained some properties of it (see also Bairamov et al. (2002)). Asadi (2006) researched the MPL function

$$\mathrm{E}(t - X_{r:n} | X_{n:n} \leq t), \ r = 1, 2, \ldots, n,$$

of components of parallel systems with n IID components. Tavangar and Asadi (2010) considered the MPL function of a $(n - k + 1)$-out-of-n system with IID components. There they studied

$$\mathrm{E}(t - X_{r:n} | X_{k:n} \leq t), \ r = 1, 2, \ldots, k,$$

and obtained several properties of it.

Recently, Poursaeed and Nematollahi (2008) considered a parallel system such that the total number of failures at time t_1 is r and at time $t_2 (t_1 < t_2)$ the system is still working or all components of the system have failed. Under these conditions (say under double monitoring), they studied the MRL and MPL functions of system in the IID case of components, that is,

$$\mathrm{E}(X_{n:n} - t_2 | X_{r:n} < t_1 < X_{r+1:n}, X_{n:n} > t_2), \tag{3.8}$$

and

$$\mathrm{E}(t_2 - X_{m:n} | X_{r:n} < t_1 < X_{r+1:n}, X_{n:n} < t_2). \tag{3.9}$$

For new further results on these topics, see Navarro and Hernandez (2008). It is remarked that Gurler and Bairamov (2009) researched the MRL function of parallel and k-out-of-n systems when the components are INID.

Along this line, this section investigates the (MRL) functions, and the MPL functions of the parallel system with INID components under the condition that, at time $t_1 (> 0)$ the total number of failures of the components is r and at time t_2 $(t_2 > t_1)$ the system is still working, or it has been failed, extending (3.8) and (3.9) to the INID case. Specifically, we obtain the representation of the MRL function of parallel systems under double monitoring, and in some special conditions several properties and mixture representations are get. Moreover, we get the representation of the MPL function of the components of parallel systems, and under some special conditions several properties and mixture representations of MPL functions are established. Also some useful examples related to several lifetime distributions and the corresponding MRL (MPL) functions are given.

The MRL and MPL functions of systems when the components of the system are nonidentical, may be expressed in terms of symmetric functions and permanents. In the following, we introduce the definitions of symmetric functions and permanents as well as some useful properties.

Definition 3.2.1 Assume that $\mathbf{x} = (x_1, \ldots, x_n) \in \Re^n$, $n \geq 1$. The jth $(1 \leq j \leq n)$ elementary symmetric functions (denoted by $\delta_j(x_1, \ldots, x_n)$), is the sum of all products of j distinct variables chosen from n variables (see Macdonald, 1979). That is

$$\delta_j(x_1, \ldots, x_n) = \delta_j(\mathbf{x}) = \sum_{1 \leq k_1 < \cdots < k_j \leq n} x_{k_1} \cdots x_{k_j}.$$

Let $\delta_j(\mathbf{x}) = 0$ for $j < 0$ and $j > n$, $\delta_j(\mathbf{x}) = 1$ for $j = 0$. Then the generating function (demoted be $G(x)$) for the elementary symmetric function is

$$G(x) = \prod_{j=1}^{n}(1 - x_i) = \sum_{j=0}^{n}(-1)^j \delta_j(\mathbf{x}). \tag{3.10}$$

A recurrence relation for the the elementary symmetric function can be obtained from (3.10)

$$\delta_j(x_1, \ldots, x_n) = \delta_j(x_1, \ldots, x_{n-1}) + x_n \delta_j(x_1, \ldots, x_{n-1}). \tag{3.11}$$

When the random variables are independent but not identically distributed, the distribution of order statistics can be expressed in terms of permanents. Consider an $n \times n$ matrix $A = (a_{ij})$, the permanent of A, denoted by $Per A$, is defined as

$$Per A = \sum_{\sigma} \prod_{i=1}^{n} a_{i\sigma(i)},$$

where the summation is over all permutations $\sigma = (\sigma(1), \ldots, \sigma(n))$ of $(1, \ldots, n)$. If $\boldsymbol{a}_1, \ldots, \boldsymbol{a}_n$ are column vectors, then we will denote by $[\boldsymbol{a}_1, \ldots, \boldsymbol{a}_n]$ the permanent of the matrix A. The permanent

$$\left[\underbrace{\boldsymbol{a}_1}_{r_1} \; \underbrace{\boldsymbol{a}_2}_{r_2} \; \underbrace{\boldsymbol{a}_n}_{r_n}\right]$$

is obtained by taking r_1 copies of $\boldsymbol{a}_1$, r_2 copies of $\boldsymbol{a}_2$, r_n copies of $\boldsymbol{a}_n$. If r_i equals 1, then we omit it in the notation above. If r_i equals 0, then $\boldsymbol{a}_i$ is not present in the permanent.

Let $X_1, X_2, \ldots, X_n$ be independent random variables, and F_i, $\bar{F}_i$, f_i are distribution, survival and density functions of X_i, $i = 1, 2, \ldots, n$. The column vectors $(F_1, \ldots, F_n)'$, $(\bar{F}_1, \ldots, \bar{F}_n)'$ and $(f_1, \ldots, f_n)'$ will be denoted simply by $\mathbf{F}(x)$, $\bar{\mathbf{F}}(x)$ and $\mathbf{f}(x)$, respectively.

It is very useful to represent the joint density functions of order statistics by the help of the theory of permanents when the underlying random variables are not identical. According to Bapat and Beg (1989), for any $x \geq 0$, the distribution function of $X_{r:n}$ may be given by

$$P(X_{r:n} \leq x) = \sum_{j=r}^{n} \frac{1}{j!(n-j)!} Per \begin{bmatrix} F_1(x) & \bar{F}_1(x) \\ \vdots & \vdots \\ \underbrace{F_n(x)}_{j} & \underbrace{\bar{F}_n(x)}_{n-j} \end{bmatrix} = \sum_{j=r}^{n} \frac{1}{j!(n-j)!} Per \left[\underbrace{F(x)}_{j} \; \underbrace{\bar{F}(x)}_{n-j} \right]$$

3.2.2 *The Residual Life of n-Out-of-n Systems*

In this section, we consider the MRL function of the parallel systems under double monitoring $(X_{r:n} < t_1 < X_{r+1:n}, X_{n:n} > t_2)$, that is, at time $t_1 (> 0)$ the total number of failures of the components is r and at time t_2 $(t_2 > t_1)$ the system is still working.

Definition 3.2.2 Given that $(X_{r:n} < t_1 < X_{r+1:n}, X_{n:n} > t_2)$, the MRL function of the parallel system having n INID components is

$$M_n^r(t_1, t_2) = \mathrm{E}(X_{n:n} - t_2 | X_{r:n} < t_1 < X_{r+1:n}, X_{n:n} > t_2).$$

Theorem 3.2.3 *Assume that* $M_n^r(t_1, t_2) = E(X_{n:n} - t_2 | X_{r:n} < t_1 < X_{r+1:n}, X_{n:n} > t_2)$ *for any* $t_2 > t_1 \geq 0$ *and* $r < n$, *then*

$$M_n^r(t_1, t_2) = \frac{\sum_{j=1}^{n-r} \binom{n-r}{j} \int_{t_2}^{\infty} Per[\underbrace{F(t_1)}_{r} \; \underbrace{\bar{F}(t_1) - \bar{F}(x)}_{n-r-j} \; \underbrace{\bar{F}(x)}_{j}]dx}{\sum_{j=1}^{n-r} \binom{n-r}{j} Per[\underbrace{F(t_1)}_{r} \; \underbrace{\bar{F}(t_1) - \bar{F}(t_2)}_{n-r-j} \; \underbrace{\bar{F}(t_2)}_{j}]}.$$

Proof For all permutations $l_1, \ldots, l_n$ of $1, 2, \ldots, n$ such that $l_1 < \cdots < l_r$, $l_{r+1} < \cdots < l_{n-j}$ and $l_{n-j+1} < \cdots < l_n$, $j = 1, \ldots, n-r$,

$$\begin{aligned} & P(X_{r:n} < t_1 < X_{r+1:n}, X_{n:n} > t_2) \\ &= \sum_{j=1}^{n-r} \sum_{l_1, \cdots, l_n} \prod_{m=1}^{r} F_{l_m}(t_1) \prod_{m=r+1}^{n-j} [\bar{F}_{l_m}(t_1) - \bar{F}_{l_m}(t_2)] \prod_{m=n-j+1}^{n} \bar{F}_{l_m}(t_2) \end{aligned}$$

$$= \sum_{j=1}^{n-r} \frac{1}{r!j!(n-r-j)!} Per \left[\underbrace{F(t_1)}_{r} \ \underbrace{\bar{F}(t_1) - \bar{F}(t_2)}_{n-r-j} \ \underbrace{\bar{F}(t_2)}_{j} \right].$$

For any $x \geq 0$, let $S_n^r(x|t_1, t_2) = P(X_{n:n} - t_2 > x | X_{r:n} < t_1 < X_{r+1:n}, X_{n:n} > t_2)$, then

$$S_n^r(x|t_1, t_2) = \frac{P(X_{n:n} - t_2 > x, X_{r:n} < t_1 < X_{r+1:n}, X_{n:n} > t_2)}{P(X_{r:n} < t_1 < X_{r+1:n}, X_{n:n} > t_2)}$$

$$= \frac{\sum_{j=1}^{n-r} \sum_{l_1 \cdots l_n} \prod_{m=1}^{r} F_{l_m}(t_1) \prod_{m=r+1}^{n-j} [\bar{F}_{l_m}(t_1) - \bar{F}_{l_m}(t_2 + x)] \prod_{m=n-j+1}^{n} \bar{F}_{l_m}(t_2 + x)}{\sum_{j=1}^{n-r} \sum_{l_1 \cdots l_n} \prod_{m=1}^{r} F_{l_m}(t_1) \prod_{m=r+1}^{n-j} [\bar{F}_{l_m}(t_1) - \bar{F}_{l_m}(t_2)] \prod_{m=n-j+1}^{n} \bar{F}_{l_m}(t_2)}$$

$$= \frac{\sum_{j=1}^{n-r} \frac{1}{r!j!(n-r-j)!} Per[\underbrace{F(t_1)}_{r} \ \underbrace{\bar{F}(t_1) - \bar{F}(t_2 + x)}_{n-r-j} \ \underbrace{\bar{F}(t_2 + x)}_{j}]}{\sum_{j=1}^{n-r} \frac{1}{r!j!(n-r-j)!} Per[\underbrace{F(t_1)}_{r} \ \underbrace{\bar{F}(t_1) - \bar{F}(t_2)}_{n-r-j} \ \underbrace{\bar{F}(t_2)}_{j}]}$$

$$= \frac{\sum_{j=1}^{n-r} \binom{n-r}{j} Per[\underbrace{F(t_1)}_{r} \ \underbrace{\bar{F}(t_1) - \bar{F}(t_2 + x)}_{n-r-j} \ \underbrace{\bar{F}(t_2 + x)}_{j}]}{\sum_{j=1}^{n-r} \binom{n-r}{j} Per[\underbrace{F(t_1)}_{r} \ \underbrace{\bar{F}(t_1) - \bar{F}(t_2)}_{n-r-j} \ \underbrace{\bar{F}(t_2)}_{j}]}.$$

Therefore the MRL function of $(X_{n:n} - t_2 | X_{r:n} < t_1 < X_{r+1:n}, X_{n:n} > t_2)$ is

$$M_n^r(t_1, t_2) = \int_0^\infty S_n^r(x|t_1, t_2)dx = \frac{\sum_{j=1}^{n-r} \binom{n-r}{j} \int_{t_2}^\infty Per[\underbrace{F(t_1)}_{r} \ \underbrace{\bar{F}(t_1) - \bar{F}(x)}_{n-r-j} \ \underbrace{\bar{F}(x)}_{j}]dx}{\sum_{j=1}^{n-r} \binom{n-r}{j} Per[\underbrace{F(t_1)}_{r} \ \underbrace{\bar{F}(t_1) - \bar{F}(t_2)}_{n-r-j} \ \underbrace{\bar{F}(t_2)}_{j}]}.$$

Example 3.2.4 Consider one 1-out-of-3 system having three INID components with lifetimes X_1, X_2, X_3. Suppose that X_i is distributed exponential with parameter λ_i for $i = 1, 2, 3$. It follows from the expressions above

$S_3^1(x|t_1, t_2)$

$$= \frac{\sum_{j=1}^{2} \sum_{l_1, l_2, l_3} \prod_{m=1}^{1} (1 - e^{-\lambda_{l_m} t_1}) \prod_{m=2}^{3-j} (e^{-\lambda_{l_m} t_1} - e^{-\lambda_{l_m} (t_2 + x)}) \prod_{m=4-j}^{3} e^{-\lambda_{l_m} (t_2 + x)}}{\sum_{j=1}^{2} \sum_{l_1, l_2, l_3} \prod_{m=1}^{1} (1 - e^{-\lambda_{l_m} t_1}) \prod_{m=2}^{3-j} (e^{-\lambda_{l_m} t_1} - e^{-\lambda_{l_m} t_2}) \prod_{m=4-j}^{3} e^{-\lambda_{l_m} t_2}}$$

$$= \frac{\sum_{j=1}^{2} \sum_{l_1, l_2, l_3} \prod_{m=1}^{1} (1 - e^{-\lambda_{l_m} t_1}) \prod_{m=2}^{3-j} (e^{-\lambda_{l_m} t_1} - e^{-\lambda_{l_m} (t_2 + x)}) \prod_{m=4-j}^{3} e^{-\lambda_{l_m} (t_2 + x)}}{\sum_{j=1}^{2} \sum_{l_1, l_2, l_3} \prod_{m=1}^{1} (1 - e^{-\lambda_{l_m} t_1}) \prod_{m=2}^{3-j} (e^{-\lambda_{l_m} t_1} - e^{-\lambda_{l_m} t_2}) \prod_{m=4-j}^{3} e^{-\lambda_{l_m} t_2}}$$

$$= \frac{\sum_{l_1, l_2, l_3} (1 - e^{-\lambda_{l_1} t_1}) e^{-\lambda_{l_2} t_1 - \lambda_{l_3} (t_2 + x)}}{\sum_{l_1, l_2, l_3} (1 - e^{-\lambda_{l_1} t_1}) e^{-\lambda_{l_2} t_1 - \lambda_{l_3} t_2}},$$

and hence

$$M_3^1(t_1,t_2)=\frac{\sum_{l_1,l_2,l_3}\int_0^{+\infty}(1-e^{-\lambda_{l_1}t_1})e^{-\lambda_{l_2}t_1-\lambda_{l_3}(t_2+x)}dx}{\sum_{l_1,l_2,l_3}(1-e^{-\lambda_{l_1}t_1})e^{-\lambda_{l_2}t_1-\lambda_{l_3}t_2}}=\frac{\sum_{l_1,l_2,l_3}(1-e^{-\lambda_{l_1}t_1})e^{-\lambda_{l_2}t_1-\lambda_{l_3}t_2}\times\frac{1}{\lambda_{l_3}}}{\sum_{l_1,l_2,l_3}(1-e^{-\lambda_{l_1}t_1})e^{-\lambda_{l_2}t_1-\lambda_{l_3}t_2}}.$$

Remark For $r=0$, that is, at time t_1, all the components is still working, then

$$S_n^0(x|t_1,t_2)=\frac{\sum_{j=1}^n\binom{n}{j}Per[\underbrace{\bar F(t_1)-\bar F(t_2+x)}_{n-j}\ \underbrace{\bar F(t_2+x)}_{j}]}{\sum_{j=1}^n\binom{n}{j}Per[\underbrace{\bar F(t_1)-\bar F(t_2)}_{n-j}\ \underbrace{\bar F(t_2)}_{j}]},$$

also when $t_1=t_2=t$, it is reduced to

$$S_n^0(x|t)=\frac{\sum_{j=1}^n\binom{n}{j}Per[\underbrace{\bar F(t)-\bar F(t+x)}_{n-j}\ \underbrace{\bar F(t+x)}_{j}]}{n!\prod_{i=1}^n\bar F_i(t)},$$

and the corresponding MRL function is (see Gurler and Bairamov (2009))

$$M_n^0(t)=\frac{1}{(n-1)!\prod_{i=1}^n\bar F_i(t)}\int_t^\infty x\,Per[\underbrace{f(x)}_{1}\ \underbrace{\bar F(x)-\bar F(t)}_{n-1}]dx-t.$$

Remark For $r=n-1$, $t_1=t_2=t\geq 0$, that is, at time t only one component is still working,

$$S_n^{n-1}(x|t)=\frac{Per[\underbrace{F(t)}_{n-1}\ \underbrace{\bar F(t+x)}_{1}]}{Per[\underbrace{F(t)}_{n-1}\ \underbrace{\bar F(t)}_{1}]}=\frac{\sum_{l_1,\ldots,l_n}\bar F_{l_n}(x|t)\prod_{i=1}^{n-1}\psi_{l_i}(t)}{\sum_{\{l_1,\ldots,l_{n-1}\}\in\{1,\ldots,n\}}\prod_{i=1}^{n-1}\psi_{l_i}(t)},$$

and hence

$$M_n^{n-1}(t)=\frac{\sum_{l_1,\ldots,l_n}\int_0^\infty\bar F_{l_n}(x|t)dx\prod_{i=1}^{n-1}\psi_{l_i}(t)}{\sum_{\{l_1,\ldots,l_{n-1}\}\in\{1,\ldots,n\}}\prod_{i=1}^{n-1}\psi_{l_i}(t)},$$

where $\psi_{l_i}(t)=\frac{F_{l_i}(t)}{\bar F_{l_i}(t)}$, $i\in\{1,2,\ldots,n\}$, and $\bar F_{l_n}(x|t)=\frac{\bar F_{l_n}(x+t)}{\bar F_{l_n}(t)}$ for $l_n\in\{1,2,\ldots,n\}$. It is clear that

$$\min\{\bar F_1(x|t),\bar F_2(x|t),\ldots,\bar F_n(x|t)\}\leq S_n^{n-1}(x|t)\leq\max\{\bar F_1(x|t),\bar F_2(x|t),\ldots,\bar F_n(x|t)\},$$

and

$$\min\{m_1(t), m_2(t), \ldots, m_n(t)\} \leq M_n^{n-1}(t) \leq \max\{m_1(t), m_2(t), \ldots, m_n(t)\}.$$

Example 3.2.5 Consider one 1-out-of-3 system with INID components with lifetimes X_1, X_2, X_3. Suppose that X_i is distributed exponential with parameter λ_i for $i = 1, 2, 3$. It follows from the expressions above

$$S_3^2(x|t) = \frac{\sum_{l_1,l_2,l_3} e^{-\lambda_{l_1}(t+x)}(1 - e^{-\lambda_{l_2}t})(1 - e^{-\lambda_{l_3}t})}{\sum_{l_1,l_2,l_3} e^{-\lambda_{l_1}t}(1 - e^{-\lambda_{l_2}t})(1 - e^{-\lambda_{l_3}t})}$$

and the MRL function is

$$\begin{aligned} M_3^2(t) &= \frac{\sum_{l_1,l_2,l_3} \frac{1}{\lambda_{l_1}} e^{-\lambda_{l_1}t}(1-e^{-\lambda_{l_2}t})(1-e^{-\lambda_{l_3}t})}{\sum_{l_1,l_2,l_3} e^{-\lambda_{l_1}t}(1 - e^{-\lambda_{l_2}t})(1 - e^{-\lambda_{l_3}t})} \\ &= \frac{(e^{\lambda_2 t} - 1)(e^{\lambda_3 t} - 1)\frac{1}{\lambda_1} + (e^{\lambda_1 t} - 1)(e^{\lambda_3 t} - 1)\frac{1}{\lambda_2} + (e^{\lambda_1 t} - 1)(e^{\lambda_2 t} - 1)\frac{1}{\lambda_3}}{(e^{\lambda_2 t} - 1)(e^{\lambda_3 t} - 1) + (e^{\lambda_1 t} - 1)(e^{\lambda_3 t} - 1) + (e^{\lambda_1 t} - 1)(e^{\lambda_2 t} - 1)}. \end{aligned}$$

In practice, it may be difficult to compute the results above, sometimes they can be checked numerically.

3.2.3 The Inactivity Time of n-Out-of-n System

In this section we investigate the MPL function of components of a parallel system whose component lifetimes are INID under the condition $(X_{r:n} < t_1 < X_{r+1:n}, X_{n:n} \leq t_2)$, that is, at time $t_1 (> 0)$ the total number of failures of the components is r and at time t_2 $(t_2 > t_1)$ the system has failed.

Definition 3.2.6 Given that $(X_{r:n} < t_1 < X_{r+1:n}, X_{n:n} \leq t_2)$, the MPL function of the parallel system having n INID components is

$$H_n^r(t_1, t_2) = \mathrm{E}(t_2 - X_{m:n}|X_{r:n} < t_1 < X_{r+1:n}, X_{n:n} \leq t_2).$$

Theorem 3.2.7 *For $r < m \leq n$, and $0 < t_1 < t_2$,*

$$H_n^r(t_1, t_2) = \frac{\sum_{j=m-r}^{n-r} \binom{n-r}{j} \int_0^{t_2} Per[\underbrace{F(t_1)}_{r}\ \underbrace{\bar{F}(t_1) - \bar{F}(t_2 - x)}_{j}\ \underbrace{\bar{F}(t_2 - x) - \bar{F}(t_2)}_{n-r-j}]dx}{Per[\underbrace{F(t_1)}_{r}\ \underbrace{\bar{F}(t_1) - \bar{F}(t_2)}_{n-r}]}.$$

Proof Let $S_n^r(x|t_1, (t_2))$ be the reliability function of $(t_2 - X_{m:n}|X_{r:n} < t_1 < X_{r+1:n}, X_{n:n} < t_2)$ for $r < m \leq n$, $x > 0$ and $0 < t_1 < t_2 - x$, then

$$S_n^r(x|t_1,(t_2))$$
$$=\frac{P(t_2-X_{m:n}>x,X_{r:n}<t_1<X_{r+1:n},X_{n:n}<t_2)}{P(X_{r:n}<t_1<X_{r+1:n},X_{n:n}<t_2)}$$
$$=\frac{\sum_{j=m-r}^{n-r}\sum_{l_1\cdots l_n}\prod_{i=1}^{r}F_{l_i}(t_1)\prod_{i=r+1}^{r+j}[\bar F_{l_i}(t_1)-\bar F_{l_i}(t_2-x)]\prod_{m=r+j+1}^{n}[\bar F_{l_i}(t_2+x)-\bar F_{l_i}(t_2)]}{\sum_{l_1\cdots l_n}\prod_{i=1}^{r}F_{l_i}(t_1)\prod_{i=r+1}^{n}[\bar F_{l_i}(t_1)-\bar F_{l_i}(t_2)]}$$
$$=\frac{\sum_{j=m-r}^{n-r}\frac{1}{r!j!(n-r-j)!}Per[\underbrace{F(t_1)}_{r}\ \underbrace{\bar F(t_1)-\bar F(t_2-x)}_{j}\ \underbrace{\bar F(t_2-x)-\bar F(t_2)}_{n-r-j}]}{\frac{1}{r!(n-r)!}Per[\underbrace{F(t_1)}_{r}\ \underbrace{\bar F(t_1)-\bar F(t_2)}_{n-r}]}$$
$$=\frac{\sum_{j=m-r}^{n-r}\binom{n-r}{j}Per[\underbrace{F(t_1)}_{r}\ \underbrace{\bar F(t_1)-\bar F(t_2-x)}_{j}\ \underbrace{\bar F(t_2-x)-\bar F(t_2)}_{n-r-j}]}{Per[\underbrace{F(t_1)}_{r}\ \underbrace{\bar F(t_1)-\bar F(t_2)}_{n-r}]}.$$

Therefore the desired result follows immediately.

Theorem 3.2.8 *For $r=n$, $t_1=t_2=t>0$, the MPL function of the parallel system is*

$$H_n^n(t)=\frac{\int_0^t\sum_{j=0}^{n}(-1)^j\sigma_j(\bar F_1(x),\ldots,\bar F_n(x))dx}{\sum_{j=0}^{n}(-1)^j\sigma_j(\bar F_1(t),\ldots,\bar F_n(t))}.$$

Proof For any $t>0$,

$$P(X_{n:n}<t)=\prod_{i=1}^{n}(1-\bar F_i(t))=\sum_{j=0}^{n}(-1)^j\sigma_j(\bar F_1(t),\ldots,\bar F_n(t)),$$

hence, for any $0<x<t$,

$$\begin{aligned}P(t-X_{n:n}>x|X_{n:n}<t)&=\frac{P(X_{n:n}<t-x)}{P(X_{n:n}<t)}\\&=\frac{\prod_{i=1}^{n}(1-\bar F_i(t-x))}{\prod_{i=1}^{n}(1-\bar F_i(t))}\\&=\frac{\sum_{j=0}^{n}(-1)^j\sigma_j(\bar F_1(t-x),\ldots,\bar F_n(t-x))}{\sum_{j=0}^{n}(-1)^j\sigma_j(\bar F_1(t),\ldots,\bar F_n(t))}.\end{aligned}$$

It follows that

$$H_n^n(t)=\int_0^t P(t-X_{n:n}>x|X_{n:n}<t)dx=\frac{\int_0^t\sum_{j=0}^{n}(-1)^j\sigma_j(\bar F_1(x),\ldots,\bar F_n(x))dx}{\sum_{j=0}^{n}(-1)^j\sigma_j(\bar F_1(t),\ldots,\bar F_n(t))}.$$

Example 3.2.9 Consider one 1-out-of-n system with INID components with lifetimes $X_1, \ldots, X_n$. Suppose that X_i is distributed exponential with parameter λ_i for $i = 1, \ldots, n$. It follows from Theorem 3.2.8, that,

$$H_n^n(t) = \frac{\sum_{j=0}^{n}(-1)^j \frac{\sum_{1\le l_1<\cdots l_j\le n}(1-e^{-\sum_{k=1}^{j}\lambda_{l_k}t})}{\sum_{k=1}^{j}\lambda_{l_i}}}{\sum_{j=0}^{n}(-1)^j \sum_{1\le l_1<\cdots l_j\le n} e^{-\sum_{k=1}^{j}\lambda_{l_k}t}},$$

and

$$\lim_{t\to\infty} H_n^n(t) = 1 - \sum_{j=0}^{n}\frac{1}{\lambda_i} + \sum_{i<j}\frac{1}{\lambda_i+\lambda_j} - \sum_{i<j<k}\frac{1}{\lambda_i+\lambda_j+\lambda_k} + \cdots + (-1)^n\frac{1}{\lambda_1+\lambda_2+\cdots+\lambda_n}.$$

When $n = 3$, the denominator of $H_3^3(t)$ is given by

$$P(X_{3:3} < t) = 1 - \sum_{j=1}^{3} e^{-\lambda_j t} + e^{-(\lambda_1+\lambda_2)t} + e^{-(\lambda_2+\lambda_3)t} + e^{-(\lambda_1+\lambda_3)t} - e^{-(\lambda_1+\lambda_2+\lambda_3)t}.$$

It can be find that $P(X_{3:3} < t)$ is the difference of 1 and the generalized mixture of the series structure.

If X_i is distributed exponential with common parameter λ for $i = 1, \ldots, n$, then the MPL of the parallel system is

$$H_n^n(t) = \frac{\sum_{j=0}^{n}(-1)^j\binom{n}{j}\frac{1-e^{-j\lambda t}}{r\lambda}}{\sum_{j=0}^{n}(-1)^j\binom{n}{j}e^{-j\lambda t}},$$

The lemma below is useful to get the next theorem.

Lemma 3.2.10 *Let $A_n(t) = \prod_{i=1}^{n} F_i(t)$ be the distribution function of parallel systems with n INID components, then*

$$A_n(t) = A_{n-1}(t)F_n(t).$$

Proof From (3.11), we have

$$\begin{aligned}
A_n(t) &= \prod_{i=1}^{n}(1-\bar{F}_i(t)) \\
&= \sum_{j=0}^{n}(-1)^j\delta_j(\bar{F}_1(t),\ldots,\bar{F}_n(t)) \\
&= \sum_{j=0}^{n}(-1)^j[\delta_j(\bar{F}_1(t),\ldots,\bar{F}_{n-1}(t)) + \bar{F}_n(t)\delta_{j-1}(\bar{F}_1(t),\ldots,\bar{F}_{n-1}(t))]
\end{aligned}$$

$$
\begin{aligned}
&= \sum_{j=0}^{n-1}(-1)^j[\delta_j(\bar{F}_1(t),\ldots,\bar{F}_{n-1}(t)) + \bar{F}_n(t)\delta_{j-1}(\bar{F}_1(t),\ldots,\bar{F}_{n-1}(t))] \\
&\quad + \bar{F}_n(t)(-1)^n\delta_{n-1}(\bar{F}_1(t),\ldots,\bar{F}_{n-1}(t)) + \bar{F}_n(t)(-1)^n\delta_{n-1}(\bar{F}_1(t),\ldots,\bar{F}_{n-1}(t)) \\
&= A_{n-1}(t) + \bar{F}_n(t)[\sum_{j=0}^{n-1}(-1)^j\delta_{j-1}(\bar{F}_1(t),\ldots,\bar{F}_{n-1}(t)) + (-1)^n\delta_{n-1}(\bar{F}_1(t),\ldots,\bar{F}_{n-1}(t))] \\
&= A_{n-1}(t)(1-\bar{F}_n(t)) \\
&= A_{n-1}(t)F_n(t).
\end{aligned}
$$

In the following theorem, we give a recurrence formula for the MPL function of the parallel system.

Theorem 3.2.11 *Let $H_k^k(t)$ be the MPL function of parallel systems with k INID components, then for any $t \geq 0$,*

$$
H_n^n(t) = \frac{1}{F_n(t)} H_{n-1}^{n-1}(t) - \frac{\int_0^t \bar{F}_n(x)A_{n-1}(x)dx}{F_n(t)A_{n-1}(t)}.
$$

Proof From Theorem 3.2.8 and Lemma 3.2.10, we have

$$
\begin{aligned}
H_n^n(t) &= \frac{\int_0^t A_n(x)dx}{A_n(t)} \\
&= \frac{\int_0^t F_n(x)A_{n-1}(x)dx}{F_n(t)A_{n-1}(t)} \\
&= \frac{\int_0^t A_{n-1}(x)dx - \int_0^t \bar{F}_n(x)A_{n-1}(x)dx}{F_n(t)A_{n-1}(t)} \\
&= \frac{1}{F_n(t)} H_{n-1}^{n-1}(t) - \frac{\int_0^t \bar{F}_n(x)A_{n-1}(x)dx}{F_n(t)A_{n-1}(t)}.
\end{aligned}
$$

This completes the proof of the theorem.

Chapter 4
The Stochastic Properties of Remaining Components

This chapter investigates the residual lives of the remaining components of an $(n - k + 1)$-out-of-n system with independent and identical components after the kth failure has been observed, and of a general coherent system after the system has failed. Some aging properties and stochastic comparisons on the residual life are obtained.

4.1 On the $(n - k + 1)$-Out-of-n System

This section focuses on the remaining components of $(n - k + 1)$-out-of-n system with independent and identical components after the kth failure has been observed. It is given some aging properties of the residual lifetimes of the remaining components of the failed system. And we stochastically compare the residual lives of the remaining items of such systems with different types of components.

4.1.1 Introduction

Let absolutely continuous random variables $X_1, X_2, \ldots, X_n$ be independent with a common distribution function F, representing the lifetimes of n components of an $(n - k + 1)$ $(1 \leq k \leq n)$ out of n system, and $X_{i,n}$ be the ith smallest order statistic among $X_1, X_2, \ldots, X_n$.

When an-$(n - k + 1)$-out-of-n system has failed, operators or designers may consider reusing the unfailed units in other systems, or constructing a new system with them. Thus it may be useful for the operator's decision making to know some aging properties of the residual lifetimes of the remaining components of the failed system,

Z. Zhang, *Conditional Order Statistics and Conditional Spacings*,
https://doi.org/10.1007/978-981-96-7151-9_4

or to stochastically compare the residual lives of the remaining items of such systems with different types of components.

As shown in Bairamov and Arnold (2008), even under the classical assumptions that the original lifetimes were independent and identically distributed (IID), the residual lifetimes of the unfailed units are exchangeable but not independent.

Let $X_1^{(k,n)}, X_2^{(k,n)}, \ldots, X_{n-k}^{(k,n)}$ be the residual lifetimes of the $(n-k)$ remaining functioning components for $k \in \{1, 2, \ldots, n\}$ after the kth failure has been observed. Under the condition that the failure of one component dose not affect the functioning of the remaining ones, Bairamov and Arnold in (2008) obtained that the joint survival function of the residual lifelengths is

$$\begin{aligned}\bar{F}_n^{(k)}(x_1, x_2, \ldots, x_{n-k}) &= P(X_1^{(k,n)} > x_1, X_2^{(k,n)} > x_2, \ldots, X_{n-k}^{(k,n)} > x_{n-k}) \\ &= \int_0^\infty \prod_{i=1}^{n-k} \bar{F}_t(x_i) dF_{k,n}(t), \end{aligned} \tag{4.1}$$

where $F_{k,n}(t)$ is the distribution function of $X_{k,n}$, and $\bar{F}_t(x_i) = \bar{F}(t + x_i)/\bar{F}(t)$ is the survival function of the residual life $X_t = (X - t|X > t)$ of a unit surviving up to time t. By Bairamov and Arnold in (2008), $X_i^{(k,n)}$ $(i = 1, 2, \ldots, n-k)$ are exchangeable and conditionally independent given $X_{k,n}$, with common marginal survival function

$$\bar{F}_n^{(k)}(x) = \int_0^\infty \bar{F}_t(x) dF_{k,n}(t). \tag{4.2}$$

4.1.2 Aging and Stochastic Properties of Remaining Components

In Bairamov and Arnold (2008), it is proved that if X is NBU (NWU), then $X_1^{(k,n)} \le_{st} X_1 (X_1^{(k,n)} \ge_{st} X_1)$. This section is devoted to further investigation of some slightly stronger or weak sufficient conditions for some stochastic order between $X_1^{(k,n)}$ and X_1. And some aging properties of $X_1^{(k,n)}$ according to the ones of X_1 are obtained. Also a sufficient condition for $\mathrm{E}X_1^{(k,n)} \le (\ge) \mathrm{E}X_1$ is get.

Theorem 4.1.1 *If X is DLR, then, $X_1^{(j,m)} \le_{lr} X_1^{(i,n)}$ for all $j \le i$, and $m - j \ge n - i$.*

Proof Let $f_m^{(j)}(x)$, $f_n^{(i)}(x)$ be the density functions of $X_1^{(j,m)}$, $X_1^{(i,n)}$, respectively. Then

$$f_n^{(i)}(x) = \int_0^\infty f_t(x) dF_{i,n}(t),$$

$$f_m^{(j)}(x) = \int_0^\infty f_t(x) dF_{j,m}(t).$$

That X is DLR implies that $X_t \leq_{lr} X_{t'}$ whenever $t \leq t'$. And from Theorem 1.C.37 of Shaked and Shanthikumar (2007), $X_{j,m} \leq_{lr} X_{i,n}$ for all $j \leq i$, and $m-j \geq n-i$, hence from Theorem 1.C.17 of Shaked and Shanthikumar (2007) again, the result holds.

Theorem 4.1.2 *If X is DFR, then, $X_1^{(j,m)} \leq_{hr} X_1^{(i,n)}$ for all $j \leq i$, and $m-j \geq n-i$.*

The proof is similar to that of the theorem above, hence is omitted.

Theorem 4.1.3 *If X is ILR (DLR), then, $X_1^{(k,n)} \leq_{lr} X_1 (X_1^{(k,n)} \geq_{lr} X_1)$.*

Proof By (4.2), the density function of $X_1^{(k,n)}$ is

$$f_n^{(k)}(x) = \int_0^\infty f_t(x) dF_{k,n}(t).$$

The ILR (DLR) property of X implies that

$$\frac{f_n^{(k)}(x)}{f(x)} = \int_0^\infty \frac{f_t(x)}{f(x)} dF_{k,n}(t) = \int_0^\infty \frac{f(t+x)}{f(x)\bar{F}(t)} dF_{k,n}(t)$$

is decreasing (increasing) in $x \geq 0$, which shows the result.

Theorem 4.1.4 *If X is IFR (DFR), then, $X_1^{(k,n)} \leq_{hr} X_1 (X_1^{(k,n)} \geq_{hr} X_1)$.*

Proof For any $x \geq 0$, the IFR (DFR) property of X implies that

$$\frac{\bar{F}_n^{(k)}(x)}{\bar{F}(x)} = \int_0^\infty \frac{\bar{F}_t(x)}{\bar{F}(x)} dF_{k,n}(t) = \int_0^\infty \frac{\bar{F}(t+x)}{\bar{F}(x)\bar{F}(t)} dF_{k,n}(t)$$

is decreasing (increasing) in $x \geq 0$, which validates the result.

Theorem 4.1.5 *If X is IRHR, then, $X_1^{(k,n)} \geq_{rh} X_1$.*

Proof For any $x \geq 0$,

$$\frac{F_n^{(k)}(x)}{F(x)} = \int_0^\infty \frac{F(t+x)-F(t)}{\bar{F}(t)F(x)} dF_{k,n}(t) = \int_0^\infty \left[\frac{F(t+x)}{\bar{F}(t)F(x)} - \frac{F(t)}{\bar{F}(t)F(x)}\right] dF_{k,n}(t).$$

The IRHR property of X implies that the first term in integrand is increasing in $x \geq 0$, and it validates the result.

Theorem 4.1.6 *If X is NBUC (NWUC), then, $X_1^{(k,n)} \leq_{icx} X_1 (X_1^{(k,n)} \geq_{icx} X_1)$.*

Proof For any $y \geq 0$,

$$\begin{aligned}
\int_y^{\infty} \bar{F}_n^{(k)}(x)dx &= \int_y^{\infty}\int_0^{\infty} \frac{\bar{F}(t+x)}{\bar{F}(t)} dF_{k,n}(t)dx \\
&= \int_0^{\infty}\int_y^{\infty} \frac{\bar{F}(t+x)}{\bar{F}(t)} dxdF_{k,n}(t) \\
&\leq \int_0^{\infty}\int_y^{\infty} \bar{F}(x)dxdF_{k,n}(t) \\
&= \int_y^{\infty} \bar{F}(x)\int_0^{\infty} dF_{k,n}(t)dx \\
&= \int_y^{\infty} \bar{F}(x)dx,
\end{aligned}$$

where the inequality follows from the NBUC of X (The case of NWUC follows by reversing the inequality above), and hence the result follows.

Theorem 4.1.7 *If X is NBU(2) (NWU(2)), then, $X_1^{(k,n)} \leq_{icv} X_1 (X_1^{(k,n)} \geq_{icx} X_1)$.*

Proof We only give the proof for the NBU(2) of X. For any $y \geq 0$,

$$\begin{aligned}
\int_0^{y} \bar{F}_n^{(k)}(x)dx &= \int_0^{y}\int_0^{\infty} \frac{\bar{F}(t+x)}{\bar{F}(t)} dF_{k,n}(t)dx \\
&= \int_0^{\infty}\int_0^{y} \frac{\bar{F}(t+x)}{\bar{F}(t)} dxdF_{k,n}(t) \\
&\leq \int_0^{\infty}\int_0^{y} \bar{F}(x)dxdF_{k,n}(t) \\
&= \int_0^{y} \bar{F}(x)\int_0^{\infty} dF_{k,n}(t)dx \\
&= \int_0^{y} \bar{F}(x)dx,
\end{aligned}$$

where the inequality follows from the NBU(2) of X, hence by the equivalent definition of the icv order, the proof of the result is complete.

Theorem 4.1.8 *If X is NBUE (NWUE), then, $E(X_1^{(k,n)}) \leq E(X_1)(E(X_1^{(k,n)}) \geq E(X_1))$.*

Proof We only give the proof for the NBUE of X. Note that

$$\begin{aligned}
\int_0^{\infty} \bar{F}_n^{(k)}(x)dx &= \int_0^{\infty}\int_0^{\infty} \frac{\bar{F}(t+x)}{\bar{F}(t)} dF_{k,n}(t)dx \\
&= \int_0^{\infty}\int_0^{\infty} \frac{\bar{F}(t+x)}{\bar{F}(t)} dxdF_{k,n}(t)
\end{aligned}$$

$$
\begin{aligned}
&\leq \int_0^\infty \int_0^\infty \bar{F}(x)dx dF_{k,n}(t) \\
&= \int_0^\infty \bar{F}(x) \int_0^\infty dF_{k,n}(t)dx \\
&= \int_0^\infty \bar{F}(x)dx,
\end{aligned}
$$

where the inequality follows from the NBUE of that X, thus the result is right.

In reliability theory, it is very important to compare stochastic behavior of the variability of coherent systems so as to design more reliable systems. This section conducts stochastic comparisons based on residual lives of the remaining components of two $(n-k+1)$ out of n systems $\mathbb{S}_1$ and $\mathbb{S}_2$ with two sets of components.

Let X, Y be the lifetimes of the identical components of $\mathbb{S}_1$ and $\mathbb{S}_2$, respectively, with corresponding reliability functions $\bar{F}$ and $\bar{G}$. Denoted by $X_i^{(k)}$ $(i = 1, 2, \ldots, n-k)$, and $Y_i^{(k)}$ $(i = 1, 2, \ldots, n-k)$ respective residual lives of the remaining components of systems $\mathbb{S}_1$ and $\mathbb{S}_2$.

Theorem 4.1.9 *Let* $\mathbf{X}_{n-k} = (X_1^{(k)}, \ldots, X_{n-k}^{(k)})$, $\mathbf{Y}_{n-k} = (Y_1^{(k)}, \ldots, Y_{n-k}^{(k)})$. *If* $X \leq_{hr} Y$, *and either* X *or* Y *is DFR, then,* $\mathbf{X}_{n-k} \leq_{uo} \mathbf{Y}_{n-k}$.

Proof The proof of the DFR of X is given below, the case of Y is similar manner. By (4.1), the reliability functions of $(X_1^{(k)}, X_2^{(k)}, \ldots, X_{n-k}^{(k)})$ and $(Y_1^{(k)}, Y_1^{(k)}, \ldots, Y_{n-k}^{(k)})$ are, respectively,

$$\bar{F}_n^{(k)}(x_1, x_2, \ldots, x_{n-k}) = \int_0^\infty \prod_{i=1}^{n-k} \bar{F}_t(x_i) dF_{k,n}(t),$$

$$\bar{G}_n^{(k)}(x_1, x_2, \ldots, x_{n-k}) = \int_0^\infty \prod_{i=1}^{n-k} \bar{G}_t(x_i) dG_{k,n}(t).$$

Then

$$
\begin{aligned}
&\bar{G}_n^{(k)}(x_1, x_2, \ldots, x_{n-k}) - \bar{F}_n^{(k)}(x_1, x_2, \ldots, x_{n-k}) \\
&= \int_0^\infty \prod_{i=1}^{n-k} \bar{G}_t(x_i) dG_{k,n}(t) - \int_0^\infty \prod_{i=1}^{n-k} \bar{F}_t(x_i) dF_{k,n}(t) \\
&= \int_0^\infty \prod_{i=1}^{n-k} [\bar{G}_t(x_i) - \bar{F}_t(x_i)] dG_{k,n}(t) + \int_0^\infty \prod_{i=1}^{n-k} \bar{F}_t(x_i) d[G_{k,n}(t) - F_{k,n}(t)].
\end{aligned}
\tag{4.3}
$$

$X \leq_{hr} Y$ implies that $\bar{G}_t(x_i) - \bar{F}_t(x_i)$ is nonnegative for all $t \geq 0$ and $x_i, i = 1, 2, \ldots, n-k$, hence the first term of above equality is nonnegative for $x_i, i = 1, 2, \ldots, n-k$. Furthermore, by $X \leq_{hr} Y$,

$$\int_y^\infty d[G_{k,n}(t) - F_{k,n}(t)] = \bar{G}_t(y) - \bar{F}_t(y)$$

is nonnegative for any $y \geq 0$, and the DFR of X implies that $\prod_{i=1}^{n-k} \bar{F}_t(x_i)$ is increasing in $t \geq 0$. It follows from Lemma 1.b of page 120 in Barlow and Proschan (1975), that the second term of above equality is nonnegative for $x_i, i = 1, 2, \ldots, n-k$. Thus the expression desired result follows immediately.

The following corollary can be easily obtained by the theorem above.

Corollary 4.1.10 *If $X \leq_{hr} Y$, then $X_1^{(k)} \leq_{st} Y_1^{(k)}$.*

Lemma 4.1.11 (Müller and Styan 2002) *Let Θ be a subset of the real line $\Re$, and let X be a nonnegative random variable having a distribution belonging to the family $\mathcal{P} = \{H(\cdot|\theta),\ \theta \in \Theta\}$ which satisfies that*

$$H(\cdot|\theta_2) \geq_{st} (\leq_{st}) H(\cdot|\theta_1) \quad whenever\ \theta_1,\ \theta_2 \in \Theta,\ and\ \theta_2 > \theta_1 > 0.$$

Let $\Psi(x,\ \theta)$ be a real valued function defined on $\Re \times \Theta$, and be measurable in x for each θ such that $\mathbb{E}_\theta[\Psi(x,\ \theta)]$ exists. If $\Psi(x,\ \theta)$ is increasing in θ and increasing (decreasing) in x, then $\mathbb{E}_\theta[\Psi(X,\ \theta)]$ is increasing in θ.

Theorem 4.1.12 *If $X \leq_{lr} Y$, and either X or Y is DLR, then $X_1^{(k)} \leq_{lr} Y_1^{(k)}$.*

Proof We only address the proof for the case when X is of DLR, the proof for the case of Y is DLR can be completed in a similar manner. By (4.2), for any $x \geq 0$, the density function of $Y_1^{(k)}$ is

$$g_n^{(k)}(x) = k\binom{n}{k} \int_0^\infty \bar{G}^{n-k-1}(t) g(t+x) G^{k-1}(t) g(t) dt.$$

It is noted that

$$\begin{aligned}\frac{g_n^{(k)}(x)}{f_n^{(k)}(x)} &= \frac{\int_0^\infty \bar{G}^{n-k-1}(t) g(t+x) G^{k-1}(t) g(t) dt}{\int_0^\infty \bar{F}^{n-k-1}(t) f(t+x) F^{k-1}(t) f(t) dt} \\ &= E[\psi(T, x)],\end{aligned}$$

where

$$\psi(t, x) = \frac{\bar{G}^{n-k-1}(t) g(t+x) G^{k-1}(t) g(t)}{\bar{F}^{n-k-1}(t) f(t+x) F^{k-1}(t) f(t)},$$

and the nonnegative random variable T has a distribution function belonging to the family $P_1 = \{H(\cdot|x) : x \geq 0\}$ with corresponding density function given by

$$h(t|x) = c(x) \bar{F}^{n-k-1}(t) f(t+x) F^{k-1}(t) f(t), \quad t \geq 0,$$

here $c(x)$ is the normalizing constant. It follows from that X is DLR, and (4.2), that $H(\cdot|x_1) \le_{lr} H(\cdot|x_2)$ whenever $x_2 \ge x_1 \ge 0$. It is seen from $X \le_{lr} Y$ that, $\psi(t,x)$ is increasing in $t \ge 0$ and $x \ge 0$. Thus by Lemma 4.1.11, $E[\psi(T,x)]$ is increasing in $x \ge 0$, and hence the result holds.

Theorem 4.1.13 *If $X \le_{lr\uparrow} Y$, and either X or Y is DLR, then $X_1^{(k)} \le_{lr\uparrow} Y_1^{(k)}$.*

Proof We only address the proof for the case when X is of DLR. Note that, for any $x \ge 0,\ y \ge 0$,

$$\begin{aligned}\frac{g_n^{(k)}(x)}{f_n^{(k)}(x+y)} &= \frac{\int_0^\infty \bar{G}^{n-k-1}(t)g(t+x)G^{k-1}(t)g(t)dt}{\int_0^\infty \bar{F}^{n-k-1}(t)f(t+x+y)F^{k-1}(t)f(t)dt}\\ &= E[\psi(T,x)],\end{aligned}$$

where

$$\psi(t,x) = \frac{\bar{G}^{n-k-1}(t)g(t+x)G^{k-1}(t)g(t)}{\bar{F}^{n-k-1}(t)f(t+x+y)F^{k-1}(t)f(t)},$$

and the nonnegative random variable T has a distribution function belonging to the family $P_1 = \{H_1(\cdot|x) : x \ge 0\}$ with corresponding density function given by

$$h_1(t|x) = c(x)\bar{F}^{n-k-1}(t)f(t+x+y)F^{k-1}(t)f(t), \quad t \ge 0,$$

here $c(x)$ is the normalizing constant. The rest proof is completely similar as that of Theorem 4.1.12 above, and hence is omitted.

Theorem 4.1.14 *If $X \le_{lr} Y$, and either X or Y is DFR, then $X_1^{(k)} \le_{hr} Y_1^{(k)}$.*

Proof We only give the proof for the case when X is of DFR, the proof for the case of Y is DFR can be completed in a similar manner. By (4.2), for any $x \ge 0$, the survival function of $Y_1^{(k)}$ is

$$\bar{G}_n^{(k)}(x) = k\binom{n}{k}\int_0^\infty G^{k-1}(t)\bar{G}^{n-k}(t)\bar{G}(t+x)g(t)dt.$$

It is noted that

$$\begin{aligned}\frac{\bar{G}_n^{(k)}(x)}{\bar{F}_n^{(k)}(x)} &= \frac{k\binom{n}{k}\int_0^\infty G^{k-1}(t)\bar{G}^{n-k}(t)\bar{G}(t+x)g(t)dt}{k\binom{n}{k}\int_0^\infty F^{k-1}(t)\bar{F}^{n-k}(t)\bar{F}(t+x)f(t)dt}\\ &= E[\psi(T,x)],\end{aligned}$$

where

$$\psi(t,x) = \frac{G^{k-1}(t)\bar{G}^{n-k}(t)\bar{G}(t+x)g(t)}{F^{k-1}(t)\bar{F}^{n-k}(t)\bar{F}(t+x)f(t)},$$

and the nonnegative random variable T has a distribution function belonging to the family $P_1 = \{H_2(\cdot|x) : x \ge 0\}$ with corresponding density function given by

$$h_2(t|x) = c(x)F^{k-1}(t)\bar{F}^{n-k}(t)\bar{F}(t+x)f(t), \quad t \geq 0,$$

here $c(x)$ is the normalizing constant. It follows from that X is DFR, that $H_2(\cdot|x_1) \leq_{lr} H_2(\cdot|x_2)$ whenever $x_2 \geq x_1 \geq 0$. It is seen from $X \leq_{lr} Y$ that $\psi(t, x)$ is increasing in $t \geq 0$ and in $x \geq 0$. Thus by Lemma 4.1.11, that $E[\psi(T, x)]$ is increasing in $x \geq 0$, and hence the result holds.

4.2 On Surviving Components upon Failure of a Coherent System

This section considers a coherent system consisting of n components with independent and identically distributed components. We obtain a mixture representation of the reliability function of the residual lifetime of living components of a coherent system when the system has failed. Based on the concept of signature, stochastic comparisons on the residual life of two systems with stochastically ordered signatures are conducted.

4.2.1 *Introduction*

In practice, many coherent systems with n components have the property that some components would never fail even the system has failed. Such systems have the signature vector as follows:

$$s = (s_1, \ldots, s_i, 0, \ldots, 0) \tag{4.4}$$

with $s_i > 0$ for $1 < i < n$. The components with lifetimes $X_{j:n} (j = i+1, \ldots, n)$ must be alive even when the system failed. Hence, after the failure of the system, the surviving components may be removed from the system and then can be used for other systems or testing purposes. Thus, it could be important for system designers to get some useful information on the reliability properties of the surviving components in the used coherent system. Therefore, many authors investigated the stochastic properties of surviving components when the system has failed at some time. Suppose, for example, consider a coherent system with 4 exchangeable components, whose signature is $s = (\frac{1}{2}, \frac{1}{2}, 0, 0)$. Then the system would fail upon the second component failure, but clearly there are still two surviving components.

Balakrishnan and Asadi (2012) have discussed the residual lifetime of the living components of system having signature as (4.4), under the condition that the system is working at time t; that is, $(X_{k:n} - t \mid T > t), k = i+1, \ldots, n$. Kelkin et al. (2013) have considered another case, given that the system exactly failed at time t, that is, $(X_{k:n} - t \mid T = t), k = i+1, \ldots, n$.

Recently, Goliforushani et al. (2012) have demonstrated a kind of conditional random variable $(X_{k:n} - t \mid T < t < X_{k:n}), k = i + 1, \ldots, n$, which represents the residual lifetimes of the components of the system with signature vector of (4.4), under the condition that the system has failed by time t.

In fact, the event "$(T < t < X_{k:n})$" and event "$(T < t)$" have the same meaning for the coherent systems with the signature form in (4.4). Of course, their representations of the reliability functions are different and hence distribution and density functions. Based on this, this paper considers the conditional random variable $(X_{k:n} - t \mid T < t), k = i + 1, \ldots, n$, which shows in fact the residual lifetime of the surviving components with lifetime $X_{k:n}, k = i + 1, \ldots, n$ when the system has failed before time t.

4.2.2 *Main Results of Residual Lifetimes of Living Components*

Let T be the lifetime of a coherent system with n $i.i.d$ components, and let $X_1, \ldots, X_n$ be the component lifetimes having a common continuous distribution F. Furthermore, we assume that the signature of the system has the form of (4.4); that is,

$$s = (s_1, \ldots, s_i, 0, \ldots, 0), \qquad s_i > 0, \quad 1 \le i < n.$$

Firstly, we give the reliability function of conditional random variables $(X_{k:n} - t \mid T < t)$ in the following theorem.

Theorem 4.2.1 *For $1 \le i \le k \le n$ and $0 \le x \le t$,*

$$P(X_{k:n} - t > x \mid T < t) = \sum_{j=1}^{i} s_j(t) P(X_{k:n} - t > x \mid X_{j:n} < t). \tag{4.5}$$

Proof For all $0 \le x \le t$ and $1 \le i < k \le n$,

$$\begin{aligned}
P(X_{k:n} - t > x \mid T < t) &= \frac{P(X_{k:n} - t > x, T < t)}{P(T < t)} \\
&= \frac{\sum_{j=1}^{i} P(T = X_{j:n}) P(X_{k:n} - t > x, T < t \mid T = X_{j:n})}{\sum_{m=1}^{i} P(T = X_{m:n}) P(T < t \mid T = X_{m:n})} \\
&= \frac{\sum_{j=1}^{i} s_j P(X_{k:n} - t > x, X_{j:n} < t)}{\sum_{m=1}^{i} s_m P(X_{m:n} < t)} \\
&= \frac{\sum_{j=1}^{i} s_j P(X_{k:n} - t > x \mid X_{j:n} < t) P(X_{j:n} < t)}{\sum_{m=1}^{i} s_m P(X_{m:n} < t)} \\
&= \sum_{j=1}^{i} s_j(t) P(X_{k:n} - t > x \mid X_{j:n} < t),
\end{aligned}$$

where the function $s_j(t) = s_j P(X_{j:n} < t)/\sum_{m=1}^{i} s_m P(X_{m:n} < t)$ for $j = 1, \ldots, i$.

Remark Expression in (4.5) indicates that, the residual lifetime of the kth component, given that the system failed at time t, can be represented as a mixture of the residual lifetime of the kth component lifetime given that $n - j + 1$-out-of-n system has failed at time t with weight coefficients $s_j(t)$ for $j = 1, \ldots, i$. For instance, consider the system with lifetime $T = \min(X_{2:3}, X_4)$, whose signature is $\boldsymbol{s} = (1/4, 3/4, 0, 0)$. Given that the system has failed at some time t, then, by some computations, the coefficient vector $\boldsymbol{s}(t) = (s_1(t), \ldots, s_n(t))$ is given by

$$\boldsymbol{s}(t) = \left(\frac{1 - \frac{3}{2}F(t) + F^2(t) - \frac{1}{4}F^3(t)}{1 + 3F(t) - 5F^2(t) + 2F^3(t)}, \frac{\frac{9}{2}F(t) - 6F^2(t) + \frac{9}{4}F^3(t)}{1 + 3F(t) - 5F^2(t) + 2F^3(t)}, 0, 0 \right),$$

where $F(t)$ is the distribution of components lifetime. Hence, from Theorem 4.1 we have

$$\begin{aligned} P(X_{3:4} - t > x \mid T < t) &= \sum_{j=1}^{2} s_j(t) P(X_{3:4} - t > x \mid X_{j:4} < t) \\ &= \frac{1 - \frac{3}{2}F(t) + F^2(t) - \frac{1}{4}F^3(t)}{1 + 3F(t) - 5F^2(t) + 2F^3(t)}) P(X_{3:4} - t > x \mid X_{1:4} < t) \\ &\quad + \frac{\frac{9}{2}F(t) - 6F^2(t) + \frac{9}{4}F^3(t)}{1 + 3F(t) - 5F^2(t) + 2F^3(t)} P(X_{3:4} - t > x \mid X_{2:4} < t). \end{aligned}$$

Furthermore, it is noted that $\lim_{t\to\infty} \boldsymbol{s}(t) = (1/4, 3/4, 0, 0)$ and $\boldsymbol{s}(0) = (1, 0, 0, 0)$.

Theorem 4.2.2 *Let $\boldsymbol{s}(t) = (s_1(t), \ldots, s_n(t))$ be the coefficient vector in* (4.5) *of a coherent system with n IID components. Then, for all* $0 \le t_1 \le t_2$, $\boldsymbol{s}(t_1) \le_{lr} \boldsymbol{s}(t_2)$.

Proof It is enough to prove that, for all $j = 1, 2, \ldots, n$ and $0 \le t_1 \le t_2$,

$$\frac{\sum_{l=j}^{n} s_l P(X_{l:n} < t_1)}{\sum_{m=1}^{n} s_m P(X_{m:n} < t_1)} \le \frac{\sum_{l=j}^{n} s_l P(X_{l:n} < t_2)}{\sum_{m=1}^{n} s_m P(X_{m:n} < t_2)}.$$

It is equivalent to

$$\sum_{l=j}^{n} \sum_{m=1}^{n} s_l s_m \left[F_{X_{l:n}}(t_2) F_{X_{m:n}}(t_1) - F_{X_{l:n}}(t_1) F_{X_{l:n}}(t_2) \right] \ge 0.$$

After some algebraic computations, the above inequality can be written as

$$\sum_{l=j}^{n} \sum_{m=1}^{j-1} s_l s_m \left[F_{X_{l:n}}(t_2) F_{X_{m:n}}(t_1) - F_{X_{l:n}}(t_1) F_{X_{l:n}}(t_2) \right] \ge 0.$$

It is well known that $X_{l:n} \leq_{rh} X_{m:n}$ and for all $l \leq m$, that is $F_{X_{l:n}}(t)/F_{X_{m:n}}(t)$ is decreasing in $t \geq 0$, where $F_{X_{l:n}}$ and $F_{X_{m:n}}$ are respective distribution functions of the order statistics $X_{l:n}$ and $X_{m:n}$. So, the bracket in summation above is nonnegative. Hence the result follows immediately.

Theorem 4.2.3 *If the distribution F is absolutely continuous, then for $1 \leq j < k \leq n$ and $t \geq 0$,*

$$(X_{k:n} - t \mid X_{j-1:n} < t) \geq_{lr} (X_{k:n} - t \mid X_{j:n} < t).$$

Proof Let $X_{j-1,k,n,(t)} = (X_{k:n} - t \mid X_{j-1:n} < t)$, $X_{j,k,n,(t)} = (X_{k:n} - t \mid X_{j:n} < t)$, $a_j(t) = 1/P(X_{j:n} < t)$. Denote $f_{X_{j-1,k,n,(t)}}(x)$ and $f_{X_{j,k,n,(t)}}(x)$ be respective density function. Then,

$$\begin{aligned}
&P(X_{k:n} - t > x \mid X_{j:n} < t) \\
&= a_j(t) P(X_{k:n} - t > x, X_{j:n} < t) \\
&= a_j(t) \sum_{p=j}^{n} \binom{n}{p} F(t)^p \sum_{q=0}^{k-j-1} \binom{n-p}{q} (\bar{F}(t) - \bar{F}(t+x))^q \bar{F}(t+x)^{n-p-q} \\
&= a_j(t) \sum_{p=j}^{n} \binom{n}{p} F(t)^p \bar{F}(t)^{n-p} \sum_{q=0}^{k-j-1} \binom{n-p}{q} (1 - \bar{F}_t(x))^q \bar{F}_t(x)^{n-p-q} \\
&= a_j(t) \sum_{p=j}^{n} \binom{n}{p} F(t)^p \bar{F}(t)^{n-p} \int_{1-\bar{F}_t(x)}^{1} (k-j-1) \binom{n-p}{k-j-1} u^{k-j-2} (1-u)^{n-p-k+j+1} du.
\end{aligned}$$

From the above expression, it is shown that

$$\begin{aligned}
&f_{X_{j,k,n,(t)}}(x) \\
&= a_j(t) \sum_{p=j}^{n} \binom{n}{p} F(t)^p \bar{F}(t)^{n-p} \frac{f(t+x)}{1-F(t)} (k-j-1) \binom{n-p}{k-j-1} (1 - \bar{F}_t(x))^{k-j-2} \bar{F}_t(x)^{n-p-k+j+1} \\
&= a_j(t) f(t+x) (1 - \bar{F}_t(x))^{k-j-2} \sum_{p=j}^{n} (k-j-1) \binom{n}{p} \binom{n-p}{k-j-1} F(t)^p \bar{F}(t)^{n-p-1} \bar{F}_t(x)^{n-p-k+j+1} \\
&= a_j(t)(k-j-1) f(t+x) (\bar{F}(t) - \bar{F}(t+x))^{k-j-2} \sum_{p=j}^{n} \binom{n}{p} \binom{n-p}{k-j-1} F(t)^p (\bar{F}(t+x))^{n-p-k+j+1}.
\end{aligned}$$

In a similar manner,

$$\begin{aligned}
&f_{X_{j-1,k,n,(t)}}(x) \\
&= a_{j-1}(t)(k-j) f(t+x) (\bar{F}(t) - \bar{F}(t+x))^{k-j-1} \sum_{p=j-1}^{n} \binom{n}{p} \binom{n-p}{k-j} F(t)^p (\bar{F}(t+x))^{n-p-k+j}.
\end{aligned}$$

Hence, we readily have

$$\frac{f_{X_{j,k,n,(t)}}(x)}{f_{X_{j-1,k,n,(t)}}(x)}$$

$$
\begin{aligned}
&= \frac{a_j(t)(k-j-1)f(t+x)(\bar{F}(t)-\bar{F}(t+x))^{k-j-2}\sum_{p=j}^{n}\binom{n}{p}\binom{n-p}{k-j-1}F(t)^p(\bar{F}(t+x))^{n-p-k+j+1}}{a_{j-1}(t)(k-j)f(t+x)(\bar{F}(t)-\bar{F}(t+x))^{k-j-1}\sum_{p=j-1}^{n}\binom{n}{p}\binom{n-p}{k-j}F(t)^p(\bar{F}(t+x))^{n-p-k+j}}\\
&= \frac{a_j(t)(k-j-1)}{a_{j-1}(t)(k-j)}\left(\frac{\bar{F}(t+x)}{\bar{F}(t)-\bar{F}(t+x)}\right)\frac{\sum_{p=j}^{n}\binom{n}{p}\binom{n-p}{k-j-1}\left(\frac{F(t)}{\bar{F}(t+x)}\right)^p}{\sum_{p=j-1}^{n}\binom{n}{p}\binom{n-p}{k-j}\left(\frac{F(t)}{\bar{F}(t+x)}\right)^p}\\
&= \frac{a_j(t)(k-j-1)}{a_{j-1}(t)(k-j)}H(x\mid t).
\end{aligned}
$$

Now, we only need to prove that $H(x \mid t)$ is a decreasing function of x. Let

$$H(x \mid t) = B(t,x)\Phi(A(t,x)),$$

where

$$\Phi(u) = \frac{\sum_{p=j}^{n}\binom{n}{p}\binom{n-p}{k-j-1}u^p}{\sum_{p=j-1}^{n}\binom{n}{p}\binom{n-p}{k-j}u^p}, \qquad A(t,x) = \frac{F(t)}{\bar{F}(t+x)}, \qquad B(t,x) = \frac{\bar{F}(t+x)}{\bar{F}(t)-\bar{F}(t+x)}.$$

Then,

$$
\begin{aligned}
\frac{\partial}{\partial u}\Phi(u) = &\frac{p_1\sum_{p_1=j}^{n}\binom{n}{p_1}\binom{n-p_1}{k-j-1}u^{p_1-1}\left(\sum_{p_2=j}^{n}\binom{n}{p_2}\binom{n-p_2}{k-j}u^{p_2}+\binom{n}{j-1}\binom{n-j+1}{k-j+1}u^{j-1}\right)}{\left(\sum_{p=j-1}^{n}\binom{n}{p}\binom{n-p}{k-j}u^p\right)^2}\\
&-\frac{\sum_{p_1=j}^{n}\binom{n}{p_1}\binom{n-p_1}{k-j-1}u^{p_1}\left(p_2\sum_{p_2=j}^{n}\binom{n}{p_2}\binom{n-p_2}{k-j}u^{p_2-1}(j-1)\binom{n}{j-1}\binom{n-j+1}{k-j+1}u^{j-2}\right)}{\left(\sum_{p=j-1}^{n}\binom{n}{p}\binom{n-p}{k-j}u^p\right)^2}.
\end{aligned}
$$

The numerator of the above fraction is equal to

$$
\begin{aligned}
&\sum_{p_1=j}^{n}\sum_{p_2=j}^{n}\binom{n}{p_1}\binom{n-p_1}{k-j-1}\binom{n}{p_2}\binom{n-p_2}{k-j}u^{p_1+p_2-1}(p_1-p_2)\\
&+\sum_{p_1=j}^{n}\binom{n}{p_1}\binom{n-p_1}{k-j-1}\binom{n}{j-1}\binom{n-j+1}{k-j+1}u^{p_1+j-2}(p_1-j+1).
\end{aligned}
$$

The second term is obviously non-negative. Hence, we just need to prove that the first term is non-negative. It is noted that,

$$
\begin{aligned}
&\sum_{p_1=j}^{n}\sum_{p_2=j}^{n}\binom{n}{p_1}\binom{n}{p_2}\binom{n-p_1}{k-j-1}\binom{n-p_2}{k-j}u^{p_1+p_2-1}(p_1-p_2)\\
&=\sum_{p_1=j}^{n}\sum_{p_2=j}^{p_1}\binom{n}{p_1}\binom{n}{p_2}\binom{n-p_1}{k-j-1}\binom{n-p_2}{k-j}u^{p_1+p_2-1}(p_1-p_2)\\
&+\sum_{p_1=j}^{n}\sum_{p_2=p_1}^{n}\binom{n}{p_1}\binom{n}{p_2}\binom{n-p_1}{k-j-1}\binom{n-p_2}{k-j}u^{p_1+p_2-1}(p_1-p_2)
\end{aligned}
$$

$$
\begin{aligned}
&= \sum_{p_1=j}^{n} \sum_{p_2=j}^{p_1} \binom{n}{p_1}\binom{n}{p_2}\binom{n-p_1}{k-j-1}\binom{n-p_2}{k-j} u^{p_1+p_2-1}(p_1-p_2) \\
&\quad + \sum_{p_2=j}^{n} \sum_{p_1=j}^{p_2} \binom{n}{p_1}\binom{n}{p_2}\binom{n-p_1}{k-j-1}\binom{n-p_2}{k-j} u^{p_1+p_2-1}(p_1-p_2) \\
&= \sum_{p_1=j}^{n} \sum_{p_2=j}^{p_1} \binom{n}{p_1}\binom{n}{p_2}\binom{n-p_1}{k-j-1}\binom{n-p_2}{k-j} u^{p_1+p_2-1}(p_1-p_2) \\
&\quad + \sum_{p_1=j}^{n} \sum_{p_2=j}^{p_1} \binom{n}{p_1}\binom{n}{p_2}\binom{n-p_1}{k-j}\binom{n-p_2}{k-j-1} u^{p_1+p_2-1}(p_2-p_1) \\
&= \sum_{p_1=j}^{n} \sum_{p_2=j}^{p_1} \binom{n}{p_1}\binom{n}{p_2} u^{p_1+p_2-1}(p_1-p_2)\left[\binom{n-p_1}{k-j-1}\binom{n-p_2}{k-j} - \binom{n-p_1}{k-j}\binom{n-p_2}{k-j-1}\right],
\end{aligned}
$$

where

$$
\begin{aligned}
&\binom{n-p_1}{k-j-1}\binom{n-p_2}{k-j} - \binom{n-p_1}{k-j}\binom{n-p_2}{k-j-1} \\
&= \frac{(n-p_1)!(n-p_2)!}{(k-j-1)!(k-j)!}\left[\frac{1}{(n-p_1-k+j+1)!(n-p_2-k+j)!} - \frac{1}{(n-p_2-k+j+1)!(n-p_1-k+j)!}\right].
\end{aligned}
$$

From $p_1 \leq p_2$, we have

$$
\frac{(n-p_2-k+j+1)!}{(n-p_1-k+j+1)!} \geq \frac{(n-p_2-k+j)!}{(n-p_1-k+j)!}.
$$

So the first term is also non-negative, thus $\Phi(u)$ is an increasing function of u. On the other hand, $A(t, x) = F(t)/\bar{F}(t+x)$ is a decreasing function in x, and so $\Phi(A(t, x))$ is a decreasing function in x. Also

$$
\begin{aligned}
\frac{\partial}{\partial x} B(t, x) &= \frac{-f(t+x)(F(t+x) - F(t) - f(t+x)(1 - F(t+x)))}{(\bar{F}(t) - \bar{F}(t+x))^2} \\
&= \frac{-f(t+x)(1 - F(t))}{(\bar{F}(t) - \bar{F}(t+x))^2} \\
&\leq 0.
\end{aligned}
$$

Thus, $B(t, x)$ is a decreasing function in x, and then $H(x \mid t) = B(t, x)\Phi(A(t, x))$ is an increasing function in x. This completes the proof of the theorem.

The following lemma is useful in proving Theorem 4.2.5.

Lemma 4.2.4 (Shaked and Shanthikumar 2007) *Let α and β be real valued function such that α is nonnegative and α, β/α are decreasing. Assume X_i have the distribution function of F_i, $i = 1, 2$. If $X_1 \leq_{rh} X_2$, then*

$$\frac{\int_{-\infty}^{\infty}\beta(x)dF_1(x)}{\int_{-\infty}^{\infty}\alpha(x)dF_1(x)} \geq \frac{\int_{-\infty}^{\infty}\beta(x)dF_2(x)}{\int_{-\infty}^{\infty}\alpha(x)dF_2(x)}.$$

Theorem 4.2.5 *Let T_1, T_2 be the lifetimes of two coherent structure of size n, both based on IID components lifetimes $X_1, \ldots, X_n$ with common distribution function F. Assume that the two systems have signatures $s_1(t) = (s_{1,1}, \ldots, s_{1,j}, 0, \ldots, 0)$ and $s_2(t) = (s_{2,1}, \ldots, s_{2,j}, 0, \ldots, 0)$ with $s_{1,j}(t) > 0$, $s_{2,j}(t) > 0$. Let $s_1(t)$ and $s_2(t)$ be corresponding coefficient vectors as in Theorem 4.1.*
(a) If $s_1(t) \leq_{st} s_2(t)$, then $(X_{k:n} - t \mid T_1 < t) \geq_{st} (X_{k:n} - t \mid T_2 < t)$;
(b) If $s_1(t) \leq_{rh} s_2(t)$, then $(X_{k:n} - t \mid T_1 < t) \geq_{hr} (X_{k:n} - t \mid T_2 < t)$;
(c) If $s_1(t) \leq_{lr} s_2(t)$, then $(X_{k:n} - t \mid T_1 < t) \geq_{lr} (X_{k:n} - t \mid T_2 < t)$.

Proof (a) By Theorem 4.2.3, $P(X_{k:n} - t \mid X_{j:n} < t)$ is decreasing in j, $j = 1, \ldots, i$. According to (4.4), for any $t \geq 0$ and $x \geq 0$,

$$\begin{aligned} P(X_{k:n} - t > x \mid T_1 < t) &= \sum_{j=1}^{i} s_{1,j}(t)P(X_{k:n} - t > x \mid X_{j:n} < t) \\ &\geq \sum_{j=1}^{i} s_{2,j}(t)P(X_{k:n} - t > x \mid X_{j:n} < t) \\ &= P(X_{k:n} - t > x \mid T_2 < t). \end{aligned}$$

The last inequality follows from (1.A.7) of Shaked and Shanthikumar (2007). This completes the proof of part (a).

(b) Let $\bar{H}_{j,k,n,(t)}(x)$ $(1 \leq j < k \leq n)$ be the reliability function of the random variable $(X_{k:n} - t \mid X_{j:n} < t)$. Then, by the proof of (a),

$$P(X_{k:n} - t > x \mid T_1 < t) = \sum_{j=1}^{i} s_{1,j}(t)\bar{H}_{j,k,n,(t)}(x);$$

$$P(X_{k:n} - t > x \mid T_2 < t) = \sum_{j=1}^{i} s_{2,j}(t)\bar{H}_{j,k,n,(t)}(x).$$

In order to obtain the result, we only prove

$$\frac{P(X_{k:n} - t > x \mid T_1 < t)}{P(X_{k:n} - t > x \mid T_2 < t)} = \frac{\sum_{j=1}^{i} s_{1,j}(t)\bar{H}_{j,k,n,(t)}(x)}{\sum_{j=1}^{i} s_{2,j}(t)\bar{H}_{j,k,n,(t)}(x)}$$

is increasing in x. That is, for all $x_2 \geq x_1 \geq 0$

$$\frac{\sum_{j=1}^{i} s_{1,j}(t)\bar{H}_{j,k,n,(t)}(x_2)}{\sum_{j=1}^{i} s_{2,j}(t)\bar{H}_{j,k,n,(t)}(x_2)} \geq \frac{\sum_{j=1}^{i} s_{1,j}(t)\bar{H}_{j,k,n,(t)}(x_1)}{\sum_{j=1}^{i} s_{2,j}(t)\bar{H}_{j,k,n,(t)}(x_1)}.$$

Let $\alpha(j) = \bar{H}_{j,k,n,(t)}(x_1)$ and $\beta(j) = \bar{H}_{j,k,n,(t)}(x_2)$. Then by Theorem 4.2.3, $\alpha(j)$, $\alpha(j)/\beta(j)$ both are decreasing in j for all $t \geq 0$. Hence from Lemma 4.2.4, the desired result follows.

(c) Let $f_{X_{j,k,n,(t)}}(x)$ be the density function of the random variable $(X_{k:n} - t \mid X_{j:n} < t)$, $\sum_{j=1}^{i} s_{1,j}(t) f_{X_{j,k,n,(t)}}(x)$ and $\sum_{j=1}^{i} s_{2,j}(t) f_{X_{j,k,n,(t)}}(x)$ be the density functions of $(X_{k:n} - t \mid T_1 < t)$ and $(X_{k:n} - t \mid T_2 < t)$, respectively. We only need to prove that

$$\frac{\sum_{j=1}^{i} s_{1,j}(t) f_{X_{j,k,n,(t)}}(x)}{\sum_{j=1}^{i} s_{2,j}(t) f_{X_{j,k,n,(t)}}(x)}$$

is increasing in x. Obviously, this is equivalent to, for all $x_2 \geq x_1 \geq 0$

$$g(x_1, x_2) = \sum_{j=1}^{i}\sum_{l=1}^{i} s_{1,j}s_{2,l}\left[f_{X_{j,k,n,(t)}}(x_2) f_{X_{l,k,n,(t)}}(x_1) - f_{X_{j,k,n,(t)}}(x_1) f_{X_{l,k,n,(t)}}(x_2)\right] \geq 0.$$

Since $s_1(t) \leq_{lr} s_2(t)$ and $(X_{k:n} - t \mid X_{j-1:n} < t) \geq_{lr} (X_{k:n} - t \mid X_{j:n} < t)$,

$$\begin{aligned}
g(x_1, x_2) &= \sum_{j=1}^{i}\sum_{l=1}^{i} s_{1,j}s_{2,l}\left[f_{X_{j,k,n,(t)}}(x_2) f_{X_{l,k,n,(t)}}(x_1) - f_{X_{j,k,n,(t)}}(x_1) f_{X_{l,k,n,(t)}}(x_2)\right] \\
&= \sum_{j=1}^{i}\sum_{l=j}^{i} s_{1,j}s_{2,l}\left[f_{X_{j,k,n,(t)}}(x_2) f_{X_{l,k,n,(t)}}(x_1) - f_{X_{j,k,n,(t)}}(x_1) f_{X_{l,k,n,(t)}}(x_2)\right] \\
&\quad + \sum_{j=1}^{i}\sum_{l=1}^{j} s_{1,j}s_{2,l}\left[f_{X_{j,k,n,(t)}}(x_2) f_{X_{l,k,n,(t)}}(x_1) - f_{X_{j,k,n,(t)}}(x_1) f_{X_{l,k,n,(t)}}(x_2)\right] \\
&= \sum_{j=1}^{i}\sum_{l=j}^{i} s_{1,j}s_{2,l}\left[f_{X_{j,k,n,(t)}}(x_2) f_{X_{l,k,n,(t)}}(x_1) - f_{X_{j,k,n,(t)}}(x_1) f_{X_{l,k,n,(t)}}(x_2)\right] \\
&\quad + \sum_{j=1}^{i}\sum_{l=j}^{i} s_{1,l}s_{2,j}\left[f_{X_{l,k,n,(t)}}(x_2) f_{X_{j,k,n,(t)}}(x_1) - f_{X_{l,k,n,(t)}}(x_1) f_{X_{j,k,n,(t)}}(x_2)\right] \\
&= \sum_{j=1}^{i}\sum_{l=j}^{i} s_{1,j}s_{2,l}\left(\frac{s_{2,l}}{s_{1,l}} - \frac{s_{2,j}}{s_{1,j}}\right)\left[f_{X_{j,k,n,(t)}}(x_2) f_{X_{l,k,n,(t)}}(x_1) - f_{X_{j,k,n,(t)}}(x_1) f_{X_{l,k,n,(t)}}(x_2)\right] \\
&\geq 0.
\end{aligned}$$

This validates the desired result.

Now, we show if the residual lifetimes of the components of the system are IFR (increasing failure rate), then $P(X_{k:n} - t > x \mid T < t)$ is a decreasing function of time t. In order to get this result, we need the following result.

Theorem 4.2.6 *For all* $0 \le x \le t$ *and* $1 \le j < k \le n$,

$$P(X_{k:n} - t > x \mid X_{j:n} < t) = \frac{\sum_{l=j}^{n} \binom{n}{l} \phi_X(t)^l P(X^t_{k-l:n-l} > x)}{\sum_{m=j}^{n} \binom{n}{m} \phi_X(t)^m},$$

where $\phi_X(t) = F(t)/\bar{F}(t)$ *and* $X^t_{k-l:n-l}$ *denotes the* $(k-l)$*th order statistic among* $(n-l)$ *IID random variables with common distribution function* $\bar{F}_t(x) = \bar{F}(t + x)/\bar{F}(t)$.

Proof For $0 \le x \le t$ and $1 \le j < k \le n$,

$$\begin{aligned}
P(X_{k:n} - t > x \mid X_{j:n} < t) &= \frac{P(X_{k:n} - t > x, X_{j:n} < t)}{P(X_{j:n} < t)} \\
&= \frac{\sum_{l=j}^{n} P(X_{k:n} - t > x, X_{l:n} < t < X_{l+1:n})}{P(X_{j:n} < t)} \\
&= \sum_{l=j}^{n} P(X_{k:n} - t > x \mid X_{l:n} < t < X_{l+1:n}) \frac{P(X_{l:n} < t < X_{l+1:n})}{P(X_{j:n} < t)} \\
&= \sum_{l=j}^{n} \bar{H}_{k,l,l,(t)}(x) M^n_{l,j}(t),
\end{aligned}$$

where $\bar{H}_{k,l,l,(t)}(x) = P(X_{k:n} - t > x \mid X_{l:n} < t < X_{l+1:n})$, $M^n_{l,j}(t) = \frac{P(X_{l:n} < t < X_{l+1:n})}{P(X_{j:n} < t)}$. It is noted,

$$\begin{aligned}
\bar{H}_{k,l,l,(t)}(x) &= P(X_{k:n} - t > x \mid X_{l:n} < t < X_{l+1:n}) \\
&= \frac{\binom{n}{l} F(t)^l \sum_{m=n-k+1}^{n-l} \binom{n-l}{m} [F(t+x) - F(t)]^{n-l-m} (1 - F(t+x))^m}{\binom{n}{l} F(t)^l \bar{F}(t)^{n-l}} \\
&= \sum_{m=n-k+1}^{n-l} \binom{n-l}{m} \bar{F}_t(x)^m (1 - \bar{F}_t(x))^{n-l-m} \\
&= P(X^t_{k-l:n-l} > x),
\end{aligned}$$

and

$$\begin{aligned}
M^n_{l,j}(t) &= \frac{P(X_{l:n} < t < X_{l+1:n})}{P(X_{j:n} < t)} \\
&= \frac{\binom{n}{l} F(t)^l \bar{F}(t)^{n-l}}{\sum_{m=j}^{n} \binom{n}{m} F(t)^m \bar{F}(t)^{n-m}} \\
&= \frac{\binom{n}{l} (\phi_X(t))^l}{\sum_{m=j}^{n} \binom{n}{m} (\phi_X(t))^m}.
\end{aligned}$$

So, we have

$$P(X_{k:n} - t > x \mid X_{j:n} < t) = \frac{\sum_{l=j}^{n} \binom{n}{l} \phi_X(t)^l P(X^t_{k-l:n-l} > x)}{\sum_{m=j}^{n} \binom{n}{m} \phi_X(t)^m}.$$

This completes the desired result.

These two lemmas are very useful later.

Lemma 4.2.7 (Kelkin and Asadi 2014) *If X is IFR, then $P(X^t_{k-l:n-l} > x)$ is stochastically decreasing in $t > 0$, where $X^t_{k-l:n-l}$ denotes the $(k-l)$th order statistic among $(n-l)$ IID random variables with common distribution function $\bar{F}_t(x) = \bar{F}(t+x)/\bar{F}(t)$.*

Lemma 4.2.8 (Kelkin and Asadi 2014) *For $k > l \geq m$, we have*

$$P(X^t_{k-l:n-l} > x) \leq_{lr} P(X^t_{k-m:n-m} > x),$$

where $X^t_{k-m:n-m}$ denotes the $(k-m)$th order statistic among $(n-m)$ IID random variables with common distribution function $\bar{F}_t(x) = \bar{F}(t+x)/\bar{F}(t)$.

Theorem 4.2.9 *If X is IFR, then $P(X_{k:n} - t > x \mid X_{j:n} < t)(j < k)$ is stochastically decreasing in $t > 0$.*

Proof Note that $\frac{\partial \phi_X(t)}{\partial t} = \frac{r(t)}{\bar{F}(t)}$,

$$\begin{aligned}
\frac{\partial \bar{H}_{j,k,n,(t)}(x)}{\partial t} &= \frac{\frac{r(t)}{\bar{F}(t)} l \sum_{l=j}^{n} \binom{n}{l} \phi_X(t)^{l-1} P(X^t_{k-l:n-l} > x) \sum_{m=j}^{n} \binom{n}{m} \phi_X(t)^m}{(\sum_{m=j}^{n} \binom{n}{m} \phi_X(t)^m)^2} \\
&\quad + \frac{\sum_{l=j}^{n} \binom{n}{l} \phi_X(t)^l \frac{\partial}{\partial t} P(X^t_{k-l:n-l} > x) \sum_{m=j}^{n} \binom{n}{m} \phi_X(t)^m}{(\sum_{m=j}^{n} \binom{n}{m} \phi_X(t)^m)^2} \\
&\quad - \frac{\frac{r(t)}{\bar{F}(t)} \sum_{l=j}^{n} \binom{n}{l} \phi_X(t)^l P(X^t_{k-l:n-l} > x) m \sum_{m=j}^{n} \binom{n}{m} \phi_X(t)^{m-1}}{(\sum_{m=j}^{n} \binom{n}{m} \phi_X(t)^m)^2}.
\end{aligned}$$

The numerator of the above fraction is rewritten as

$$\frac{r(t)}{\bar{F}(t)}(C + D), \tag{4.6}$$

where

$$\begin{aligned}
C = l \sum_{l=j}^{n} \binom{n}{l} \phi_X(t)^{l-1} P(X^t_{k-l:n-l} > x) \sum_{m=j}^{n} \binom{n}{m} \phi_X(t)^m \\
- \sum_{l=j}^{n} \binom{n}{l} \phi_X(t)^l P(X^t_{k-l:n-l} > x) m \sum_{m=j}^{n} \binom{n}{m} \phi_X(t)^{m-1},
\end{aligned}$$

and

$$D = \frac{\bar{F}(t)}{r(t)} \sum_{l=j}^{n} \binom{n}{l} \phi_X(t)^l \frac{\partial}{\partial t} P(X^t_{k-l:n-l} > x) \sum_{m=j}^{n} \binom{n}{m} \phi_X(t)^m.$$

Then it suffices to prove that the term C and D in (4.6) are both negative. By Lemma 4.2.7 and the condition X is IFR, $P(X^t_{k-l:n-l} > x)$ is decreasing in $t > 0$. Thus, we have $\frac{\partial}{\partial t} P(X^t_{k-l:n-l} > x) \leq 0$ for all $t > 0$, and hence $D \leq 0$.

Next, we need to prove C is also negative. It can be shown that,

$$\begin{aligned} C &= l \sum_{l=j}^{n} \binom{n}{l} \phi_X(t)^{l-1} P(X^t_{k-l:n-l} > x) \sum_{m=j}^{n} \binom{n}{m} \phi_X(t)^m \\ &\quad - \sum_{m=j}^{n} \binom{n}{m} \phi_X(t)^m P(X^t_{k-m:n-m} > x) l \sum_{l=j}^{n} \binom{n}{l} \phi_X(t)^{l-1} \\ &= \sum_{l=j}^{n} \sum_{m=j}^{n} \binom{n}{l} \binom{n}{m} l \phi_X(t)^{l+m-1} \left[P(X^t_{k-l:n-l} > x) - P(X^t_{k-m:n-m} > x) \right] \\ &= \sum_{l=j}^{n} \sum_{m=j}^{l} \binom{n}{l} \binom{n}{m} l \phi_X(t)^{l+m-1} \left[P(X^t_{k-l:n-l} > x) - P(X^t_{k-m:n-m} > x) \right] \\ &\quad + \sum_{l=j}^{n} \sum_{m=l}^{n} \binom{n}{l} \binom{n}{m} l \phi_X(t)^{l+m-1} \left[P(X^t_{k-l:n-l} > x) - P(X^t_{k-m:n-m} > x) \right]. \end{aligned}$$

By interchanging the order of summations, the second term can be represented as

$$\begin{aligned} &\sum_{m=j}^{n} \sum_{l=j}^{m} \binom{n}{l} \binom{n}{m} m \phi_X(t)^{l+m-1} \left[P(X^t_{k-m:n-m} > x) - P(X^t_{k-l:n-l} > x) \right] \\ &= \sum_{l=j}^{n} \sum_{m=j}^{l} \binom{n}{l} \binom{n}{m} m \phi_X(t)^{l+m-1} \left[P(X^t_{k-m:n-m} > x) - P(X^t_{k-l:n-l} > x) \right]. \end{aligned}$$

It follows that

$$C = \sum_{l=j}^{n} \sum_{m=j}^{l} \binom{n}{l} \binom{n}{m} (l-m) \phi_X(t)^{l+m-1} \left[P(X^t_{k-l:n-l} > x) - P(X^t_{k-m:n-m} > x) \right].$$

By Lemma, for all $l \geq m$,

$$P(X^t_{k-l:n-l} > x) - P(X^t_{k-m:n-m} > x) \leq 0.$$

This the desired result follows.

Theorem 4.2.10 *If X is IFR, then $(X_{k:n} - t \mid T < t)$ is stochastically decreasing in $t > 0$.*

Proof From Theorem 4.2.3, we know that for all $x > 0$ and $t > 0$, $P(X_{k:n} - t > x \mid X_{j:n} < t)$ is a decreasing function of $t > 0$, and so

$$P(X_{k:n} - t > x \mid T < t) = \sum_{j=1}^{i} s_j(t) P(X_{k:n} - t > x \mid X_{j:n} < t)$$

is a decreasing function of t.

In what follows, we consider two coherent systems having the same structure with n IID components with their lifetimes denoted by $X_1, \ldots, X_n$ and $Y_1, \ldots, Y_n$. Let F and G be continuous distribution functions of X_1 and Y_1, respectively. Suppose T_1 and T_2 are the corresponding lifetimes of the two coherent systems. We wish to show that if $X_1 \leq_{hr} Y_1$, then the remaining components of the system with lifetime T_1 are more reliable than the remaining components of the system with lifetime T_2. For establishing this result, we first need the following lemma.

Lemma 4.2.11 *If $X_1 \leq_{hr} Y_1$ and $1 < l < k < n$, then*

$$P(X^t_{k-l:n-l} > x) \leq_{st} P(Y^t_{k-l:n-l} > x).$$

Proof The distribution function of $P(X^t_{k-l:n-l} > x)$ can be rephrased as

$$P(X^t_{k-l:n-l} > x) = \int_{1-\bar{F}_t(x)}^{1} (k-l)\binom{n-l}{k-l} u^{k-l-1}(1-u)^{n-k+l} du.$$

The assumption $X_1 \leq_{hr} Y_1$ implies that $\bar{F}_t(x) < \bar{G}_t(x)$, so $1 - \bar{F}_t(x) > 1 - \bar{G}_t(x)$. Hence, $P(X^t_{k-l:n-l} > x) \leq P(Y^t_{k-l:n-l} > x)$. This proof is finished.

Theorem 4.2.12 *If $X_1 \leq_{hr} Y_1$, then*

$$(X_{k:n} - t \mid X_{j:n} < t) \leq_{st} (Y_{k:n} - t \mid Y_{j:n} < t).$$

Proof For $1 \leq j < k \leq n$ and $0 \leq t \leq x$, we have

$$\begin{aligned}
& P(X_{k:n} - t > x \mid X_{j:n} < t) - P(Y_{k:n} - t > x \mid Y_{j:n} < t) \\
&= \frac{\sum_{l=j}^{n} \binom{n}{l} \phi_X(t)^l P(X^t_{k-l:n-l} > x)}{\sum_{m=j}^{n} \binom{n}{m} \phi_X(t)^m} - \frac{\sum_{l=j}^{n} \binom{n}{l} \phi_Y(t)^l P(X^t_{k-l:n-l} > x)}{\sum_{m=j}^{n} \binom{n}{m} \phi_Y(t)^m} \\
&= \frac{\sum_{l=j}^{n} \binom{n}{l} \phi_X(t)^l \sum_{m=j}^{n} \binom{n}{m} \phi_Y(t)^l P(X^t_{k-l:n-l} > x) - \sum_{m=j}^{n} \binom{n}{m} \phi_X(t)^m \sum_{l=j}^{n} \binom{n}{l} \phi_Y(t)^l P(Y^t_{k-l:n-l} > x)}{\sum_{m=j}^{n} \binom{n}{m} \phi_X(t)^m \sum_{m=j}^{n} \binom{n}{m} \phi_Y(t)^m} \\
&= \frac{\sum_{l=j}^{n} \binom{n}{l} \phi_X(t)^l \sum_{m=j}^{n} \binom{n}{m} \phi_Y(t)^l (P(X^t_{k-l:n-l} > x) - P(Y^t_{k-m:n-m} > x))}{\sum_{m=j}^{n} \binom{n}{m} \phi_X(t)^m \sum_{m=j}^{n} \binom{n}{m} \phi_Y(t)^m}.
\end{aligned}$$

By Lemmas 4.2.7 and 4.2.11, we know

$$P(X^t_{k-l:n-l} > x) - P(Y^t_{k-m:n-m} > x) \leq 0.$$

Hence, the last term is negative. This completes the proof.

Theorem 4.2.13 *Let $X_1, \ldots, X_n$ be the independent and identically distributed random variable with a common distribution function F and let $Y_1, \ldots, Y_n$ be the independent and identically distributed random variable with a common distribution function G. Denote $T_1 = \tau_1(X_1, \ldots, X_n)$ and $T_2 = \tau_2(Y_1, \ldots, Y_n)$ are the lifetimes of two coherent systems with coefficient vectors $s_1(t) = (s_{1,1}(t), \ldots, s_{1,j}(t), 0, \ldots, 0)$ and $s_2(t) = (s_{2,1}(t), \ldots, s_{2,j}(t), 0, \ldots, 0)$. If $X_1 \leq_{hr} Y_1$, then*

$$(X_{k:n} - t \mid T_1 < t) \leq_{st} (Y_{k:n} - t \mid T_2 < t).$$

Proof If $X_1 \leq_{hr} Y_1$, then from Theorem 4.2.12, we get that

$$(X_{k:n} - t \mid X_{j:n} < t) \leq_{st} (Y_{k:n} - t \mid Y_{j:n} < t).$$

On the other hand, the two systems have the same signature while the coefficient vectors as in Theorem 4.5 are different because of different components. Therefore, it is noted that

$$\begin{aligned} P(X_{k:n} - t > x \mid T_1 < t) &= \sum_{j=1}^{i} s_{1,j}(t) P(X_{k:n} - t > x \mid X_{j:n} < t) \\ &\leq \sum_{j=1}^{i} s_{1,j}(t) P(Y_{k:n} - t > x \mid Y_{j:n} < t) \\ &= \sum_{j=1}^{i} s_{2,j}(t) P(Y_{k:n} - t > x \mid Y_{j:n} < t) \\ &= P(Y_{k:n} - t > x \mid T_2 < t). \end{aligned}$$

This completes the proof of the theorem.

4.3 On Surviving Components upon Surviving Coherent Systems

In this section, we discuss some ordering and aging properties of the residual lives of surviving components, given that a system has survived till some time. First, we study the stochastic behavior of residual lifetimes of the surviving components of a general coherent system with an arbitrary structure when it is still working at some time. The obtained results are generalizations of results for some specific conditional systems known in the literature to the case of general coherent systems.

4.3.1 Introduction

In real life, many scrapped cars, refrigerators, televisions and other electrical appliances are thrown away even they don't fail. Some of components from these products still have given performance, so they can be used in other systems or for some other testing purposes. Based on this, many scholars have discussed the aging and stochastic behaviors of the remaining life of these unfailed components given that a system survives a certain time.

Suppose a coherent system of size n has its signature vector to be

$$s = (s_1, s_2, \ldots, s_r, 0, \ldots, 0)\ ,\ r = 1, \ldots, n-1, \text{with } s_r > 0. \tag{4.7}$$

Observe that the system with its signature as in (4.7) has the property that the components with lifetimes $X_{j:n}$ $(j = r+1, \ldots, n)$ would not be observed even the system has failed. For the case of a system with its signature vector as in (4.7), Balakrishnan and Asadi (2012) studied some stochastic properties of the unfailed components of the system under the condition that the system has survived till time t; that is,

$$(X_{j:n} - t|T > t) \ \text{ for } \ r < j < n.$$

It should be noted that for a general coherent system with arbitrary signature

$$\mathbf{s} = (s_1, s_2, \ldots, s_n),$$

some components or even all the components may be alive if the system is working at time t. So, we can consider the residual lives of the surviving components in the system when the system and surviving components have survived till time t, that is,

$$(X_{j:n} - t|X_{j:n} > t, T > t), \ \text{ for } \ j = 1, \ldots, n. \tag{4.8}$$

So we study here the residual lives of surviving components of coherent systems when the system and the surviving components have survived till time t. A mixture representation for the residual lives of the surviving components of used coherent systems is first obtained. We then utilize it to establish some ordering, preservation and aging properties of the residual lives of the surviving components of coherent systems.

4.3.2 Residual Lives of Surviving Components in General Coherent Systems

More generally, we now focus on the residual lives of possibly surviving components in a general coherent system with signature $\mathbf{s} = (s_1, s_2, \ldots, s_n)$, given that the system and components have survived till time t, that is, $(X_{j:n} - t|X_{j:n} - t, T > t)$, for $j = 1, \ldots, n$. We first present corresponding survival function in the following theorem.

Theorem 4.3.1 *For any* $j \in \{1, 2, \ldots, n\}$, *a given* $t > 0$, *and* $x > 0$,

$$P(X_{j:n} - t > x \mid X_{j:n} > t, T > t) = \sum_{i=1}^{n} s_i(t) P(X_{j:n} - t > x | X_{j:n} > t, X_{i:n} > t),$$

where $s_i(t) = s_i P(X_{i:n} > t, X_{j:n} > t / P(X_{j:n} > t, T > t)$.

Proof For a given $t > 0$, and $x > 0$, we find

$$\begin{aligned}
&P(X_{j:n} - t > x \mid X_{j:n} > t, T > t) \\
&= \sum_{i=1}^{n} P(X_{j:n} - t > x, T = X_{i:n} | X_{j:n} > t, T > t) \\
&= \sum_{i=1}^{n} P(T = X_{i:n} | X_{j:n} > t, T > t) P(X_{j:n} - t > x | T = X_{i:n}, X_{j:n} > t, T > t) \\
&= \sum_{i=1}^{n} P(T = X_{i:n}) \frac{P(X_{j:n} > t, T > t | T = X_{i:n})}{P(X_{j:n} > t, T > t)} P(X_{j:n} - t > x | X_{j:n} > t, X_{i:n} > t) \\
&= \sum_{i=1}^{n} s_i(t) P(X_{j:n} - t > x | X_{i:n} > t, X_{j:n} > t).
\end{aligned}$$

Hence, the theorem.

Remark In fact, for $i = 1, \ldots, n$, $s_i(t) = P(T = X_{i:n} | X_{j:n} > t, T > t)$ is the conditional signature, as stated in Navarro et al. (2008). Moreover, $s_1(t)$ is decreasing in time t, while $s_n(t)$ is increasing in time t. Also, we observe that the survival function of the residual lives of surviving components, given that a general coherent system has survived till time t, can be represented as a mixture of those of the residual lives of possibly surviving components, given that an i-out-of-n system survives up to time t, with the conditional signature as mixture coefficients.

As noted by Balakrishnan and Asadi (2012), we can then state the following result.

Theorem 4.3.2 *If the common distribution function of the component lifetimes within the system is IFR, then* $P(X_{j:n} - t > x \mid X_{j:n} > t, T > t)$ *decreases in time* t.

Remark Balakrishnan and Asadi (2012) presented a counterexample to show that this result is not true for the case when the components of the system are DFR for a coherent system having its signature as in (4.8). By using the same example, we can state that the above result doesn't hold for the case when the components of the system are DFR for a general coherent system having any signature.

Theorem 4.3.3 *Let $T_1 = \phi_1(X_1, \cdots, X_n)$ and $T_2 = \phi_2(X_1, \cdots, X_n)$ be the lifetimes of two coherent systems with respective system signatures $\mathbf{s}_1 = (s_{11}, \ldots, s_{1i}, \ldots, s_{1n})$ and $\mathbf{s}_2 = (s_{21}, \ldots, s_{2i}, \ldots, s_{2n})$, both based on n components with IID lifetimes $X_1, \ldots, X_n$ having a common continuous distribution function F. Let, for $t \geq 0$, $\mathbf{s}_1(t) = (s_{11}(t), \ldots, s_{1i}(t), \ldots, s_{1n}(t))$ and $\mathbf{s}_2(t) = (s_{21}(t), \ldots, s_{2i}(t), \ldots, s_{2n}(t))$ denote the respective coefficient vectors (i.e., the conditional signature elements), with $s_{1i}(t) = s_{1i} P(X_{i:n} > t, X_{j:n} > t / P(X_{j:n} > t, T_1 > t)$ and $s_{2i}(t) = s_{2i} P(X_{i:n} > t, X_{j:n} > t / P(X_{j:n} > t, T_2 > t)$, for $i = 1, \ldots, n$. If $\mathbf{s}_1(t) \leq_{st} \mathbf{s}_2(t)$, then $(X_{j:n} - t \mid X_{j:n} > t, T_1 > t) \geq_{st} (X_{j:n} - t \mid X_{j:n} > t, T_2 > t)$.*

Proof According to Theorem 4.3.1, for a given $t > 0$, and $x > 0$, we have

$$P(X_{j:n} - t > x \mid X_{j:n} > t, T_1 > t) = \sum_{i=1}^{n} s_{1i}(t) P(X_{j:n} - t > x | X_{j:n} > t, X_{i:n} > t),$$

$$P(X_{j:n} - t > x \mid X_{j:n} > t, T_2 > t) = \sum_{i=1}^{n} s_{2i}(t) P(X_{j:n} - t > x | X_{j:n} > t, X_{i:n} > t).$$

It is clear that $P(X_{j:n} - t > x | X_{j:n} > t, X_{i:n} > t)$ is decreasing in $i = 1, 2, \ldots, n$, hence, by the condition $\mathbf{s}_1(t) \leq_{st} \mathbf{s}_2(t)$ and Shaked and Shanthikumar (2007), the desired result holds.

Theorem 4.3.4 *Suppose there are two coherent systems having the same structure (and so the same signature $\mathbf{s} = (s_1, \ldots, s_i, \ldots, s_n)$) with two sets of IID component lifetimes $X_1, \ldots, X_n$ and $Y_1, \ldots, Y_n$ having distributions F and G, respectively, with respective coefficient vectors $\mathbf{s}_1(t) = (s_{11}(t), \ldots, s_{1i}(t), \ldots, s_{1n}(t))$ and $\mathbf{s}_2(t) = (s_{21}(t), \ldots, s_{2i}(t), \ldots, s_{2n}(t))$, with $s_{1i}(t) = s_i P(X_{j:n} > t, X_{i:n} > t) / P(X_{j:n} > t, T_1 > t)$ and $s_{2i}(t) = s_i P(Y_{j:n} > t, Y_{i:n} > t) / P(Y_{j:n} > t, T_2 > t)$, for $i = 1, \ldots, n$, where T_1 and T_2 denote the corresponding lifetimes of the two systems. If $F \geq_{st} G$ and $\mathbf{s}_1(t) \leq_{st} \mathbf{s}_2(t)$, then, for fixed $t \geq 0$, any $x \geq 0$, and $j = 1, \ldots, n$,*

$$P(X_{j:n} - t > x \mid X_{j:n} > t, T_1 > t) \geq P(Y_{j:n} - t > x \mid Y_{j:n} > t, T_2 > t).$$

Proof According to Theorem 4.3.1, for a given $t > 0$, and $x > 0$, we have

$$\begin{aligned} P(X_{j:n} - t > x \mid X_{j:n} > t, T_1 > t) &= \sum_{i=1}^{n} s_{1i}(t) P(X_{j:n} - t > x | X_{j:n} > t, X_{i:n} > t) \\ &\geq \sum_{i=1}^{n} s_{2i}(t) P(X_{j:n} - t > x | X_{j:n} > t, X_{i:n} > t) \end{aligned}$$

$$\geq \sum_{i=1}^{n} s_{2i}(t)P(Y_{j:n} - t > x|Y_{j:n} > t, Y_{i:n} > t)$$

$$= P(Y_{j:n} - t > x \mid Y_{j:n} > t, T_2 > t),$$

where the first inequality follows from that $P(X_{j:n} - t > x|X_{j:n} > t, X_{i:n} > t)$ is decreasing in i and $\mathbf{s}_1(t) \leq_{st} \mathbf{s}_2(t)$, and the second one holds due to Theorem 4.3.3 and $F \geq_{st} G$.

4.3.3 Conclusion

This section primarily discusses some stochastic properties of residual lives of surviving components, given that a general coherent system and components have survived till some time. First, a mixture representation for the residual lives of possibly surviving components of a coherent system with a general structure is derived. Then, this result is then utilized to carry out some stochastic comparisons of different systems having different components or different signatures. The results established here generalize some existing results in the literature for some specific conditional systems.

4.4 Surviving Components of Coherent System Under Double Monitoring

The section is mainly for the residual lives of surviving components of coherent systems with n independent and identically distributed components, given that before time t_1 $(t_1 > 0)$, exactly r $(r < n)$ components have failed and the system failed at time t_2 $(t_2 > t_1)$. A mixture representation of the residual life of surviving components of a failed coherent system can be represented as the mixture of the residual lives of $X_{k,n}$. Also some aging properties and preservation results of the residual lives of the surviving components of failed systems are conducted.

4.4.1 Introduction and Preliminaries

There are many coherent systems whose signatures have the form

$$\mathbf{s} = (s_1, s_2, \ldots, s_m, 0, \ldots, 0). \tag{4.9}$$

Table 4.1 Signature vectors of coherent systems

System	$T = \phi(X_1, X_2, X_3)$	**p**
1	$\min\{X_1, X_2, X_3\}$	$(1, 0, 0)$
2	$\min\{X_1, \max\{X_2, X_3\}\}$	$(\frac{1}{3}, \frac{2}{3}, 0)$
3	$X_{2,3}$(2-out-of-3)	$(0, 1, 0)$
4	$\max\{X_1, \min\{X_2, X_3\}\}$	$(0, \frac{2}{3}, \frac{1}{3})$
5	$\max\{X_1, X_2, X_3\}$	$(0, 0, 1)$

For example, from Table 4.1 , the signatures of Systems 1, 2 and 3 (three of five systems with three IID components), have the form as that in Eq. (4.9). A coherent system with a signature of the form (4.9) has the property that the component with lifetime $X_{k,n}$ for $k = m + 1, \dots, n$, would never fail in the system even if the system has failed. Hence, after the failure of the system, the surviving components may be removed from the system and then can be used for other systems or testing purposes. Therefore, the study of the reliability properties such as aging properties of such surviving components may be of interest to engineers and system designers.

We consider a coherent system (with lifetime T having n IID components) with a signature vector of form (4.9) for $1 < m < n$, and we are interested in the conditional residual life

$$(X_{k,n} - t_2 | X_{r,n} \le t_1 < X_{r+1,n}, T = t_2) \tag{4.10}$$

for $0 \le r < m$ (assume $X_{0,n} = 0$) and $k = m + 1, \dots, n$. In fact the random variable in Eq. (4.10) is the residual life of a surviving component at time t_2 ($t_2 > t_1 > 0$) in the system, given that before time t_1, exactly r ($r < n$) components have failed and the system is just failed at time t_2.

In next section, at first we obtain a mixture representation of residual lifetime of surviving components in a failed coherent system with IID components, given that before time t_1 ($t_1 > 0$), exactly r ($r < n$) components have failed and the system is failed at time t_2 ($t_2 > t_1$). Secondly we present some aging properties and preservation results of the residual lives of the surviving components of the failed coherent system. In addition, we obtain some stochastic properties of the mixture distribution and the residual life of the surviving components of a failed k-out-of-n systems.

4.4.2 *Aging Properties and Stochastic Comparisons of Remaining Components*

We consider a system with lifetime T having n original components with IID lifetimes $X_1, \dots, X_n$ according to the continuous distribution function F. Suppose that the system has the signature of the form $\mathbf{s} = (s_1, \dots, s_m, 0, \dots, 0)$. For $0 \le r < m < k \le n$ and $0 < t_1 < t_2$, the reliability function of the conditional residual life $(X_{k,n} -$

$t_2|X_{r,n} \le t_1 < X_{r+1,n}, T = t_2)$ of the surviving components can be represented as follows:

$$\begin{aligned}
&\Pr(X_{k,n} - t_2 > x|X_{r,n} \le t_1 < X_{r+1,n}, T = t_2) \\
&= \sum_{i=r+1}^{m} \Pr(X_{k,n} - t_2 > x, T = X_{i,n}|A_r(t_1), T = t_2) \\
&= \sum_{i=r+1}^{m} \Pr(X_{k,n} - t_2 > x \mid A_r(t_1), T = t_2, T = X_{i,n}) \times \Pr(T = X_{i,n}|A_r(t_1), T = t_2) \\
&= \sum_{i=r+1}^{m} \Pr(X_{k,n} - t_2 > x \mid A_r(t_1), X_{i,n} = t_2) \times \Pr(T = X_{i,n} \mid A_r(t_1), T = t_2) \\
&= \sum_{i=r+1}^{m} p_i(t_1, t_2) \Pr(X_{k,n} - t_2 > x|A_r(t_1), X_{i,n} = t_2),
\end{aligned} \tag{4.11}$$

where the event $A_r(t_1) = \{X_{r,n} \le t_1 < X_{r+1,n}\}$, and for $i = r+1, \ldots, m$, $p_i(t_1, t_2) = \Pr(T = X_{i,n} \mid A_r(t_1), T = t_2)$ can be regarded as the conditional signature (or dynamic signature) of the system, which is the probability that the unit with lifetime $X_{i,n}$ $(i = r+1, \ldots, m)$ would cause the system to fail given that the system just failed at time t_2. The third equality holds because the events $\{T = X_{i,n}\}$ and $\{A_r(t_1), X_{i,n} = t_2\}$, $\{T = X_{i,n}\}$ and $\{X_{k,n} - t_2, A_r(t_1), X_{i,n} = t_2\}$ are independent, respectively. The vector

$$\mathbf{p}(t_1, t_2) = (0, \ldots, 0, p_{r+1}(t_1, t_2), \ldots, p_m(t_1, t_2), 0, \ldots, 0) \tag{4.12}$$

is called the conditional signature vector, and this signature vector depends on the underlying distribution (see Theorem 5.3.1 in the later).

Remark It can be found that, from Eq. (4.12), the residual life of surviving components in a failed coherent system can be represented as the mixture of the residual lives of the surviving components in a $(n-i+1)$ $(i < k)$-out-of-n system given that before time t_1 $(t_1 > 0)$, exactly r $(r < n)$ components have failed, and at time t_2 it has just failed. The residual life of surviving components in a failed coherent system can also be expressed as the mixture of the residual life of a $(n-k+1)$-out-of-n system given that i $(i < k)$ components have just failed at time t_2, and the mixture distribution is the conditional signature vector $\mathbf{p}(t_1, t_2) = (0, \ldots, 0, p_{r+1}(t_1, t_2), \ldots, p_m(t_1, t_2), 0, \ldots, 0)$.

For example, consider the system with lifetime $T = \min\{X_1, \max\{X_2, X_3, X_4\}\}$, whose signature is $\mathbf{s} = (\frac{1}{4}, \frac{1}{4}, \frac{1}{2}, 0)$, where X_1, X_2, X_3, X_4 are IID with survival function

$$\bar{F}(x) = \frac{1}{(x+1)^3},$$

for $x \ge 0$. Suppose that at time t_1, exactly one component has failed and at time $t_2 > t_1$ the system just failed. Now we consider the residual life of the fourth component that was surely surviving at the time that the system failed. By some computations, the coefficients (conditional signature) are given by

$$p_2(t_1, t_2) = \frac{1}{2\frac{(1+t_2)^3}{(1+t_1)^3} - 1}, \; p_3(t_1, t_2) = \frac{2\frac{(1+t_2)^3}{(1+t_1)^3} - 2}{2\frac{(1+t_2)^3}{(1+t_1)^3} - 1}.$$

Hence from (4.11), we have

$$\begin{aligned}\Pr(X_{4,4} - t_2 > x \mid A_r(t_1), T = t_2) = {} & \frac{1}{2\frac{(1+t_2)^3}{(1+t_1)^3} - 1} \Pr(X_{4,4} - t_2 > x \mid A_r(t_1), X_{2,4} = t_2) \\ & + \frac{2\frac{(1+t_2)^3}{(1+t_1)^3} - 2}{2\frac{(1+t_2)^3}{(1+t_1)^3} - 1} \Pr(X_{4,4} - t_2 > x \mid A_r(t_1), X_{3,4} = t_2),\end{aligned}$$

where $A_r(t_1) = \{X_{1,4} \le t_1 < X_{2,4}\}$.

Remark The conditional signature $p_i(t_1, t_2)$ can be explained using Bayesian point of view. Suppose that the probability vector $\mathbf{s} = (s_1, s_2, \ldots, s_m, 0, \ldots, 0)$ is a prior distribution where s_i, $i = 1, 2, \ldots, m$, denotes the probability that the system lifetime T equals the ordered lifetime $X_{i,n}$ before the operation of the system, which can be regarded as the prior belief of the system designer about the failure time probability of the system when the system has not yet been used. Now assume that the system is put in operation at original time 0, and the designer has the information (data) that, at time $t_1 > 0$, the number of failed components is just r and the system is still working, at time $t_2 > t_1 > 0$, the system just failed. Under these circumstances, the probability vector $(0, \ldots, 0, p_{r+1}(t_1, t_2), \ldots, p_m(t_1, t_2), 0, \ldots, 0)$, can be interpreted as the posterior belief (posterior distribution) of the designer on the probability of failure time of the system given that the system is just failed at time t_2.

In the following we assume that the conditional signature vectors of coherent systems takes the form as in (4.9), implying that at time t_1, exactly r components have failed while the system was still operating, and at time $t_2(> t_1)$, the system just failed. Some properties of the conditional signature vector $\mathbf{p}(t_1, t_2)$ are given first. The following theorem shows that $p_k(t_1, t_2)$ can be represented as the function of the ratio $\bar{F}(t_1)/\bar{F}(t_2)$.

Theorem 4.4.1 *For $0 \le r < m < k \le n$ and $0 < t_1 < t_2$,*

$$p_k(t_1, t_2) = \frac{s_k(\phi(t_1, t_2) - 1)^k}{\sum_{j=r+1}^{m} s_j \phi((t_1, t_2) - 1)^j},$$

where $\phi(t_1, t_2) = \bar{F}(t_1)/\bar{F}(t_2)$.

Proof For $0 \le r < m < k < n$ and $0 < t_1 < t_2$,

$$p_k(t_1, t_2) = \Pr(T = X_{k,n} \mid A_r(t_1), T = t_2)$$

$$
\begin{aligned}
&= \frac{\Pr(T = X_{k,n}, A_r(t_1), T = t_2)}{\Pr(A_r(t_1), T = t_2)} \\
&= \frac{\Pr(A_r(t_1), T = t_2 | T = X_{k,n}) \Pr(T = X_{k,n})}{\sum_{j=r+1}^{m} \Pr(A_r(t_1), T = t_2, T = X_{j,n})} \\
&= \frac{\Pr(A_r(t_1), X_{k,n} = t_2 | T = X_{k,n}) \Pr(T = X_{k,n})}{\sum_{j=r+1}^{m} \Pr(A_r(t_1), T = t_2 | T = X_{j,n}) \Pr(T = X_{j,n})} \\
&= \frac{s_k \Pr(A_r(t_1), X_{k,n} = t_2)}{\sum_{j=r+1}^{m} s_j \Pr(A_r(t_1), X_{j,n} = t_2)} \\
&= \frac{s_k \binom{n}{r} (\bar{F}(t_1) - \bar{F}(t_2))^{k-r-1} \bar{F}^{n-k}(t_2)}{\sum_{j=r+1}^{m} s_j \binom{n}{r} (\bar{F}(t_1) - \bar{F}(t_2))^{j-r-1} \bar{F}^{n-j}(t_2)} \\
&= \frac{s_k \left[\frac{\bar{F}(t_1) - \bar{F}(t_2)}{\bar{F}(t_2)} \right]^k}{\sum_{j=r+1}^{m} s_j \left[\frac{\bar{F}(t_1) - \bar{F}(t_2)}{\bar{F}(t_2)} \right]^j} \\
&= \frac{s_k (\phi(t_1, t_2) - 1)^k}{\sum_{j=r+1}^{m} s_j (\phi(t_1, t_2) - 1)^j},
\end{aligned}
$$

where the fifth equality follows from the fact that the event $\{A_r(t_1), X_{k,n} = t_2\}$ and $\{T = X_{k,n}\}$ are independent.

Theorem 4.4.2 *(a) Given a fixed t_1, $p_{r+1}(t_1, t_2)$ is decreasing in t_2, and $\lim_{t_2 \to \infty} p_{r+1}(t_1, t_2) = 0$; Given a fixed t_2, $p_{r+1}(t_1, t_2)$ is increasing in t_1; (b) Given a fixed t_1, $p_m(t_1, t_2)$ is increasing in t_2, and $\lim_{t_2 \to \infty} p_m(t_1, t_2) = \infty$; Given a fixed t_2, $p_m(t_1, t_2)$ is decreasing in t_1.*

Proof For $j = r + 1, \ldots, m$, and $t_2 > t_1 > 0$,

$$
\begin{aligned}
p_j(t_1, t_2) &= \frac{s_j (\phi(t_1, t_2) - 1)^j}{\sum_{i=r+1}^{m} s_i (\phi(t_1, t_2) - 1)^i} \\
&= \frac{s_j}{\sum_{i=r+1}^{m} s_i (\phi(t_1, t_2) - 1)^{i-j}}.
\end{aligned}
$$

It is clear that $\phi(t_1, t_2) - 1$ is increasing in t_2 and hence $(\phi(t_1, t_2) - 1)^{i-j}$ is increasing in t_2 when $j = r + 1$. Thus $p_{r+1}(t_1, t_2)$ is decreasing in t_2 and $\lim_{t_2 \to \infty} p_{r+1}(t_1, t_2) = 0$. The proof of the remaining parts of (a) and (b) are similar and hence are omitted.

Theorem 4.4.3 *For $0 < t_1 < t_2 < t_3$, let $\boldsymbol{p}(t_1, t_2) = (0, \ldots, 0, p_{r+1}(t_1, t_2), \ldots, p_m(t_1, t_2), 0 \ldots, 0)$, and $\boldsymbol{p}(t_1, t_3) = (0, \ldots, 0, p_{r+1}(t_1, t_3), \ldots, p_m(t_1, t_3), 0, \ldots, 0)$. Then $\boldsymbol{p}(t_1, t_2) \leq_{st} \boldsymbol{p}(t_1, t_3)$. That is, $\boldsymbol{p}(t_1, t_2)$ is stochastically increasing in time t_2.*

Proof It is sufficient to show that for all $i = r + 1, \ldots, m$,

$$\sum_{j=i}^{m} p_j(t_1, t_2) \le \sum_{j=i}^{m} p_j(t_1, t_3).$$

That is,

$$\frac{\sum_{j=i}^{m} s_j(\phi(t_1, t_2) - 1)^j}{\sum_{k=r+1}^{m} s_k(\phi(t_1, t_2) - 1)^k} \le \frac{\sum_{j=i}^{m} s_j(\phi(t_1, t_3) - 1)^j}{\sum_{k=r+1}^{m} s_k(\phi(t_1, t_3) - 1)^k}. \tag{4.13}$$

Equation (4.13) is equivalent to

$$\sum_{j=i}^{m} \sum_{k=r+1}^{m} s_j s_k[(\phi(t_1, t_3) - 1)^j (\phi(t_1, t_2) - 1)^k \\ -(\phi(t_1, t_3) - 1)^k (\phi(t_1, t_2) - 1)^j] \ge 0.$$

The left side in the inequality above can be represented as

$$\begin{aligned}
&\sum_{j=i}^{m} \sum_{k=r+1}^{m} s_j s_k[(\phi(t_1, t_3) - 1)^j (\phi(t_1, t_2) - 1)^k - (\phi(t_1, t_3) - 1)^k (\phi(t_1, t_2) - 1)^j] \\
&= \sum_{j=i}^{m} \sum_{k=r+1}^{i-1} s_j s_k[(\phi(t_1, t_3) - 1)^j (\phi(t_1, t_2) - 1)^k - (\phi(t_1, t_3) - 1)^k (\phi(t_1, t_2) - 1)^j] \\
&\quad + \sum_{j=i}^{m} \sum_{k=i}^{m} s_j s_k[(\phi(t_1, t_3) - 1)^j (\phi(t_1, t_2) - 1)^k - (\phi(t_1, t_3) - 1)^k (\phi(t_1, t_2) - 1)^j] \\
&= \sum_{j=i}^{m} \sum_{k=r+1}^{i-1} s_j s_k[(\phi(t_1, t_3) - 1)^j (\phi(t_1, t_2) - 1)^k - (\phi(t_1, t_3) - 1)^k (\phi(t_1, t_2) - 1)^j] \\
&= \sum_{j=i}^{m} \sum_{k=r+1}^{i-1} s_j s_k(\phi(t_1, t_3) - 1)^k (\phi(t_1, t_2) - 1)^k [\phi(t_1, t_3) - 1)^{j-k} - (\phi(t_1, t_2) - 1)^{j-k}].
\end{aligned}$$

It is clear that the last term is nonnegative for $t_1 < t_2 < t_3$, and hence the result follows.

For example, for the case of the example, the corresponding conditional signature vector is

$$\mathbf{p}(t_1, t_2) = \left(0, \frac{1}{2\frac{(1+t_2)^3}{(1+t_1)^3} - 1}, \frac{2\frac{(1+t_2)^3}{(1+t_1)^3} - 2}{2\frac{(1+t_2)^3}{(1+t_1)^3} - 1}, 0\right).$$

It can be easily checked that $\mathbf{p}(t_1, t_2) \le_{st} \mathbf{p}(t_1, t_3)$ for $t_2 \le t_3$.

The following proposition indicates that the stronger the signature vector of a system is in the likelihood ratio order, the stronger is the corresponding conditional signature vector of the system in the usual stochastic order.

Theorem 4.4.4 *Suppose that* $\mathbf{s}^{(1)} = (s_1^{(1)}, \ldots, s_m^{(1)}, 0, \ldots, 0)$ *and* $\mathbf{s}^{(2)} = (s_1^{(2)}, \ldots, s_m^{(2)}, 0 \ldots, 0)$ *are signatures of two coherent systems with a set of IID components,* *respectively. Let* $\mathbf{p}^{(1)}(t_1, t_2) = (0, \ldots, 0, p_{r+1}^{(1)}(t_1, t_2), \ldots, p_m^{(1)}(t_1, t_2), 0, \ldots, 0)$ *and* $\mathbf{p}^{(2)}(t_1, t_2) = (0, \ldots, 0, p_{r+1}^{(2)}(t_1, t_2), \ldots, p_m^{(2)}(t_1, t_2), 0, \ldots, 0)$ *be the corresponding conditional signature vectors for* $0 < t_1 < t_2$. *If* $\mathbf{s}^{(1)} \leq_{lr} \mathbf{s}^{(2)}$. *Then* $\mathbf{p}^{(1)}(t_1, t_2) \leq_{st} \mathbf{p}^{(2)}(t_1, t_2)$.

Proof It is sufficient to show that, for all $i = r + 1, \ldots, m$,

$$\frac{\sum_{j=i}^{m} s_j^{(1)}(\phi(t_1, t_2) - 1)^j}{\sum_{k=r+1}^{m} s_k^{(1)}(\phi(t_1, t_2) - 1)^k} \leq \frac{\sum_{j=i}^{m} s_j^{(2)}(\phi(t_1, t_2) - 1)^j}{\sum_{k=r+1}^{m} s_k^{(2)}(\phi(t_1, t_2) - 1)^k}. \tag{4.14}$$

The inequality in Eq. (4.14) is equivalent to

$$\sum_{j=i}^{m} \sum_{k=r+1}^{m} (\phi(t_1, t_2) - 1)^j (\phi(t_1, t_2) - 1)^k \times (s_j^{(2)} s_k^{(1)} - s_j^{(1)} s_k^{(2)}) \geq 0.$$

That is,

$$\sum_{j=i}^{m} \sum_{k=r+1}^{i-1} (\phi(t_1, t_2) - 1)^{k+j} (s_j^{(2)} s_k^{(1)} - s_j^{(1)} s_k^{(2)}) \geq 0.$$

The condition $\mathbf{s}^{(1)} \leq_{lr} \mathbf{s}^{(2)}$ implies that the inequality in above holds and hence the result follows.

Theorem 4.4.5 *For* $i = r + 1, \ldots, m$, *and* $n \geq k > m$, $(X_{k,n} - t_2 \mid A_r(t_1), X_{i,n} = t_2) \stackrel{d}{=} Z_{k-i,n-i}^{t_2}$, *where* $\stackrel{d}{=}$ *means equality in distribution, and* $Z_{k-i,n-i}^{t_2}$ *is the* $(k - i)$*th order statistic among* $n - i$ *random variables that are IID with the same distribution as* $(X - t_2 | X > t_2)$.

Proof For any $x \geq 0$, and $0 < t_1 < t_2$,

$$\begin{aligned}
\Pr(X_{k,n} - t_2 > x \mid A_r(t_1), X_{i,n} = t_2) &= \frac{\Pr(X_{k,n} - t_2 > x, A_r(t_1), X_{i,n} = t_2)}{\Pr(A_r(t_1), X_{i,n} = t_2)} \\
&= \frac{\sum_{j=n-k+1}^{n-i} \binom{n-i}{j} \bar{F}^j(t_2 + x)(\bar{F}(t_2) - \bar{F}(t_2 + x))^{n-i-j}}{\bar{F}(t_2)^{n-i}} \\
&= \sum_{j=n-k+1}^{n-i} \binom{n-i}{j} [\bar{F}_{t_2}(x)]^j [1 - \bar{F}_{t_2}(x)]^{n-i-j} \\
&= \Pr(Z_{k-i,n-i}^{t_2} > x).
\end{aligned}$$

where $\bar{F}_{t_2}(x) = \bar{F}(t_2 + x)/\bar{F}(t_2)$. Thus the result holds.

Remark From the result in Theorem 4.4.5, it can also be found that $(X_{k,n} - t_2 \mid A_r(t_1), X_{i,n} = t_2) \stackrel{d}{=} (X_{k-i,n-i} - t_2 | X_{1,n-i} > t_2)$.

The following result follows from Theorem 1.C.31 of Shaked and Shantikumar (2007).

Theorem 4.4.6 *For $n \geq k$, $Z^{t_2}_{k-i,n-i} \leq_{lr} Z^{t_2}_{k-j,n-j}$ when $i > j$. That is, $(X_{k,n} - t_2 \mid A_r(t_1), X_{i,n} = t_2)$ is decreasing in i in the likelihood ratio order.*

Remark From Theorem 1.C.37 and Corollaries 1.C.38 and 1.C.39 of Shaked and Shanthikumar (2007), it is clear that, for $i \leq j$ and $m \geq n$,

$$Z^{t_2}_{k-j,m-j} \leq_{lr} Z^{t_2}_{k-i,n-i},$$

and hence, for $i \leq j$ and $m \geq n$, $(X_{k,m} - t_2 \mid A_r(t_1), X_{j,m} = t_2) \leq_{lr} (X_{k,n} - t_2 \mid A_r(t_1), X_{i,n} = t_2)$.

In the following, we show that if the components of the system are IFR then $\Pr(X_{k,n} - t_2 > x \mid A_r(t_1), T = t_2)$ is decreasing in t_2, and if $\Pr(X_{k,n} - t_2 > x \mid A_r(t_1), T = t_2)$ is increasing in t_2 then the components of the system are DFR. In order to get this result, we need the following lemma.

Lemma 4.4.7 *For $0 \leq r < k \leq n$ and given a fixed $t_1 > 0$, if F is IFR, then $\Pr(X_{k,n} - t_2 > x \mid A_r(t_1), X_{i,n} = t_2)$ is decreasing in t_2 ($> t_1$).*

Proof For any $x \geq 0$, and $0 < t_1 < t_2$,

$$\begin{aligned}\Pr(X_{k,n} - t_2 > x \mid A_r(t_1), X_{i,n} = t_2) &= \sum_{j=n-k+1}^{n-i} \binom{n-i}{j} \left[\bar{F}_{t_2}(x)\right]^j \left[1 - \bar{F}_{t_2}(x)\right]^{n-i-j} \\ &= \int_{1-\frac{\bar{F}(t_2+x)}{\bar{F}(t_2)}}^{1} (k-i)\binom{n-i}{k-i} u^{k-i-1}(1-u)^{n-k+i} du.\end{aligned}$$

Note that F is IFR if and only if $1 - \bar{F}(t_2 + x)/\bar{F}(t_2)$ is increasing in t_2. Therefore the result follows immediately.

Theorem 4.4.8 *(a) If F is IFR, then $\Pr(X_{k,n} - t_2 > x \mid A_r(t_1), T = t_2)$ is decreasing in t_2 for $0 \leq r < m < k \leq n$ and $0 < t_1 < t_2$; (b) If $\Pr(X_{k,n} - t_2 > x \mid A_r(t_1), T = t_2)$ is increasing in t_2 for $0 \leq r < m < k \leq n$ and $0 < t_1 < t_2$, then F is DFR.*

Proof (a) For any $x \geq 0$, and $0 \leq t_1 < t_2$,

$$\begin{aligned}&\frac{\partial}{\partial t_2}\Big[\sum_{i=r+1}^{m} p_i(t_1, t_2) \times \Pr(X_{k,n} - t_2 > x | A_r(t_1), X_{i,n} = t_2)\Big] \\ &= \sum_{i=r+1}^{m} \frac{\partial p_i(t_1, t_2)}{\partial t_2} \Pr(X_{k,n} - t_2 > x | A_r(t_1), X_{i,n} = t_2) \\ &+ \sum_{i=r+1}^{m} p_i(t_1, t_2) \frac{\partial \Pr(X_{k,n} - t_2 > x \mid A_r(t_1), X_{i,n} = t_2)}{\partial t_2}.\end{aligned}$$

From Lemma 4.4.7 above, that F is IFR means that $\partial \Pr(X_{k,n} - t_2 > x \mid A_r(t_1), X_{i,n} = t_2)/\partial t_2$ is negative. Hence it is sufficient to show that $\partial p_i(t_1, t_2)/\partial t_2$ is negative. Noting that

$$\begin{aligned}\frac{\partial p_i(t_1,t_2)}{\partial t_2} &= \frac{\partial \frac{s_i(\phi(t_1,t_2)-1)^i}{\sum_{j=r+1}^m s_j(\phi(t_1,t_2)-1)^j}}{\partial t_2} \\ &= \sum_{i=r+1}^{m} s_i s_k (\phi(t_1,t_2)-1)^{k+i-1}(i-k)\frac{\partial \phi(t_1,t_2)}{\partial t_2}\end{aligned}$$

is negative for $k > m$. It follows that the proof of (a) is complete.

(b) It is noted that for $t_1 < t_2 < t_3$ and any $x \geq 0$,

$$\begin{aligned}\sum_{i=r+1}^{m} p_i(t_1,t_2)\Pr(X_{k,n}-t_2 > x \mid A_r(t_1), X_{i,n}=t_2) &\leq \sum_{i=r+1}^{m} p_i(t_1,t_3)\Pr(X_{k,n}-t_3 > x \mid A_r(t_1), X_{i,n}=t_3) \\ &\leq \sum_{i=r+1}^{m} p_i(t_1,t_2)\Pr(X_{k,n}-t_3 > x \mid A_r(t_1), X_{i,n}=t_3)\end{aligned}$$

where the first inequality follows because $\Pr(X_{k,n} - t_2 > x \mid A_r(t_1), T = t_2)$ is increasing in t_2, and the second inequality follows from Theorem 4.4.6 and Eq. (1.A.7) on p. 4 in Shaked and Shantikumar (2007). Hence

$$\sum_{i=r+1}^{m} p_i(t_1,t_2)\left[\Pr(X_{k,n}-t_3 > x \mid A_r(t_1), X_{i,n}=t_3) - \Pr(X_{k,n}-t_2 > x \mid A_r(t_1), X_{i,n}=t_2)\right] \geq 0.$$

That the inequality holds implies that there exists at least one i $(r+1 \leq i \leq m)$ such that

$$\Pr(X_{k,n}-t_3 > x \mid A_r(t_1), X_{i,n}=t_3) - \Pr(X_{k,n}-t_2 > x \mid A_r(t_1), X_{i,n}=t_2) \geq 0.$$

Also for all i $(r+1 \leq i \leq m)$, $p_i(t_1, t_2) \geq 0$, which in turn implies that from (4.14),

$$\frac{\bar{F}(t_2+x)}{\bar{F}(t_2)} \leq \frac{\bar{F}(t_3+x)}{\bar{F}(t_3)},$$

and hence F is DFR.

In the following we give some sufficient conditions for stochastic orders of two systems having a set of IID components but with different structures.

Theorem 4.4.9 *Let $T_1 = \tau_1(X_1, \ldots, X_n)$ and $T_2 = \tau_2(X_1, \ldots, X_n)$ be the lifetimes of two coherent systems, both based on n components with IID lifetimes distributed according to the common continuous distribution F, having respective signatures $\mathbf{s}^{(1)} = (s_1^{(1)}, \ldots, s_m^{(1)}, 0, \ldots, 0)$, and $\mathbf{s}^{(2)} = (s_1^{(2)}, \ldots, s_m^{(2)}, 0, \ldots, 0)$. In addition, for*

all $t_2 > t_1$, *their corresponding conditional signature vectors are*
$\boldsymbol{p}^{(1)}(t_1, t_2)$
$= (0, \ldots, 0, p_{r+1}^{(1)}(t_1, t_2), \ldots, p_m^{(1)}(t_1, t_2), 0, \ldots, 0)$ *and*
$\boldsymbol{p}^{(2)}(t_1, t_2)$
$= (0, \ldots, 0, p_{r+1}^{(2)}(t_1, t_2), \ldots, p_m^{(2)}(t_1, t_2), 0, \ldots, 0)$.

(a) If $\boldsymbol{p}^{(1)}(t_1, t_2) \leq_{st} \boldsymbol{p}^{(2)}(t_1, t_2)$, *then* $(X_{k,n} - t_2|A_r(t_1), T_1 = t_2) \geq_{st} (X_{k,n} - t_2|A_r(t_1), T_2 = t_2)$;

(b) If $\boldsymbol{p}^{(1)}(t_1, t_2) \leq_{hr} \boldsymbol{p}^{(2)}(t_1, t_2)$, *then* $(X_{k,n} - t_2|A_r(t_1), T_1 = t_2) \geq_{rh} (X_{k,n} - t_2|A_r(t_1), T_2 = t_2)$;

(c) If $\boldsymbol{p}^{(1)}(t_1, t_2) \leq_{rh} \boldsymbol{p}^{(2)}(t_1, t_2)$, *then* $(X_{k,n} - t_2|A_r(t_1), T_1 = t_2) \geq_{hr} (X_{k,n} - t_2|A_r(t_1), T_2 = t_2)$;

(d) If $\boldsymbol{p}^{(1)}(t_1, t_2) \leq_{lr} \boldsymbol{p}^{(2)}(t_1, t_2)$, *then* $(X_{k,n} - t_2|A_r(t_1), T_1 = t_2) \geq_{lr} (X_{k,n} - t_2|A_r(t_1), T_2 = t_2)$.

Proof (a) By Eq. (4.11), for any $x \geq 0$,

$$\begin{aligned}\Pr(X_{k,n} - t_2 > x|A_r(t_1), T_1 = t_2) &= \sum_{i=r+1}^{m} p_i^{(1)}(t_1, t_2) \Pr(X_{k,n} - t_2 > x \mid A_r(t_1), X_{i,n} = t_2)\\ &= \sum_{i=r+1}^{m} p_i^{(1)}(t_1, t_2) \Pr(Z_{k-i,n-i}^{t_2} > x) = \mathrm{E}\left[\Pr(Z_{k-I,n-I}^{t_2} > x)\right],\end{aligned}$$

where the random variable I is distributed as $\mathbf{p}^{(1)}(t_1, t_2)$. Also

$$\Pr(X_{k,n} - t_2 > x|A_r(t_1), T_2 = t_2) = \mathrm{E}[\Pr(Z_{k-J,n-J}^{t_2} > x)],$$

where the random variable J is distributed as $\mathbf{p}^{(2)}(t_1, t_2)$. By the Proposition 6, $\Pr(Z_{k-i,n-i}^{t_2} > x)$ is decreasing in i. Therefore the result of (a) follows from the Eq. (1.A.7) of p. 4 in Shaked and Shantikumar (2007).

By the conditions and Theorem 1.B.50. and Theorem 1.B.12. of Shaked and Shantikumar (2007), the results (b) and (c) can be easily obtained respectively.

(d) Let $f_{T_1}(x)$, $f_{T_2}(x)$ be the density functions of $(X_{k,n} - t_2|A_r(t_1), T_1 = t_2)$ and $(X_{k,n} - t_2|A_r(t_1), T_2 = t_2)$, respectively. To prove the result we need only to show that

$$f_{T_1}(x)/f_{T_2}(x)$$

is increasing in $x \geq 0$. That is, we need to show that

$$\frac{\sum_{i=r+1}^{m} p_i^{(1)}(t_1, t_2) f_{Z_{k-i,n-i}^{t_2}}(x)}{\sum_{i=r+1}^{m} p_i^{(2)}(t_1, t_2) f_{Z_{k-i,n-i}^{t_2}}(x)} \text{ is increasing in } x \geq 0, \tag{4.15}$$

where $f_{Z_{k-i,n-i}^{t_2}}(x)$ denotes the density function of the random variable $Z_{k-i,n-i}^{t_2}$, i.e. $(X_{k,n} - t_2|A_r(t_1), X_{i,n} = t_2)$. Then, for $x_1 \leq x_2$, Eq. (4.15) is equivalent to

$$\frac{\sum_{i=r+1}^{m} p_i^{(1)}(t_1,t_2) f_{Z^{t_2}_{k-i,n-i}}(x_1)}{\sum_{i=r+1}^{m} p_i^{(2)}(t_1,t_2) f_{Z^{t_2}_{k-i,n-i}}(x_1)} \le \frac{\sum_{i=r+1}^{m} p_i^{(1)}(t_1,t_2) f_{Z^{t_2}_{k-i,n-i}}(x_2)}{\sum_{i=r+1}^{m} p_i^{(2)}(t_1,t_2) f_{Z^{t_2}_{k-i,n-i}}(x_2)}. \tag{4.16}$$

After some computations it can be shown that the inequality in (4.16) is equivalent to

$$\sum_{i=r+1}^{m}\sum_{j=r+1}^{i}\left(p_i^{(1)}(t_1,t_2)p_j^{(2)}(t_1,t_2) - p_i^{(2)}(t_1,t_2)p_j^{(1)}(t_1,t_2)\right)$$
$$\cdot\left(f_{Z^{t_2}_{k-i,n-i}}(x_2) f_{Z^{t_2}_{k-j,n-j}}(x_1) - f_{Z^{t_2}_{k-i,n-i}}(x_1) f_{Z^{t_2}_{k-j,n-j}}(x_2)\right)$$

is nonnegative.

The condition $\mathbf{p}^{(1)}(t_1,t_2) \le_{lr} \mathbf{p}^{(2)}(t_1,t_2)$ implies that $p_i^{(1)}(t_1,t_2)p_j^{(2)}(t_1,t_2) - p_i^{(2)}(t_1,t_2)p_j^{(1)}(t_1,t_2) \le 0$, and from Theorem 4.4.6, that, $f_{Z^{t_2}_{k-i,n-i}}(x_2) f_{Z^{t_2}_{k-j,n-j}}(x_1) - f_{Z^{t_2}_{k-i,n-i}}(x_1) f_{Z^{t_2}_{k-j,n-j}}(x_2) \le 0$, for $x_1 \le x_2$. It follows that the inequality in (4.16) holds and hence the proof is complete.

For example, consider two coherent systems in Fig. 4.1, the signature of system above is $\mathbf{p} = (\frac{1}{2}, \frac{1}{2}, 0, 0)$, the signature of one below is $\mathbf{q} = (\frac{1}{4}, \frac{1}{4}, \frac{1}{2}, 0)$. Suppose that at time t_1, the number of failed components is exactly 1, and at time t_2 ($t_2 > t_1$), the systems are just failed. It is computed that the corresponding signature vectors are $\mathbf{p}(t_1,t_2) = (0, 1, 0, 0)$ and

$$\mathbf{q}(t_1,t_2) = \left(0, \frac{1}{2\frac{(1+t_2)^3}{(1+t_1)^3} - 1}, \frac{2\frac{(1+t_2)^3}{(1+t_1)^3} - 2}{2\frac{(1+t_2)^3}{(1+t_1)^3} - 1}, 0\right),$$

respectively. It is easily shown that $\mathbf{p}(t_1,t_2) \le_{lr} \mathbf{q}(t_1,t_2)$, hence $\mathbf{p}(t_1,t_2) \le_{hr} (\le_{rh})(\le_{st})\mathbf{q}(t_1,t_2)$. By the results of Theorem 4.4.9, the residual life of the last remaining component in the system above is stronger than that of the last remaining component in the one below in the sense of in usual stochastic order, (reversed) hazard rate order and likelihood ratio order.

The following theorem gives some sufficient conditions for stochastic orders of two systems having two different sets of IID components but with common structure.

Theorem 4.4.10 *Let $T_1 = \tau_1(X_1, \ldots, X_n)$ and $T_2 = \tau_2(Y_1, \ldots, Y_n)$ be the lifetimes of two coherent systems having the same structures with common signature $s = (s_1, \ldots, s_m, 0, \ldots, 0)$. Assume that $X_1, \ldots, X_n$ are independent and identically distributed as F, $Y_1, \ldots, Y_n$ are independent and identically distributed as G. If $F \le_{hr} G$, then for $t_1 < t_2$,*

$$\Pr(X_{k,n} - t_2 > x | A_r(t_1), T_1 = t_2) \le \Pr(Y_{k,n} - t_2 > x | A_r(t_1), T_2 = t_2),$$

where $A_r(t_1) = \{X_{r,n} \le t_1 < X_{r+1,n}\}$, $A_r(t_1) = \{Y_{r,n} \le t_1 < Y_{r+1,n}\}$.

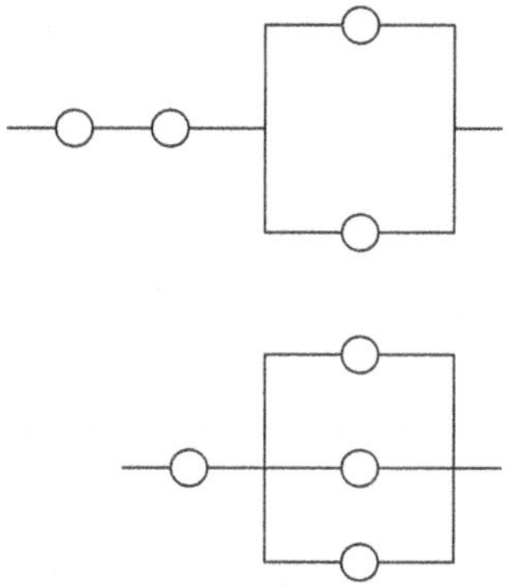

Fig. 4.1 Two coherent systems with likelihood ratio-ordered conditional signature

Proof For $i = r+1, \ldots, m$, and $0 < t_1 < t_2$, let

$$p_i^X(t_1, t_2) = \frac{s_i \varphi_X^i(t_1, t_2)}{\sum_{j=r+1}^m s_j \varphi_X^j(t_1, t_2)},$$

and

$$p_i^Y(t_1, t_2) = \frac{s_i \varphi_Y^i(t_1, t_2)}{\sum_{j=r+1}^m s_j \varphi_Y^j(t_1, t_2)},$$

where $\varphi_X(t_1, t_2) = \bar{F}(t_1)/\bar{F}(t_2) - 1$ and $\varphi_Y(t_1, t_2) = \bar{G}(t_1)/\bar{G}(t_2) - 1$. By Theorem 4.4.5, for any $x \geq 0$,

$$\Pr(X_{k,n} - t_2 > x | A_r(t_1), T_1 = t_2) = \sum_{i=r+1}^m p_i^X(t_1, t_2) \Pr(Z_{k-i,n-i}^{t_2,X} > x),$$

and

$$\Pr(Y_{k,n} - t_2 > x | A_r(t_1), T_2 = t_2) = \sum_{i=r+1}^m p_i^Y(t_1, t_2) \Pr(Z_{k-i,n-i}^{t_2,Y} > x),$$

where $Z_{k-i,n-i}^{t_2,X}$ is the $(k-i)$th order statistic among $n-i$ IID random variables that are distributed as $(X - t_2 | X > t_2)$, $Z_{k-i,n-i}^{t_2,Y}$ is the $(k-i)$th order statistic among $n-i$ random variables that are IID with the same distribution as $(Y - t_2 | Y > t_2)$.

To prove the result, we need only to show that

$$\sum_{i=r+1}^m p_i^X(t_1, t_2) \Pr(Z_{k-i,n-i}^{t_2,X} > x) \leq \sum_{i=r+1}^m p_i^Y(t_1, t_2) \Pr(Z_{k-i,n-i}^{t_2,Y} > x).$$

This is equivalent to

$$\frac{\sum_{i=r+1}^{m} s_i \varphi_X^i(t_1, t_2)}{\sum_{j=r+1}^{m} s_j \varphi_X^j(t_1, t_2)} \Pr(Z_{k-i,n-i}^{t_2,X} > x) \leq \frac{\sum_{i=r+1}^{m} s_i \varphi_Y^i(t_1, t_2)}{\sum_{j=r+1}^{m} s_j \varphi_Y^j(t_1, t_2)} \Pr(Z_{k-i,n-i}^{t_2,Y} > x).$$

That is,

$$\sum_{i=r+1}^{m} \sum_{j=r+1}^{m} s_i s_j \varphi_X^i(t_1, t_2) \varphi_Y^j(t_1, t_2) \Pr(Z_{k-i,n-i}^{t_2,X} > x) - \sum_{i=r+1}^{m} \sum_{j=r+1}^{m} s_i s_j \varphi_X^j(t_1, t_2) \varphi_Y^i(t_1, t_2) \Pr(Z_{k-i,n-i}^{t_2,Y} > x)$$

is nonpositive, which is equivalent to

$$\sum_{i=r+1}^{m} \sum_{j=r+1}^{m} s_i s_j \varphi_X^i(t_1, t_2) \varphi_Y^j(t_1, t_2) \times \left[\Pr(Z_{k-i,n-i}^{t_2,X} > x) - \Pr(Z_{k-i,n-i}^{t_2,Y} > x)\right] + \sum_{i=r+1}^{m} \sum_{j=r+1}^{m} s_i s_j \Pr(Z_{k-i,n-i}^{t_2,Y} > x) \times \left[\varphi_X^i(t_1, t_2) \varphi_Y^j(t_1, t_2) - \varphi_Y^i(t_1, t_2) \varphi_X^j(t_1, t_2)\right] \tag{4.17}$$

is nonpositive. That $F \leq_{hr} G$ holds if and only if $\frac{\bar{F}(t_2+x)}{\bar{F}(t_2)} \leq \frac{\bar{G}(t_2+x)}{\bar{G}(t_2)}$, and hence from (4.15),

$$\Pr(Z_{k-i,n-i}^{t_2,X} > x) - \Pr(Z_{k-i,n-i}^{t_2,Y} > x) \leq 0,$$

which in turn implies that the first term in Eq. (4.17) is nonpositive. Therefore in order to obtain the required result, it is sufficient to prove that the second term in Eq. (4.17) is nonpositive. Noting that

$$\begin{aligned} &\sum_{i=r+1}^{m} \sum_{j=r+1}^{m} s_i s_j \Pr(Z_{k-i,n-i}^{t_2,Y} > x) \varphi_X^i(t_1, t_2) \varphi_Y^j(t_1, t_2) \\ &\quad - s_i s_j \Pr(Z_{k-i,n-i}^{t_2,Y} > x) \varphi_X^j(t_1, t_2) \varphi_Y^i(t_1, t_2) \\ &\quad = \sum_{i=r+1}^{m} \sum_{j=i}^{m} s_i s_j \left[\varphi_X^i(t_1, t_2) \varphi_Y^j(t_1, t_2) \right. \\ &\quad \left. - \varphi_X^j(t_1, t_2) \varphi_Y^i(t_1, t_2)\right] \Pr(Z_{k-i,n-i}^{t_2,Y} > x) \\ &\quad + \sum_{i=r+1}^{m} \sum_{j=r+1}^{i-1} s_i s_j \left[\varphi_X^i(t_1, t_2) \varphi_Y^j(t_1, t_2) \right. \\ &\quad \left. - \varphi_X^j(t_1, t_2) \varphi_Y^i(t_1, t_2)\right] \Pr(Z_{k-i,n-i}^{t_2,Y} > x) \end{aligned}$$

$$
\begin{aligned}
&= \sum_{i=r+1}^{m}\sum_{j=i}^{m} s_i s_j \Big[\varphi_X^i(t_1,t_2)\varphi_Y^j(t_1,t_2) \\
&\quad - \varphi_X^j(t_1,t_2)\varphi_Y^i(t_1,t_2)\Big] \Pr(Z_{k-i,n-i}^{t_2,Y} > x) \\
&\quad + \sum_{j=r+1}^{m}\sum_{i=j}^{m} s_i s_j \Big[\varphi_X^i(t_1,t_2)\varphi_Y^j(t_1,t_2) \\
&\quad - \varphi_X^j(t_1,t_2)\varphi_Y^i(t_1,t_2)\Big] \Pr(Z_{k-i,n-i}^{t_2,Y} > x) \\
&= \sum_{i=r+1}^{m}\sum_{j=i}^{m} s_i s_j \Big[\varphi_X^i(t_1,t_2)\varphi_Y^j(t_1,t_2) \\
&\quad - \varphi_X^j(t_1,t_2)\varphi_Y^i(t_1,t_2)\Big] \Pr(Z_{k-i,n-i}^{t_2,Y} > x) \\
&\quad + \sum_{i=r+1}^{m}\sum_{j=i}^{m} s_i s_j \Big[\varphi_X^j(t_1,t_2)\varphi_Y^i(t_1,t_2) \\
&\quad - \varphi_X^i(t_1,t_2)\varphi_Y^j(t_1,t_2)\Big] \Pr(Z_{k-j,n-j}^{t_2,Y} > x) \\
&= \sum_{i=r+1}^{m}\sum_{j=i}^{m} \Big[\Pr(Z_{k-i,n-i}^{t_2,Y} > x) - \Pr(Z_{k-j,n-j}^{t_2,Y} > x)\Big] \\
&\quad \times s_i s_j \Big[\varphi_X^i(t_1,t_2)\varphi_Y^j(t_1,t_2) - \varphi_X^j(t_1,t_2)\varphi_Y^i(t_1,t_2)\Big].
\end{aligned}
$$

It follows from Theorem 4.4.6 that, $\Pr(Z_{k-i,n-i}^{t_2,Y} > x) - \Pr(Z_{k-j,n-j}^{t_2,Y} > x)$ is non-negative for any $x \geq 0$ and $i < j$. Thus we need only to show that

$$
\varphi_X^i(t_1,t_2)\varphi_Y^j(t_1,t_2) - \varphi_X^j(t_1,t_2)\varphi_Y^i(t_1,t_2)
$$

is negative. This can be easily proven, from the condition $F \leq_{hr} G$. The result follows directly.

4.4.3 Conclusions

The part provides results about the residual lives of surviving components of coherent systems with n independent and identically distributed components, given that before time t_1 ($t_1 > 0$), exactly r ($r < n$) components have failed and the system failed at time t_2 ($t_2 > t_1$). It is shown that the residual life of surviving components of a failed coherent system can be represented as the mixture of the residual lives of $X_{k,n}$ in a $(n-i+1)$ $(i < k)$ out of n system that failed at time t_2. By the mixture representation some aging properties and preservation results of the residual lives of the surviving components of failed systems are conducted. Also we present some independent stochastic properties of the residual life of $X_{k,n}$ in a $(n-i+1)$ $(i < k)$ out of n system and conditional signature vectors.

Chapter 5
The Number of Failed Components of Coherent Systems

This chapter is about the number of failed components in an $(n-k+1)$-out-of-n system and a general coherent system given the systems are still working at time t. The reliabilities and some stochastic properties are obtained.

5.1 The Number of Failed Components of $(n-k+1)$-Out-of-n Systems

This section is about the number of failed components in an $(n-k+1)$-out-of-n system, given that at least m $(m < k \le n)$ components have failed before time t, and the system is still working at time t. In this case, we compute the probability that there are exactly i working components. Moreover, the reliability and several stochastic properties are get.

5.1.1 Introduction

For an $(n-k+1)$-out-of-n system with n independent and identically distributed component lifetimes, there are still some components surviving given that at least m $(m < k \le n)$ components have failed before time t, and the system is still working at time t. It is noted that the probabilities

$$p_t(i, m, k, n) = P(N_t = i | X_{m:n} \le t, X_{k:n} > t), \quad i = m, m+1, \ldots, k-1, \quad (5.1)$$

where $N_t = i$ denotes the number of failed components in the system. In the next part, some properties of $p_t(i, m, k, n)$ are studied. Some different results and examples are provided.

Z. Zhang, *Conditional Order Statistics and Conditional Spacings*,
https://doi.org/10.1007/978-981-96-7151-9_5

5.1.2 Ordered Properties of Number of Failed Components of $(n-k+1)$-Out-of-n Systems

In $(n-k+1)$-out-of-n system, let $X_1, X_2, \ldots, X_n$ be lifetimes of n IID components with distribution F. Denote $X_{i:n}$ as the ith smallest order statistic of $X_1, X_2, \ldots, X_n$. The N_t denotes the number of failed components in the system on $[0, t]$. Now assume that at least m $(m < k \le n)$ components have failed before time t, and the system is still working at time t. Then, we consider the probability that there are exactly i failures

$$p_t(i, m, k, n) = P(N_t = i | X_{m:n} \le t, X_{k:n} > t), \quad i = m, m+1, \ldots, k-1. \tag{5.2}$$

To compute $p_t(i, m, k, n)$, note that the event $\{N_t = i\}$ is equivalent to the event $\{X_{i:n} \le t < X_{i+1:n}\}$. We can write

$$\begin{aligned} p_t(i, m, k, n) &= P(N_t = i | X_{m:n} \le t, X_{k:n} > t) \\ &= P(X_{i:n} < t < X_{i+1:n} | X_{m:n} \le t, X_{k:n} > t) \\ &= P(X_{i:n} < t | X_{m:n} \le t, X_{k:n} > t) - P(X_{i+1:n} < t | X_{m:n} \le t, X_{k:n} > t). \end{aligned}$$

Now we have

$$\begin{aligned} P(X_{i:n} < t | X_{m:n} \le t, X_{k:n} > t) &= \frac{P(X_{i:n} < t, X_{m:n} \le t, X_{k:n} > t)}{P(X_{m:n} \le t, X_{k:n} > t)} \\ &= \frac{P(X_{i:n} < t, X_{k:n} > t)}{P(X_{m:n} \le t, X_{k:n} > t)} = \frac{\sum_{j=i}^{k-1} \binom{n}{j} F^j(t) \bar{F}^{n-j}(t)}{\sum_{j=m}^{k-1} \binom{n}{j} F^j(t) \bar{F}^{n-j}(t)} \\ &= \frac{\sum_{j=i}^{k-1} \binom{n}{j} \phi^j(t)}{\sum_{j=m}^{k-1} \binom{n}{j} \phi^j(t)}, \end{aligned}$$

where $\phi(t) = \frac{F(t)}{\bar{F}(t)}$. The $p_t(i, m, k, n)$ can be rewritten as

$$p_t(i, m, k, n) = \frac{\binom{n}{i} \phi^i(t)}{\sum_{j=m}^{k-1} \binom{n}{j} \phi^j(t)}, \quad i = m, m+1, \ldots, k-1. \tag{5.3}$$

Remark Let υ be the median of the distribution F. Then $\phi(\upsilon) = 1$ and

$$p_\upsilon(i, m, k, n) = \frac{\binom{n}{i}}{\sum_{j=m}^{k-1} \binom{n}{j}}, \quad i = m, m+1, \ldots, k-1,$$

which shows that $p_\upsilon(i, m, k, n)$ does not depend on υ, scilicet, in an $(n-k+1)$-out-of-n system, given that at least m $(m < k \le n)$ components have failed before time v, and the system is still working at time v, the probability of having i failures in the system does not depend on the distribution function F(or v).

Remark For fixed $m > 1$, when $k = n$, i.e., the system is a parallel system, we have

$$\begin{aligned} p_v(i, m, n, n) &= \frac{\binom{n}{i}\phi^i(t)}{\sum_{j=m}^{n-1}\binom{n}{j}\phi^j(t)} \\ &= \frac{\binom{n}{i}\phi^i(t)}{(1+\phi(t))^n - \phi^n(t) - \sum_{j=0}^{m-1}\binom{n}{j}\phi^j(t)} \\ &= \frac{\binom{n}{i}}{2^n - 2 - \sum_{j=1}^{m-1}\binom{n}{j}}, \quad i = m, m+1, \ldots, n-1. \end{aligned}$$

Note that, $p_v(i, m, n, n)$ first starts to increase in i to achieve its maximum and starts to decrease.

Theorem 5.1.1 (a) For $m \le a \le k-1$, $a \in N$, we have

$$(N_t | X_{m:n} \le t, X_{k:n} > t) \le_{st} (N_t | X_{m:n} \le t, X_{k-1:n} > t);$$

(b) For $m \le b \le k-1$, $b \in N$, we have

$$(N_t | X_{m-1:n} \le t, X_{k:n} > t) \le_{st} (N_t | X_{m:n} \le t, X_{k:n} > t);$$

(c) Assume that there are two $(n-k+1)$-out-of-n systems of order n with the same structures, and the components have independent lifetimes. Let $\bar{F}$ denote the survival function of a random variable X, and $\bar{G}$ denote the survival function of a random variable Y. If $X \le_{st} Y$, then

$$(N_t | Y_{m:n} \le t, Y_{k:n} > t) \le_{lr} (N_t | X_{m:n} \le t, X_{k:n} > t).$$

Proof We prove part (a), part (b) can be established similarly. For $m \le a \le k-1$, $a \in N$, we have

$$\begin{aligned} &\frac{\sum_{i=a}^{k-1} P(N_t = i | X_{m:n} \le t, X_{k:n} > t)}{\sum_{i=a}^{k-1} P(N_t = i | X_{m:n} \le t, X_{k-1:n} > t)} \\ &= \frac{\sum_{i=a}^{k-1}\binom{n}{i}\phi^i(t)}{\sum_{j=m}^{k-1}\binom{n}{j}\phi^j(t)} \Big/ \frac{\sum_{i=a}^{k-1}\binom{n}{i}\phi^i(t)}{\sum_{j=m}^{k-2}\binom{n}{j}\phi^j(t)} \\ &= \frac{\sum_{j=m}^{k-2}\binom{n}{j}\phi^j(t)}{\sum_{j=m}^{k-1}\binom{n}{j}\phi^j(t)} < 1. \end{aligned}$$

This completes the proof.

To prove the result of (c), we have to discuss the monotonicity in t of the following formula

$$\begin{aligned}&\frac{P(N_t=i|X_{m:n}\le t, X_{k:n}>t)}{P(N_t=i|Y_{m:n}\le t, Y_{k:n}>t)}\\&=\frac{\binom{n}{i}F^i(t)\bar{F}^{n-i}(t)}{\sum_{j=m}^{k-1}\binom{n}{j}F^j(t)\bar{F}^{n-j}(t)}\Big/\frac{\binom{n}{i}G^i(t)\bar{G}^{n-i}(t)}{\sum_{j=m}^{k-1}\binom{n}{j}G^j(t)\bar{G}^{n-j}(t)}\\&\propto\left(\frac{F(t)}{G(t)}\right)^i\left(\frac{\bar{F}(t)}{\bar{G}(t)}\right)^{n-i}.\end{aligned}$$

Since $X\le_{st}Y$, then $\bar{F}(t)\le\bar{G}(t)$, $F(t)\ge G(t)$. It is increase in i.

Let us recall the log-concave function of a sequence $\{p_i\}$: Let $\{p_i, i=0,1,\ldots\}$ be the probability mass function of a random variable X taking values on $\{x_i, i=0,1,\ldots\}$. This sequence $\{p_i\}$ is said to be a log-concave function if and only if

$$p_i^2\ge p_{i-1}p_{i+1},\quad i=0,1,\ldots,$$

where $p_{-1}=0$ by convention. It is known that the log-concavity of $\{p_i\}$ implies that its distribution function is IFR. A probability mass function $\{p_i, i=0,1,\ldots\}$ with survival function $\bar{P}_i$ is said to be IFR if $\frac{\bar{P}_i}{\bar{P}_{i-1}}$ is decreasing in i, $i=0,1,\ldots,n-1$.

Theorem 5.1.2 For fixed values of m,k and n, the function $p_t(i,m,k,n)$ is a log-concave function.

Proof From the fact that for $i=m,m+1,\ldots,k-1$,

$$\binom{n}{i}^2\ge\binom{n}{i+1}\binom{n}{i-1}$$

and from the expression of $p_t(i,m,k,n)$ in (5.3), we have $p_t^2(i,m,k,n)\ge p_t(i-1,m,k,n)p_t(i+1,m,k,n)$.

Theorem 5.1.3 The function F is uniquely determined by $p_t(i,m,k,n)$ as follows:

$$F(t)=\frac{(i+1)p_t(i+1,m,k,n)}{(i+1)p_t(i+1,m,k,n)+(n-i)p_t(i,m,k,n)},\quad i=m,m+1,\ldots,k-1. \tag{5.4}$$

Proof From (5.3), we can write for $i=m,m+1,\ldots,k-1$,

$$\frac{p_t(i+1,m,k,n)}{p_t(i,m,k,n)}=\frac{\binom{n}{i+1}}{\binom{n}{i}}\phi(t).$$

This in turn, produces a representation (5.4) after calculation.

Theorem 5.1.4 Denote by υ the median of the distribution function F. Then
(a) For $t \geq \upsilon$ and $i \leq \frac{n-1}{2}$, $p_t(i+1, m, k, n) \geq p_t(i, m, k, n)$;
(b) For $t \leq \upsilon$ and $i \geq \frac{n-1}{2}$, $p_t(i+1, m, k, n) \leq p_t(i, m, k, n)$.

Proof We prove part (a), part (b) can be established similarly. For $t \geq \upsilon$, we have $F(t) \geq \frac{1}{2}$, which in turn implies that $\phi(t) \geq 1$. On the other hand for $i \leq \frac{n-1}{2}$, we have

$$\binom{n}{i+1} \geq \binom{n}{i}.$$

Hence, we have

$$\binom{n}{i+1}\phi^{i+1}(t) \geq \binom{n}{i}\phi^{i}(t).$$

We need to show the following lemma to prove the next result.

Lemma 5.1.5 For fixed j, m and n, the function $\gamma^{j}_{m,k,n}$ defined as

$$\gamma^{j}_{m,k,n} = \frac{\sum_{l=j}^{k-1}\binom{n}{l}t^{l}}{\sum_{i=m}^{k-1}\binom{n}{i}t^{i}}$$

is an increase function of t.

Proof We need to show that for $0 < t_1 < t_2$,

$$\frac{\sum_{l=j}^{k-1}\binom{n}{l}t_1^{l}}{\sum_{i=m}^{k-1}\binom{n}{i}t_1^{i}} \leq \frac{\sum_{l=j}^{k-1}\binom{n}{l}t_2^{l}}{\sum_{i=m}^{k-1}\binom{n}{i}t_2^{i}}. \tag{5.5}$$

The inequality in (5.5) holds if and only if

$$\sum_{l=j}^{k-1}\sum_{i=m}^{j-1}\binom{n}{l}\binom{n}{i}(t_1^{l}t_2^{i} - t_1^{i}t_2^{l}) \leq 0, \tag{5.6}$$

as

$$\sum_{l=j}^{k-1}\sum_{i=j}^{k-1}\binom{n}{l}\binom{n}{i}(t_1^{l}t_2^{i} - t_1^{i}t_2^{l}) = 0.$$

Since $0 \leq t_1 \leq t_2$, it's easy to see that (5.6) holds. This proof is completed.

In the following, we assume that $\bar{P}_t(j, m, k, n)$ denotes the survival function of $p_t(i, m, k, n)$, i.e. $\bar{P}_t(j, m, k, n) = \sum_{i=j}^{k-1} p_t(i, m, k, n)$.

Theorem 5.1.6 Assume there are two coherent systems of order n with the same structures and the components have independent lifetimes, with survival functions $\bar{F}$ and $\bar{G}$, respectively. If for all t, $\bar{F} \leq \bar{G}$, then $\bar{P}_t^{F}(j, m, k, n) \geq \bar{P}_t^{G}(j, m, k, n)$.

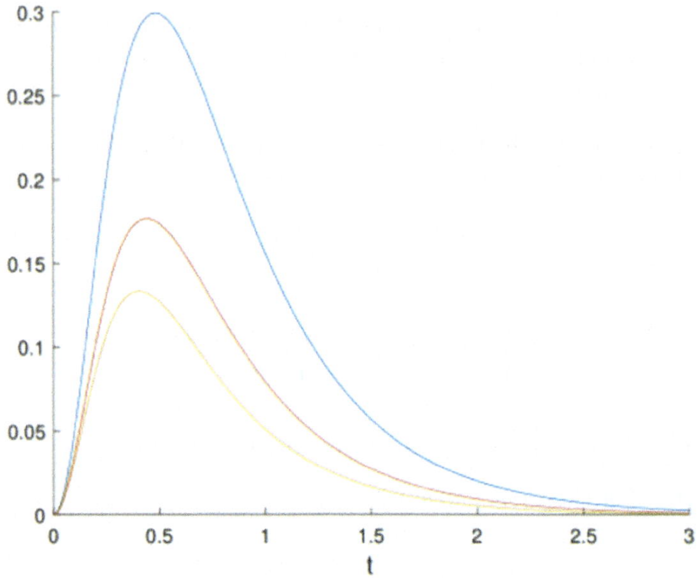

Fig. 5.1 plot of $p_t(i, m, k, n)$, $i = 3, m = 1, k = 6, n = 8, 9, 10$, component lifetime$\sim$ $Exp(1)$

Proof The $\bar{F} \leq \bar{G}$ implies that $\phi^F(t) \leq \phi^G(t)$. Now, the result follows from the fact that $\bar{P}_t^F(j, m, k, n) = \gamma^j_{m,k,n}(\phi^F(t))$ and $\bar{P}_t^G(j, m, k, n) = \gamma^j_{m,k,n}(\phi^G(t))$ and that, based on Lemma 5.1.5, $\gamma^j_{m,k,n}$ is increase in t.

Remark Figure 5.1 displays the $p_t(3, 1, 6, 8)$, $p_t(3, 1, 6, 9)$ and $p_t(3, 1, 6, 10)$. For fixed i, m and k, it clearly shows that for n, the $p_t(i, m, k, n)$ is not a monotone function. It first starts to increase in t to achieve its maximum and then starts to decrease. Assume that components of the system have exponential distribution with mean 1.

5.2 The Number of Failed Components of Coherent Systems

This section is about the number of failed components in a coherent system, given that at least m $(m < k \leq n)$ components have failed before time t, and the system is still working at time t. The probability that there are exactly i working components is computed. Furthermore, the reliability and several stochastic properties are obtained.

5.2.1 *Ordered Properties of Number of Failed Components of Coherent Systems*

For a coherent system with signature vector $\mathbf{s} = (s_1, s_2, \ldots, s_n)$, assume that $\bar{S}_j = \sum_{i=j+1}^{n} s_i$, $0 \le j \le n-1$. If $p_t^c(i, m, n)$ denotes the probability of having i failures in the system, given that at least m $(m < n)$ components have failed before time t, and the system is still working at time t, we have

$$
\begin{aligned}
p_t^c(i, m, n) &= P(X_{i:n} < t < X_{i+1:n} | X_{m:n} \le t, T > t) \\
&= \frac{P(X_{i:n} < t < X_{i+1:n}, X_{m:n} \le t, T > t)}{P(X_{m:n} \le t, T > t)} \\
&= \frac{\sum_{k=i+1}^{n} P(X_{i:n} < t < X_{i+1:n}, X_{m:n} \le t, X_{k:n} > t, T = X_{k:n})}{\sum_{l=m+1}^{n} P(X_{m:n} \le t, X_{l:n} > t, T = X_{l:n})} \\
&= \frac{\sum_{k=i+1}^{n} P(X_{i:n} < t < X_{i+1:n}, T = X_{k:n})}{\sum_{l=m+1}^{n} P(X_{m:n} \le t, X_{l:n} > t, T = X_{l:n})} \\
&= \frac{\sum_{k=i+1}^{n} P(X_{i:n} < t < X_{i+1:n}) P(T = X_{k:n})}{\sum_{l=m+1}^{n} P(X_{m:n} \le t, X_{l:n} > t) P(T = X_{l:n})} \\
&= \frac{\sum_{k=i+1}^{n} s_k \binom{n}{i} F^i(t) \bar{F}^{n-i}(t)}{\sum_{l=m+1}^{n} s_l \sum_{j=m}^{l-1} \binom{n}{j} F^j(t) \bar{F}^{n-j}(t)}, \quad i = m, m+1, \ldots, n-1,
\end{aligned}
$$

where the fourth equation comes from the fact that the event $\{X_{i:n} < t < X_{i+1:n}, X_{m:n} \le t, X_{k:n} > t\} (\equiv \{X_{i:n} < t < X_{i+1:n}\})$ and $\{T = X_{k:n}\}$ are independent, and $p_t^c(i, m, n)$ can be written as

$$
p_t^c(i, m, n) = \frac{\bar{S}_i \binom{n}{i} \phi^i(t)}{\sum_{l=m+1}^{n} s_l \sum_{j=m}^{l-1} \binom{n}{j} \phi^j(t)}, \quad i = m, m+1, \ldots, n-1. \tag{5.7}
$$

For example, the structure function of a coherent system is (Fig. 5.2)

$$
T = \min\{\max(X_1, X_2), \max(X_3, \min(X_4, X_5))\},
$$

where the $X_i's, i = 1, 2, 3, 4, 5$ are assumed to be independent and identical exponential distribution with mean 1. Let $m = 1$. Then $\phi(t) = e^t - 1$ and

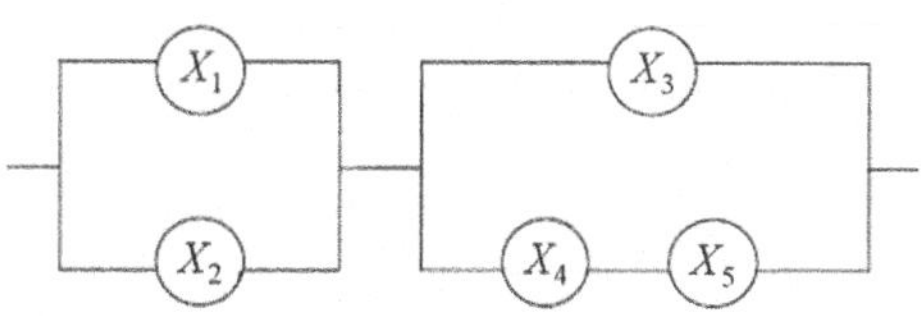

Fig. 5.2 The structure function$T = \min\{\max(X_1, X_2), \max(X_3, \min(X_4, X_5))\}$

$$p_t^c(1, 1, 5) = \frac{5}{5 + 7(e^t - 1) + 2(e^t - 1)^2},$$
$$p_t^c(2, 1, 5) = \frac{7(e^t - 1)}{5 + 7(e^t - 1) + 2(e^t - 1)^2},$$
$$p_t^c(3, 1, 5) = \frac{2(e^t -)^2}{5 + 7(e^t - 1) + 2(e^t - 1)^2}.$$

Note that the $p_t^c(4, 1, 5) = 0$ in this system, because, $P(T = X_{4:5}) = 0$, the $X_{4:5}$ is the structure of system for the component's lifetime, i.e., the lifetime $X_{4:5}$ of component will never cause the system failure.

Theorem 5.2.1 (a) For $m \leq a \leq n - 1, a \in N$, we have

$$(N_t|X_{m-1:n} \leq t, T > t) \leq_{st} (N_t|X_{m:n} \leq t, T > t);$$

(b) Assume that there are two coherent systems with lifetimes T_1 and T_2 of order n with the same structures, the components have independent lifetimes, with survival function $\bar{F}$ and $\bar{G}$, respectively. If $X \leq_{st} Y$. Then

$$(N_t|Y_{m:n} \leq t, T_2 > t) \leq_{lr} (N_t|X_{m:n} \leq t, T_1 > t).$$

Proof (a) For $m \leq a \leq n - 1, a \in N$, we have

$$\frac{\sum_{i=a}^{n-1} P(N_t = i|X_{m:n} \leq t, T > t)}{\sum_{i=a}^{n-1} P(N_t = i|X_{m-1:n} \leq t, T > t)}$$
$$= \frac{\sum_{l=m}^{n} s_l \sum_{j=m-1}^{l-1} \binom{n}{j}\phi^j(t)}{\sum_{l=m+1}^{n} s_l \sum_{j=m}^{l-1} \binom{n}{j}\phi^j(t)} > 1.$$

This proof is completed.

As for (b), one can use the same steps as the proof of Theorem 5.1.1 (c) to establish the result.

Theorem 5.2.2 *If the structure vector* $\mathbf{s}$ *is IFR, then the function* $p_t^c(i, m, n)$ *is a log-concave function and hence IFR.*

Proof To prove this result, we have to show that

$$(p_t^c(i, m, n))^2 \geq p_t^c(i - 1, m, n)p_t^c(i + 1, m, n), \quad i = m, m + 1, \ldots, n - 1,$$

where by convention $p_t^c(n, m, n) = 0$. This is equivalent to showing that

$$(\bar{S}_i\binom{n}{i})^2 \geq \bar{S}_{i-1}\bar{S}_{i+1}\binom{n}{i - 1}\binom{n}{i + 1}.$$

The result follows from the condition that $\mathbf{s}$ is IFR and Theorem 5.1.1.

The signature of a system with exchangeable components can be computed by finding the number of path sets of the system containing exactly i working components. Let $r_i(n)$ be the number of path sets of the system containing exactly i working components. Define

$$a_i(n) = \frac{r_i(n)}{\binom{n}{i}}, \quad i = 1, 2, \ldots, n. \tag{5.8}$$

Then the quantities

$$s_i(n) = a_{n-i+1}(n) - a_{n-i}(n), \quad i = 1, 2, \ldots, n, \tag{5.9}$$

or equivalently

$$a_i(n) = \sum_{j=n-i+1}^{n} s_j(n), \quad i = 1, 2, \ldots, n, \tag{5.10}$$

give the signature of a system, where $a_0(n) = 0$.

Theorem 5.2.3 *Let $T = \phi(T_1, T_2, \ldots, T_n)$ be the lifetime of a coherent system with signature $\mathbf{s} = (s_1, s_2, \ldots, s_n)$. If $T_1, T_2, \ldots, T_n$ have an absolutely continuous exchangeable joint distribution, then for $m \le j \le n-1$*

$$P\{N_t = j | T_{m:n} \le t, T > t\} = \frac{P\{T_{j+1:n} > t\} - P\{T_{j:n} > t\}}{P\{T_{m:n} \le t, T > t\}} \sum_{i=j+1}^{n} s_i,$$

where $T_{i:n}$ is the ith smallest among $T_1, T_2, \ldots, T_n$.

Proof We can write

$$P\{N_t = j | T_{m:n} \le t, T > t\} = \frac{P\{T > t | T_{m:n} \le t, N_t = j\} P\{T_{m:n} \le t, N_t = j\}}{P\{T_{m:n} \le t, T > t\}}.$$

It is clear that

$$P\{T > t | T_{m:n} \le t, N_t = j\} = P\{T > t | N_t = j\} = \frac{r_{n-j}(n)}{\binom{n}{n-j}}.$$

We obtain

$$P\{N_t = j | T_{m:n} \le t, T > t\} = \frac{r_{n-j}(n)}{\binom{n}{n-j}} \frac{P\{T_{j+1:n} > t\} - P\{T_{j:n} > t\}}{P\{T_{m:n} \le t, T > t\}}.$$

With (5.8) and (5.9), we have

$$\frac{r_{n-j}(n)}{\binom{n}{n-j}} = a_{n-j}(n) = \sum_{i=j+1}^{n} s_i.$$

This completes the proof.

Remark The conditional probability given in Theorem 5.2.3 can also be written as

$$P\{N_t = j | T_{m:n} \le t, T > t\} = \frac{P\{N_t = j\}}{P\{T_{m:n} \le t, T > t\}} \sum_{i=j+1}^{n} s_i.$$

Corollary 5.2.4 *As a direct consequence of above remark we have*

$$E\{N_t | T_{m:n} \le t, T > t\} = \frac{E(g(N_t))}{P\{T_{m:n} \le t, T > t\}},$$

where $g(j) = j \sum_{i=j+1}^{n} s_i$.

Chapter 6
Stochastic Properties and Parameter Estimation of Standby Systems

This chapter mainly investigates a general load-sharing parallel system with two components. Some comparisons among different kinds of standby systems are constructed. Under some specific conditions, that load-sharing parallel system is better than any one of its two components under the full load condition in the usual stochastic order, is shown. When the lifetimes of the components are exponential distributions, we prove that load-sharing parallel system is better than the usual parallel system with the same two components. Moreover, we discuss the stochastic comparison of the residual lifetimes between two load-sharing parallel systems. Finally, at the end of this chapter, we deal with the parameter estimation for the load-sharing parallel system when the underlying distribution of component lifetime is Exponential Pareto.

6.1 Results of Warm Standby Systems

This section has a study on the warm standby system. The stochastic comparison on the total lifetime of the warm standby system is conducted. Also, the optimal allocation of a general standby component in a series system with two independent components is discussed. Several examples are also presented to justify the main results, which in fact form nice generalizations of some existing conclusions in literature.

6.1.1 Introduction

The primary aim of reliability engineering is to increase the reliability or the availability of a system. A commonly used technique to reach this goal is to allocate a

Z. Zhang, *Conditional Order Statistics and Conditional Spacings*,
https://doi.org/10.1007/978-981-96-7151-9_6

redundancy. The key problem is to determine to which component the redundancy should be allocated so as to optimize the performance of the system, this issue has been studied through conducting stochastic comparison on system's lifetimes with various redundancy allocation in past decades.

Generally, there are three types of standby redundancy, namely, *cold* standby, *hot* standby and *warm* standby. In a cold standby redundancy, the redundant component is inactive, that is, it has zero failure rate and hence never fail in standby state; While, in a system with hot standby redundancy, the redundant component also works in the usual environment and thus may fail when it is in standby state. That is, the active component and the redundant one form a parallel system. The warm standby redundancy works in a milder environment in the standby state and hence the failure rate is nonzero and smaller than that in the usual environment, it is just an intermediate one between the cold and the hot ones. These three kinds of standby redundancy are firstly unified as the so-called *general standby* in Cha et al. (2008). For more on standby redundancy see, for example, She and Pecht (1992) and Finkelstein (2007).

Let X be the lifetime of a redundancy in the usual environment with distribution function $F(t)$ and survival function $\bar{F}(t) = 1 - F(t)$. Since the component operates in a milder environment in the standby state, the lifetime in the standby state is usually larger than that in the usual environment. In the light of the *accelerate life model* (see Bagdonavicius and Nikulin 2001) and the *decelerated life model* (see Singpurwalla 1995), it is reasonable to assume the stochastic order (see the definition) between them. This mathematically implies that the lifetime of the redundancy in standby state, say X^*, has the distribution function $F(\gamma(t))$ for some real function $\gamma(t)$ such that (i) $\gamma(0) = 0$, (ii) $\gamma(t) \le t$ for all $t \ge 0$, and (iii) $\gamma(t)$ is increasing with respect to $t \ge 0$. On the other hand, suppose that a standby redundancy having been in standby state before time $t > 0$ is just activated at this epoch and will operate in the usual environment. In comparison with the calender age t, let $\omega(t)$ be the *virtual age* (see Kijima 1989) of the standby component in the usual environment. It is reasonable to assume that (i) $\omega(0) = 0$, (ii) $\omega(t) \le t$ for all $t \ge 0$, and (iii) $\omega(t)$ is increasing with respect to $t \ge 0$. Let Y be the lifetime of an active component with distribution function $G(t)$. Assume that X and Y are independent. Then, the total lifetime of a warm standby system composed of the active component with lifetime Y and the standby component with lifetime X is

$$T_{Y|X} = \begin{cases} Y, & \text{if } Y \ge X^*, \\ Y + X_{\omega(Y)}, & \text{if } Y < X^*, \end{cases} \tag{6.1}$$

where $X_{\omega(Y)} = X - \omega(Y)|X > \omega(Y)$ is the *residual life at random time*. One may refer to Yue and Cao (2000), Li and Zuo (2004) and references therewith for more on residual life at random time. Cha et al. (2008) is among the first to bring the decelerated life model and the virtual age model together in the above *warm* standby model, they derives the reliability function for the lifetime $T_{Y|X}$ of the warm standby system in terms of the failure rate, it may be rephrased as below,

$$\bar{H}(t) = \bar{G}(t) + \int_0^t \frac{\bar{F}(t - \delta(u))}{\bar{F}(\omega(u))} \cdot \bar{F}(\gamma(t))\, \mathrm{d}G(u), \tag{6.2}$$

for any $t \geq 0$, where $\delta(u) \equiv u - \omega(u)$ for all $u \geq 0$.

This part has a further discussion on the above-mentioned general standby system. Some stochastic comparisons on the lifetimes of general standby systems with either different active components or different redundancies are conducted. Moreover, it is obtained how the variation of the distribution of the active or redundant component has effect on the whole lifetime of the system. In the end, the two-component series system with one general redundant component is considered. The optimal allocation of the standby redundancy is derived. Some numerical examples related to the main results are also presented.

6.1.2 Stochastic Comparison of Standby Systems

Cha et al. (2008) proves that, if $\gamma(t) = \omega(t)$ for all $t \geq 0$, then $X_1 \leq_{st} X_2$ implies $T_{Y|X_1} \leq_{st} T_{Y|X_2}$. That is, the lifetime of the standby system stochastically increases as the lifetime of the standby redundancy increases in the sense of the usual stochastic order. As the first main result, Theorem 6.1.1 derives the stochastic monotone property without any restriction on $\omega(t)$ and $\gamma(t)$.

Theorem 6.1.1 *If* $X_1 \leq_{hr} X_2$, *then* $T_{Y|X_1} \leq_{st} T_{Y|X_2}$.

Proof Denote $\bar{H}_1(t)$, $\bar{H}_2(t)$ the reliability functions of $T_{Y|X_1}$ and $T_{Y|X_2}$, respectively. According to (6.2), for any $t \geq 0$,

$$\bar{H}_i(t) = \bar{G}(t) + \int_0^t \frac{\bar{F}_i(t - u + \omega(u))}{\bar{F}_i(\omega(u))} \bar{F}_i(\gamma(u))\, \mathrm{d}G(u), \quad i = 1, 2.$$

Note that $X_1 \leq_{hr} X_2$ if and only if $[X_1 - t|X_1 > t] \leq_{st} [X_2 - t|X_2 > t]$ for any $t \geq 0$ (see Shaked and Shanthikumar 2007), we have, for any $0 \leq u \leq t$,

$$\frac{\bar{F}_1(t - u + \omega(u))}{\bar{F}_1(\omega(u))} \leq \frac{\bar{F}_2(t - u + \omega(u))}{\bar{F}_2(\omega(u))}$$

and hence

$$\frac{\bar{F}_1(t - u + \omega(u))}{\bar{F}_1(\omega(u))} \bar{F}_1(\gamma(u)) \leq \frac{\bar{F}_2(t - u + \omega(u))}{\bar{F}_2(\omega(u))} \bar{F}_2(\gamma(u)).$$

Now, the desired result follows directly from integrating on both sides of the above equation.

One may naturally wonder whether the condition $X_1 \leq_{hr} X_2$ in Theorem 6.1.1 is superfluous and thus may be reduced to the usual stochastic order. The following counterexample serves as a negative answer.

Example 6.1.2 For two standby components with respective reliability functions

$$\bar{F}_1(x) = \begin{cases} 1, \text{ if } x < 1, \\ \frac{1}{2}, \text{ if } 1 \le x < 2, \\ \frac{2}{7}, \text{ if } 2 \le x < 3, \\ 0, \text{ if } x \ge 3, \end{cases} \qquad \bar{F}_2(x) = \begin{cases} 1, \text{ if } x < 1, \\ \frac{3}{4}, \text{ if } 1 \le x < 2, \\ \frac{7}{20}, \text{ if } 2 \le x < 3, \\ 0, \text{ if } x \ge 3. \end{cases}$$

It is clear that $X_1 \le_{st} X_2$, however, $X_1 \le_{hr} X_2$ does not hold.

Let $\omega(u) = 0.8u$, $\gamma(u) = 0.2u$. For $P(Y = 2) = 1$, by some computations, we have

$$\bar{H}_1(t) = \begin{cases} 1, \text{ if } t < 2.4, \\ \frac{4}{7}, \text{ if } 2.4 \le t < 3.4, \\ 0, \text{ if } t \ge 3.4, \end{cases} \qquad \bar{H}_2(t) = \begin{cases} 1, \text{ if } t < 2.4, \\ \frac{7}{15}, \text{ if } 2.4 \le t < 3.4, \\ 0, \text{ if } t \ge 3.4. \end{cases}$$

Obviously, $\bar{H}_1(t) \ge \bar{H}_2(t)$ for $t \in [2.4, 3.4)$. This guarantees that $T_{Y|X_1} \le_{st} T_{Y|X_2}$ is invalid.

Under the assumption of $\gamma(t) = \omega(t)$ for all $t \ge 0$, Cha et al. (2008) also proves that, if $\delta(u)$ is increasing in $u \ge 0$, then $Y_1 \le_{st} Y_2$ implies $T_{Y_1|X} \le_{st} T_{Y_2|X}$. Theorem 6.1.4 below has a further discussion on this topic without the restriction of $\gamma(t) = \omega(t)$ for all $t \ge 0$. The following useful lemma due to Barlow and Proschan (p. 120) (1975) will be utilized in proving the second main result below.

Lemma 6.1.3 *Let W be a measure, not necessarily non-negative. If $\int_{-\infty}^{t} \mathrm{d}W(x) \ge 0$ for all t, then $\int_{-\infty}^{\infty} h(x)\,\mathrm{d}W(x) \ge 0$ for any non-negative and decreasing function h.*

Theorem 6.1.4 *Suppose X is absolutely continuous, $\delta(u)$ is increasing in $u \ge 0$ and $\omega'(u) \ge \gamma'(u)$ for any $u \ge 0$. If X is IFR and $Y_1 \le_{st} Y_2$, then $T_{Y_1|X} \le_{st} T_{Y_2|X}$.*

Proof Denote $\bar{H}_1(t)$, $\bar{H}_2(t)$ the reliability functions of $T_{Y_1|X}$ and $T_{Y_2|X}$, respectively. According to (6.2), for $i = 1, 2$, and any $t \ge 0$,

$$\begin{aligned} \bar{H}_i(t) =& \; \bar{G}_i(t) + \int_0^t \frac{\bar{F}(t-u+\omega(u))}{\bar{F}(\omega(u))} \bar{F}(\gamma(u))\,\mathrm{d}G_i(u) \\ =& \; 1 - \int_0^t \left[1 - \frac{\bar{F}(t-u+\omega(u))}{\bar{F}(\omega(u))} \bar{F}(\gamma(u))\right] \mathrm{d}G_i(u) \\ =& \; 1 - \int_0^\infty I(0 \le u \le t)\left[1 - \bar{F}(t-u+\omega(u)) \frac{\bar{F}(\gamma(u))}{\bar{F}(\omega(u))}\right] \mathrm{d}G_i(u). \end{aligned}$$

Then, for all $t \ge 0$,

$$\bar{H}_2(t) - \bar{H}_1(t) = \int_0^\infty I(0 \le u \le t)\left[1 - \frac{\bar{F}(t-u+\omega(u))}{\bar{F}(\omega(u))} \bar{F}(\gamma(u))\right] \mathrm{d}[G_1(u) - G_2(u)]. \tag{6.3}$$

For any $u \ge 0$,

$$\frac{\mathrm{d}}{\mathrm{d}u}\frac{\bar{F}(\gamma(u))}{\bar{F}(\omega(u))} = \frac{-f(\gamma(u))\bar{F}(\omega(u))\gamma'(u) + f(\omega(u))\bar{F}(\gamma(u))\omega'(u)}{\bar{F}^2(\omega(u))}$$
$$= \frac{\bar{F}(\gamma(u))}{\bar{F}(\omega(u))}\cdot\left[-\frac{f(\gamma(u))}{\bar{F}(\gamma(u))}\gamma'(u) + \frac{f(\omega(u))}{\bar{F}(\omega(u))}\omega'(u)\right].$$

Note that $\omega(0) = \gamma(0) = 0$, $\omega'(u) \geq \gamma'(u)$ for all $u \geq 0$ also implies $\omega(u) \geq \gamma(u)$ for all $u \geq 0$. Taking the IFR property of X into account, we have $\frac{\mathrm{d}}{\mathrm{d}u}\frac{\bar{F}(\gamma(u))}{\bar{F}(\omega(u))} \geq 0$ for all $u \geq 0$. Since $\delta(u)$ is increasing in $u \geq 0$, it holds that $\frac{\bar{F}(t-u+\omega(u))}{\bar{F}(\omega(u))}\bar{F}(\gamma(u))$ is increasing and hence

$$I(0 \leq u \leq t)\left[1 - \frac{\bar{F}(t-u+\omega(u))}{\bar{F}(\omega(u))}\bar{F}(\gamma(u))\right]$$

is nonnegative and decreasing in $u \geq 0$. On the other hand, $Y_1 \leq_{st} Y_2$ implies

$$\int_0^x \mathrm{d}[G_1(u) - G_2(u)] \geq 0, \quad \text{for any } x \geq 0. \tag{6.4}$$

Now, in combination with (6.3) and (6.4), it follows from Lemma 6.1.3 immediately that $\bar{H}_2(t) \geq \bar{H}_1(t)$ for all $t \geq 0$. This completes the proof.

For ease of reference, we also recall the following lemma, which will be used in the proof of the next theorem. For more details on this lemma, one may refer to Kochar and Kirmani (1995).

Lemma 6.1.5 *Let $\psi_1(x, u)$ and $\psi_2(x, u)$ be two positive real-valued functions such that (i) for $u_1 \leq u_2$, $\frac{\psi_1(x, u_2)}{\psi_1(x, u_1)}$ and $\frac{\psi_2(x, u_2)}{\psi_1(x, u_1)}$ are both increasing in x, and (ii) for each fixed x, $\frac{\psi_2(x, u)}{\psi_1(x, u)}$ is increasing in u. Then, $X \leq_{lr} Y$ implies that the ratio $\frac{\mathsf{E}[\psi_2(x, Y)]}{\mathsf{E}[\psi_1(x, X)]}$ is increasing in x provided that all concerned expectations exist.*

Theorem 6.1.6 below presents the stochastic comparison in terms of the reversed hazard rate order.

Theorem 6.1.6 *Suppose $\gamma(u) = w(u)$ for any $u \geq 0$. If $Y_1 \leq_{lr} Y_2$, X is DRHR and $\delta(u)$ is increasing in $u \geq 0$, then $T_{Y_1|X} \leq_{rh} T_{Y_2|X}$.*

Proof Denote $H_1(t)$, $H_2(t)$ distribution functions of $T_{Y_1|X}$ and $T_{Y_2|X}$, respectively. By (6.2), we have

$$H_1(t) = \int_0^t F(t-\delta(u))\,\mathrm{d}G_1(u) = \int_0^\infty I(0 \leq u \leq t)F(t-\delta(u))\,\mathrm{d}G_1(u)$$

and

$$H_2(t) = \int_0^t F(t-\delta(u))\,\mathrm{d}G_2(u) = \int_0^\infty I(0 \leq u \leq t)F(t-\delta(u))\,\mathrm{d}G_2(u).$$

The ratio

$$\frac{H_2(t)}{H_1(t)} = \frac{\displaystyle\int_0^\infty I(0 \le u \le t)F(t-\delta(u))\,\mathrm{d}G_2(u)}{\displaystyle\int_0^\infty I(0 \le u \le t)F(t-\delta(u))\,\mathrm{d}G_1(u)} = \frac{\mathsf{E}[\psi_2(t, Y_2)]}{\mathsf{E}[\psi_1(t, Y_1)]},$$

where $\psi_1(t, u) = I(0 \le u \le t)F(t-\delta(u)) = \psi_2(t, u)$. It is easy to see that $I(0 \le u \le t)$ is TP 2 in $(u, t) \in [0, \infty) \times [0, \infty)$. Furthermore, X is DRHR and $\delta(u)$ is increasing in u, this guarantees that $F(t-\delta(u))$ is TP 2 in $(u, t) \in [0, \infty) \times [0, \infty)$. Then,

$$\frac{\psi_1(t, u_2)}{\psi_1(t, u_1)} = \frac{I(0 \le u_2 \le t)F(t-\delta(u_2))}{I(0 \le u_1 \le t)F(t-\delta(u_1))} = \frac{\psi_2(t, u_2)}{\psi_1(t, u_1)}$$

is increasing in t for $u_1 \le u_2$. For each fixed x, $\dfrac{\psi_2(t, u)}{\psi_1(t, u)} = 1$ is increasing in u. In view of $Y_1 \le_{lr} Y_2$, the desired result follows from Lemma 6.1.5 immediately.

6.1.3 Allocation of a Warm Standby in a Series System

Suppose Y_1, Y_2 and X be independent random variables representing lifetimes of two active components of a series system and a standby one, respectively. In order to decide to which active component the standby one should be assigned, the following two methods need to be studied,

$$U_1 = \min\{T_{Y_1|X}, Y_2\} \quad \text{and} \quad U_2 = \min\{Y_1, T_{Y_2|X}\},$$

where $T_{Y_1|X}$ and $T_{Y_2|X}$, determined by (6.1), are lifetimes of the warm standby system with redundancy X and the active components Y_1, Y_2, respectively. Boland et al. (1992) established that, if the redundancy is a cold standby, then

$$Y_1 \le_{hr} Y_2 \quad \text{implies} \quad U_1 \ge_{st} U_2, \tag{6.5}$$

and if the redundancy is a hot standby, then

$$Y_1 \le_{st} Y_2 \quad \text{if and only if} \quad U_1 \ge_{st} U_2. \tag{6.6}$$

Afterwards, Singh and Misra (1994) further built that, if the redundancy is a hot standby, then

$$Y_1 \le_{st} Y_2 \quad \text{implies} \quad U_1 \ge_{sp} U_2. \tag{6.7}$$

These three conclusions in (6.5), (6.6) and (6.7) above claim that, for a series system with two independent components, it is optimal to allocate the redundant component to the weaker active one. Theorem 6.1.7 below, which serves as a generalization of (6.7), confirms the assertion once again for the general standby system in the sense of the stochastic precedence order.

Theorem 6.1.7 *If* $Y_1 \leq_{st} Y_2$*, then* $U_1 \geq_{sp} U_2$.

Proof Let G_1, G_2 be the distribution functions of Y_1, Y_2, and denote their reliability functions as $\bar{G}_1$ and $\bar{G}_2$. By the total probability, we have

$$\begin{aligned}
P(U_1 > U_2) = & \ P\left(\min\{T_{Y_1|X}, Y_2\} > \min\{Y_1, T_{Y_2|X}\}\right) \\
= & \ P\left(\min\{T_{Y_1|X}, Y_2\} > \min\{Y_1, T_{Y_2|X}\}, Y_1 \geq T_{Y_2|X}\right) \\
& +P\left(\min\{T_{Y_1|X}, Y_2\} > \min\{Y_1, T_{Y_2|X}\}, Y_1 < T_{Y_2|X}\right) \\
= & \ P\left(T_{Y_1|X} > T_{Y_2|X}, Y_2 > T_{Y_2|X}, Y_1 \geq T_{Y_2|X}\right) \\
& +P\left(T_{Y_1|X} > Y_1, Y_2 > Y_1, Y_1 < T_{Y_2|X}\right) \\
= & \ P(Y_2 > Y_1, X^* > Y_1) \\
= & \int_0^\infty P(Y_2 > y, X^* > y)\, \mathrm{d}G_1(y) \\
= & \int_0^\infty \bar{G}_2(y)\bar{F}(\gamma(y))\, \mathrm{d}G_1(y).
\end{aligned}$$

Likewise,

$$P(U_1 < U_2) = \int_0^\infty \bar{G}_1(y)\bar{F}(\gamma(y))\, \mathrm{d}G_2(y).$$

Since $Y_1 \leq_{st} Y_2$, it holds that, for any $t \geq 0$,

$$\begin{aligned}
& \int_0^t \bar{G}_2(y)\, \mathrm{d}G_1(y) - \int_0^t \bar{G}_1(y)\, \mathrm{d}G_2(y) \\
\geq & \int_0^t \bar{G}_1(y)\, \mathrm{d}G_1(y) - \int_0^t \bar{G}_2(y)\, \mathrm{d}G_2(y) \\
= & \int_0^t \bar{G}_2(y)\, \mathrm{d}\bar{G}_2(y) - \int_0^t \bar{G}_1(y)\, \mathrm{d}\bar{G}_1(y) \\
= & \ \frac{1}{2}\left[\bar{G}_2^2(t) - \bar{G}_1^2(t)\right] \\
& \geq 0.
\end{aligned}$$

Note that $\bar{F}(\gamma(y))$ is decreasing in $y \geq 0$, it follows from Lemma 6.1.3 that

$$P(U_1 > U_2) - P(U_1 < U_2) = \int_0^\infty \bar{F}(\gamma(y))\left[\bar{G}_2(y)\, \mathrm{d}G_1(y) - \bar{G}_1(y)\, \mathrm{d}G_2(y)\right] \geq 0.$$

This validates the desired result.

The following example asserts that the condition $Y_1 \leq_{st} Y_2$ in Theorem 6.1.7 is not necessary.

Example 6.1.8 Consider two active components having their respective reliability functions

$$\bar{G}_1(x) = \begin{cases} 1, & \text{if } 0 \le x < 1, \\ \frac{4}{5}, & \text{if } 1 \le x < 2, \\ \frac{3}{10}, & \text{if } 2 \le x < 3, \\ 0, & \text{if } x \ge 3, \end{cases} \quad \text{and} \quad \bar{G}_2(x) = \begin{cases} 1, & \text{if } 0 \le x < 1, \\ \frac{3}{4}, & \text{if } 1 \le x < 2, \\ \frac{1}{2}, & \text{if } 2 \le x < 3, \\ 0, & \text{if } x \ge 3. \end{cases}$$

It may be verified that $Y_1 \le_{st} Y_2$ is invalid; Suppose $\gamma(t) = \omega(t) = 0.5\,t$. It holds that

$$P(U_1 > U_2) = \frac{2}{5} \ge \frac{11}{40} = P(U_1 < U_2).$$

This yields $U_1 \ge_{sp} U_2$.

The last result provides the optimal allocation for the series system with two components in the sense of the usual stochastic order.

Theorem 6.1.9 *Suppose $\gamma(t) = w(t)$ for all $t \ge 0$. If $Y_1 \le_{hr} Y_2$, then $U_1 \ge_{st} U_2$.*

Proof According to (6.2), for all $t \ge 0$,

$$P(U_1 > t) = \bar{G}_2(t)\left[1 - \int_0^t F(t - \delta(u))\,\mathrm{d}G_1(u)\right],$$

and

$$P(U_2 > t) = \bar{G}_1(t)\left[1 - \int_0^t F(t - \delta(u))\,\mathrm{d}G_2(u)\right].$$

$Y_1 \le_{hr} Y_2$ implies, for $u_2 \ge u_1 > 0$, $\frac{\bar{G}_2(u_2)}{\bar{G}_2(u_1)} \ge \frac{\bar{G}_1(u_2)}{\bar{G}_1(u_1)}$, and hence

$$\frac{\bar{G}_2(u_2) - \bar{G}_2(u_1)}{\bar{G}_2(u_1)} \ge \frac{\bar{G}_1(u_2) - \bar{G}_1(u_1)}{\bar{G}_1(u_1)}.$$

Equivalently,

$$\frac{\bar{G}_2(u_2) - \bar{G}_2(u_1)}{\bar{G}_1(u_2) - \bar{G}_1(u_1)} \le \frac{\bar{G}_2(u_1)}{\bar{G}_1(u_1)}.$$

Note that, for $t \ge u_2$, $\frac{\bar{G}_2(u_1)}{\bar{G}_1(u_1)} \le \frac{\bar{G}_2(t)}{\bar{G}_1(t)}$, it holds that

$$\frac{\bar{G}_2(u_2) - \bar{G}_2(u_1)}{\bar{G}_1(u_2) - \bar{G}_1(u_1)} \le \frac{\bar{G}_2(t)}{\bar{G}_1(t)}.$$

Thus, for $t \ge u_2 \ge u_1 > 0$,

$$\bar{G}_1(t)\bar{G}_2(u_2) - \bar{G}_1(u_2)\bar{G}_2(t) \ge \bar{G}_1(t)\bar{G}_2(u_1) - \bar{G}_1(u_1)\bar{G}_2(t).$$

That is, $\bar{G}_1(t)\bar{G}_2(u) - \bar{G}_1(u)\bar{G}_2(t)$ is increasing in $u \in (0, t)$.

Now, we have, for all $t \geq 0$,

$$\begin{aligned}
& P(U_1 > t) - P(U_2 > t) \\
= \ & \bar{G}_1(t) \int_0^t F(t - \delta(u))\, \mathrm{d}G_2(u) - \bar{G}_2(t) \int_0^t F(t - \delta(u))\, \mathrm{d}G_1(u) + \bar{G}_2(t) - \bar{G}_1(t) \\
= \ & \int_0^t F(t - \delta(u))\, \mathrm{d}_u \left[\bar{G}_1(u)\bar{G}_2(t) - \bar{G}_1(t)\bar{G}_2(u)\right] - \int_0^t \mathrm{d}_u \left[\bar{G}_1(u)\bar{G}_2(t) - \bar{G}_1(t)\bar{G}_2(u)\right] \\
= \ & \int_0^t \bar{F}(t - \delta(u))\, \mathrm{d}_u \left[\bar{G}_1(t)\bar{G}_2(u) - \bar{G}_1(u)\bar{G}_2(t)\right] \\
\geq \ & 0.
\end{aligned}$$

This validates the desired result.

Example 6.1.10 below shows that the conclusion $U_1 \geq_{st} U_2$ in Theorem 6.1.9 can not be strengthened to the hazard rate order even under the stronger assumption of $Y_1 \leq_{lr} Y_2$.

Example 6.1.10 Let Y_1, Y_2 and X are all exponential distributions with parameters λ, μ and θ, respectively. It is clear that $Y_1 \leq_{lr} Y_2$. Set $\gamma(u) = \omega(u) = \beta u,\ u \geq 0$, $0 < \beta < 1$. It holds that, for any $t \geq 0$,

$$\begin{aligned}
P(U_1 > t) &= e^{-\mu t}\left[1 - \int_0^t \left(1 - e^{-\theta(t-u+\beta u)}\right) \lambda e^{-\lambda u}\, \mathrm{d}u\right] \\
&= e^{-\mu t}\left[e^{-\lambda t} + \frac{\lambda e^{-(\theta\beta+\lambda)t}}{\theta - \theta\beta - \lambda} - \frac{\lambda e^{-\theta t}}{\theta - \theta\beta - \lambda}\right], \\
P(U_2 > t) &= e^{-\lambda t}\left[1 - \int_0^t \left(1 - e^{-\theta(t-u+\beta u)}\right) \mu e^{-\mu u}\, \mathrm{d}u\right] \\
&= e^{-\lambda t}\left[e^{-\mu t} + \frac{\mu e^{-(\theta\beta+\mu)t}}{\theta - \theta\beta - \mu} - \frac{\mu e^{-\theta t}}{\theta - \theta\beta - \mu}\right].
\end{aligned}$$

Set $\lambda = 0.8$, $\mu = 0.5$, $\theta = 2$ and $\beta = 0.5$. For any $t \geq 0$,

$$P(U_1 > t) - P(U_2 > t) = e^{-2.8t}(1 + 3e^{0.5t} - 4e^{0.3t}) = e^{-2.8t} g(t).$$

Note that $g(0) = 0$ and $g'(t) = e^{-0.3t}(1.5e^{0.2t} - 1.2) \geq 0$ for all $t \geq 0$, it holds that $g(t) \geq 0$ for all $t \geq 0$ and hence $U_1 \geq_{st} U_2$. However, the ratio

$$h(t) = \frac{P(U_2 > t)}{P(U_1 > t)} = \frac{e^{-1.3t} + e^{-2.3t} - e^{-2.8t}}{e^{-1.3t} + 4e^{-2.3t} - 4e^{-2.5t}}$$

satisfies, for $t \geq 0$,

$$h'(t) = e^{-5.3t}(3e^{1.7t} - 4.8e^{1.5t} + 1.5e^{1.2t} - 0.8e^{0.7t} + 2e^{0.2t} - 1.1).$$

Since $h'(2) = 0.0002 > 0$, there exists a neighborhood $(2-\delta, 2+\delta)$, $\delta > 0$ is a small real number, such that $h'(t) \geq 0$ for $t \in (2-\delta, 2+\delta)$. This implies that $\frac{P(U_2>t)}{P(U_1>t)}$ increases in $t \in (2-\delta, 2+\delta)$. Therefore, $U_1 \geq_{hr} U_2$ is invalid.

Additionally, Example 6.1.11 shows that the condition $Y_1 \leq_{hr} Y_2$ in Theorem 6.1.9 may not be relaxed to the usual stochastic order.

Example 6.1.11 For two active components with reliability functions

$$\bar{G}_1(x) = \begin{cases} 1, \text{ if } x < 1, \\ \frac{1}{2}, \text{ if } 1 \leq x < 2, \\ \frac{2}{7}, \text{ if } 2 \leq x < 3, \\ 0, \text{ if } x \geq 3, \end{cases} \qquad \bar{G}_2(x) = \begin{cases} 1, \text{ if } x < 1, \\ \frac{3}{4}, \text{ if } 1 \leq x < 2, \\ \frac{7}{20}, \text{ if } 2 \leq x < 3, \\ 0, \text{ if } x \geq 3, \end{cases}$$

it is easy to verify that $Y_1 \leq_{st} Y_2$ but $Y_1 \leq_{hr} Y_2$ is not valid.

Let $P(X = 2) = 1$ and $\gamma(u) = \omega(u) = 0.5u,\ u \geq 0$. Then,

$$P(U_1 > x) = \begin{cases} 1, \text{ if } x < 1, \\ \frac{3}{4}, \text{ if } 1 \leq x < 2, \\ \frac{7}{20}, \text{ if } 2 \leq x < 2.5, \\ \frac{7}{40}, \text{ if } 2.5 \leq x < 3, \\ 0, \text{ if } x \geq 3, \end{cases} \qquad P(U_2 > x) = \begin{cases} 1, \text{ if } x < 1, \\ \frac{1}{2} \text{ if } 1 \leq x < 2, \\ \frac{2}{7} \text{ if } 2 \leq x < 2.5, \\ \frac{3}{14} \text{ if } 2.5 \leq x < 3, \\ 0, \text{ if } x \geq 3. \end{cases}$$

Obviously, $U_1 \geq_{st} U_2$ does not hold.

6.2 Stochastic Properties of Load Standby Systems

This section mainly investigates a load–sharing parallel system having two units. First, we construct some comparisons among a load standby system, a warm standby system, a hot standby system and a cold standby system. Moreover, some stochastic comparisons between the load-sharing parallel system and one of its two components are obtained in the sense of the usual stochastic order. Finally, the residual life of this system and its properties are examined.

6.2.1 Intruduction

Increasing the availability or the reliability of a system is the primary aim of reliability engineering. A commonly used technique for this purpose is to allocate a redundancy. In most cases, independence of the components is assumed within the system. However, in practice, the components of many systems share a given load, functioning dependently on each other. This is because one component failure may affect the failure rates of the remaining surviving components of the system. Many empirical studies about some systems have shown that the workload has a strong effect on the component failure rate. And in most situations, it has been proved that an increased load results in a higher component failure rate. Hence, considering stochastic dependencies among the components within the system is becoming more and more popular and necessary.

Yun and Cha (2010) introduced a general model of load-sharing parallel system (Fig. 6.1) having two components whose lifetimes are arbitrarily continuous. As described there, when the system starts operating, these two components C_1 and C_2 share the total load L, with respective load proportions α and $1-\alpha$. After one component fails, another would load the total L and continues working. In their work, L was assumed to be 1 without loss of generality. Zhang and Balakrishnan (2017) stated that the unit C_2 (C_1) may be regarded as a standby that takes excess load from the unit C_1 (C_2), and hence the unit C_2 (C_1) can be regarded as a load standby.

Suppose $F_i(\cdot)$ and $\bar{F}_i(\cdot) = 1 - F_i(\cdot)$ $(i = 1, 2)$ are the distribution and survival functions of the lifetime X_i of component C_i under full load condition. Let $X_{i,\mathcal{P}_j}$ be the lifetime of the unit C_i $(i = 1, 2)$ under partial load condition $\mathcal{P}_j$ $(j = 1, 2)$, $F_{i,\mathcal{P}_j}(\cdot)$ and $\bar{F}_{i,\mathcal{P}_j}(\cdot) = 1 - F_{i,\mathcal{P}_j}(\cdot)$ be the corresponding distribution and survival functions. Throughout the section, it is assumed that the components are *independent* under partial load conditions. According to the stochastic reasoning employed in *accelerated life model* (see Bagdonavicius and Nikulin (2001)) and *decelerated life model* (see Singpurwalla (1995)), it is naturally assumed that, for all $t \geq 0$,

$$F_{1,\mathcal{P}_1}(t) = F_1(g_1(\alpha)t), \quad F_{2,\mathcal{P}_2}(t) = F_2(g_2(1-\alpha)t), \tag{6.8}$$

where the load function $g_i(\alpha)$ $(i = 1, 2)$ is satisfied with: (i) $g_i(\alpha) = 0$ for $\alpha = 0$, $g_i(\alpha) = 1$ for $\alpha = 1$; (ii) $g_i(\alpha) \in [0, 1]$, and (iii) $g_i(\alpha)$ strictly increases in α. Clearly, the equalities of (1) show that the lifetimes of the units under full load are stochastically smaller than those of the units under partial load.

Assume that $T(X_{1,\mathcal{P}_1}, X_{2,\mathcal{P}_2})$ is the lifetime of the load-sharing parallel system in Fig. 6.2, and $R_{X_{1,\mathcal{P}_1}, X_{2,\mathcal{P}_2}}(\cdot)$ is the corresponding reliability function. Then, by Yun and Cha (2010), and Zhang and Balakrishnan (2017), for any $x \geq 0$,

$$\begin{aligned} R_{X_{1,\mathcal{P}_1}, X_{2,\mathcal{P}_2}}(x) &= \bar{F}_1(g_1(\alpha)x)\bar{F}_2(g_2(1-\alpha)x) + \int_0^x \frac{\bar{F}_1(v(u) - u + x)}{\bar{F}_1(v(u))} \bar{F}_1(g_1(\alpha)u) dF_2(g_2(1-\alpha)u) \\ &\quad + \int_0^x \frac{\bar{F}_2(w(u) - u + x)}{\bar{F}_2(w(u))} \bar{F}_2(g_2(1-\alpha)u) dF_1(g_1(\alpha)u), \end{aligned} \tag{6.9}$$

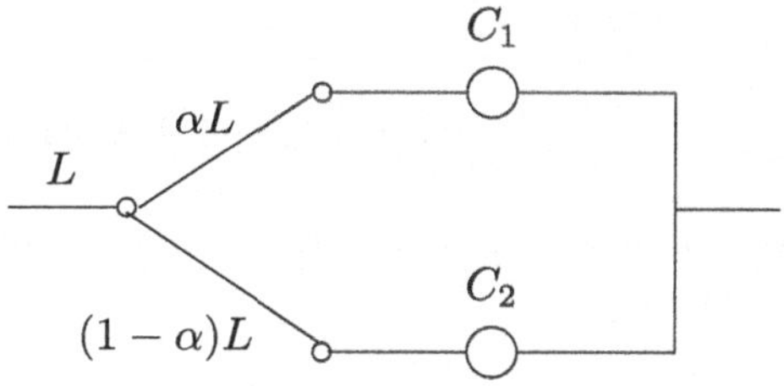

Fig. 6.1 A load-sharing parallel system

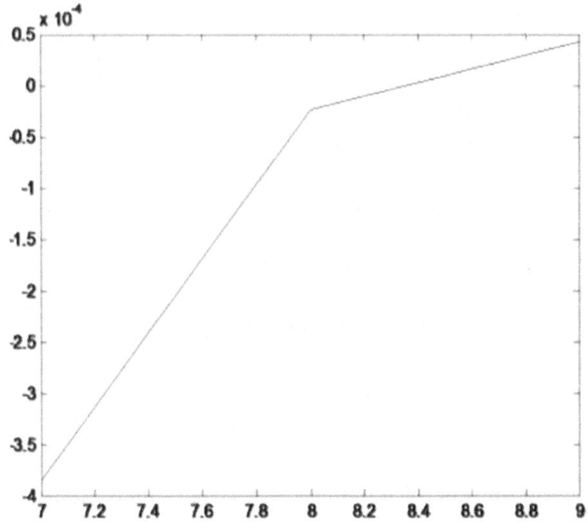

Fig. 6.2 The difference of two reliability functions

where $v(u)$ and $w(u)$ are respective virtual ages (see Finkelstein and Finkelstein (2008), Finkelstein (2004, 2007, 2009), Finkelstein and Hazra (2016)) of unit lifetimes X_1 and X_2 under full load condition, which are equivalent to age u under partial load condition. In other words, the performance of units C_1 and C_2 operating under partial load condition during $(0, u]$ is just one under the full load condition during $(0, v(u)]$ and $(0, w(u)]$, respectively. Naturally, it is assumed that (i) $v(u) = 0$, $w(u) = 0$ for $u = 0$, (ii) $v(u) \leq u$ and $w(u) \leq u$ for $u \geq 0$, (iii) $v(u)$ and $w(u)$ increase with respect to u.

If the component C_1 operates under the partial load condition $\mathcal{P}_1$ while C_2 operates under the full load condition, then the load-sharing system would be the warm standby system. In this case, its lifetime is denoted by $T(X_{1,\mathcal{P}_1}, X_2)$. According to (6.9), for $t \geq 0$, its reliability function can be represented as

$$R_{X_{1,\mathcal{P}_1},X_2}(t) = \bar{F}_2(t) + \int_0^t \frac{\bar{F}_1(t-u+v(u))}{\bar{F}_1(v(u))} \bar{F}_1(g_1(\alpha)u)\,\mathrm{d}F_2(u). \tag{6.10}$$

If both C_1 and C_2 operate under the full load condition, then the warm standby system would be a hot standby one. In this case, its lifetime is denoted by $\max\{X_1, X_2\}$, with reliability

$$R_{\max\{X_1,X_2\}}(t) = \bar{F}_2(t) + \bar{F}_1(t) - \bar{F}_2(t)\bar{F}_1(t). \tag{6.11}$$

Furthermore, the reliability function of cold standby system having C_1 and C_2 is given by

$$R_{X_1+X_2}(t) = \bar{F}_2(t) + \int_0^t \bar{F}_1(t-u)dF_2(u), \tag{6.12}$$

where $X_1 + X_2$ is the lifetime of the cold standby system with units C_1 and C_2.

In practice, it is nature of interest to design a system with high reliability in the sense of Stochastic Order. Given two components, one could use them to design various systems, for example, a load-sharing system, a warm standby system, a parallel system, or a cold standby system. It is nature of interest to compare the difference among these systems in reliability performance.

6.2.2 *Stochastic Properties of Load-Sharing Systems*

In order to investigate the impact on the system performance of the two virtual age functions, we will use $\mathcal{S}(v(\cdot), w(\cdot))$ to denote the system having two modeling functions $v(\cdot)$ and $w(\cdot)$, and $R(x; v(\cdot), w(\cdot))$ to denote the reliability function of the system. From the representation in (6.2), it is easy to observe the following facts: (i) If X_1 is IFR (DFR) and $v_1(u) \leq v_2(u)$ for all $u \geq 0$, then $R(x; v_1(\cdot), w(\cdot)) \geq (\leq) R(x; v_2(\cdot), w(\cdot))$; (ii) If X_i $(i = 1, 2)$ is IFR (DFR) and $v_1(u) \leq v_2(u)$, $w_1(u) \leq w_2(u)$ for all $u \geq 0$, then $R(x; v_1(\cdot), w_1(\cdot)) \geq (\leq) R(x; v_2(\cdot), w_2(\cdot))$.

In practice, a problem of interest is that of distributing components into specific locations within a given load-sharing parallel system in order to achieve the maximum possible system reliability. For example, suppose a reliability engineer has three components C_1, C_2 and C_3. If C_1 is fixed in one location, then he/she has two options, and can put C_2 or C_3 in another location. But which would produce higher system reliability? Theorem 6.2.2 below would show that if component C_2 is smaller than C_3 in hazard rate order, then the system with components C_1 and C_2 will be smaller than the one with C_1 and C_3 in the sense of usual stochastic order.

In order to investigate the effect on the system performance of the two components, we use $R_{i,j}(x)$ to denote the reliability function of the system lifetime $T(X_i, X_j)$ with independent component lifetimes $X_{\mathcal{P}_i}$ and $X_{\mathcal{P}_j}$ under partial load condition, whose corresponding virtual age functions are $v(u)$ and $w(u)$, and with assignment functions $g_1(\alpha)$ and $g_2(1-\alpha)$, respectively. Suppose $\bar{F}_i$ and F_i $(i = 1, 2, 3)$ are the reliability function and distribution function of component i with lifetime X_i under full load condition.

The following lemma due to Barlow and Proschan (1975) is useful to establish the main results.

Lemma 6.2.1 *Let W be a measure, not necessarily non-negative. If $\int_{-\infty}^{t} \mathrm{d}W(x) \geq 0$ for all t, then $\int_{-\infty}^{\infty} h(x)\,\mathrm{d}W(x) \geq 0$ for any non-negative and decreasing function h.*

Theorem 6.2.2 *Suppose X_1 is IFR and $g_1(\alpha)u \leq v(u)$ for all $u \geq 0$, and $u - v(u)$ is increasing in $u \geq 0$. If $X_2 \leq_{hr} X_3$, then*

$$R_{1,2}(x) \leq R_{1,3}(x), \quad x \geq 0.$$

Proof It is clear that $R_{1,2}(x)$ is just as in (6.9), and for any $x \geq 0$,

$$R_{1,3}(x) = \bar{F}_1(g_1(\alpha)x)\bar{F}_3(g_2(1-\alpha)x) + \int_0^x \frac{\bar{F}_1(v(u)-u+x)}{\bar{F}_1(v(u))}\bar{F}_1(g_1(\alpha)u)dF_3(g_2(1-\alpha)u)$$
$$+ \int_0^x \frac{\bar{F}_3(w(u)-u+x)}{\bar{F}_3(w(u))}\bar{F}_3(g_2(1-\alpha)u)dF_1(g_1(\alpha)u).$$

We thus find

$$\begin{aligned}
&R_{1,3}(x) - R_{1,2}(x)\\
&= \bar{F}_1(g_1(\alpha)x) \times [F_2(g_2(1-\alpha)x) - F_3(g_2(1-\alpha)x)]\\
&\quad + \int_0^x \frac{\bar{F}_1(v(u)-u+x)}{\bar{F}_1(v(u))}\bar{F}_1(g_1(\alpha)u)d[F_3(g_2(1-\alpha)u) - F_2(g_2(1-\alpha)u)]\\
&\quad + \int_0^x \left[\frac{\bar{F}_3(w(u)-u+x)}{\bar{F}_3(w(u))}\bar{F}_3(g_2(1-\alpha)u) - \frac{\bar{F}_2(w(u)-u+x)}{\bar{F}_2(w(u))}\bar{F}_2(g_2(1-\alpha)u)\right]dF_1(g_1(\alpha)u)\\
&= \int_0^x \left[\bar{F}_1(g_1(\alpha)x) - \frac{\bar{F}_1(v(u)-u+x)}{\bar{F}_1(v(u))}\bar{F}_1(g_1(\alpha)u)\right]d[F_2(g_2(1-\alpha)u) - F_3(g_2(1-\alpha)u)]\\
&\quad + \int_0^x \left[\frac{\bar{F}_3(w(u)-u+x)}{\bar{F}_3(w(u))}\bar{F}_3((1-\alpha)u) - \frac{\bar{F}_2(w(u)-u+x)}{\bar{F}_2(w(u))}\bar{F}_2(g_2(1-\alpha)u)\right]dF_1(g_1(\alpha)u)\\
&= \int_0^\infty I(0 \leq u \leq x)\left[\bar{F}_1(g_1(\alpha)x) - \frac{\bar{F}_1(v(u)-u+x)}{\bar{F}_1(v(u))}\bar{F}_1(g_1(\alpha)u)\right]d[F_2(g_2(1-\alpha)u) - F_3(g_2(1-\alpha)u)]\\
&\quad + \int_0^x \left[\frac{\bar{F}_3(w(u)-u+x)}{\bar{F}_3(w(u))}\bar{F}_3((1-\alpha)u) - \frac{\bar{F}_2(w(u)-u+x)}{\bar{F}_2(w(u))}\bar{F}_2(g_2(1-\alpha)u)\right]dF_1(g_1(\alpha)u). \qquad (6.13)
\end{aligned}$$

It follows that, from $X_2 \leq_{hr} X_3$, $\int_0^t d[F_2(g_2(1-\alpha)x) - F_3(g_2(1-\alpha)x)] \geq 0$ for any $t \geq 0$. Furthermore, the IFR property of X_1 and $g_1(\alpha)u \leq v(u)$ imply that

$$\frac{\bar{F}_1(g_1(\alpha)x)}{\bar{F}_1(g_1(\alpha)u)} = \frac{\bar{F}_1(g_1(\alpha)u + g_1(\alpha)(x-u))}{\bar{F}_1(g_1(\alpha)u)} \geq \frac{\bar{F}_1(g_1(\alpha)u + x - u)}{\bar{F}_1(g_1(\alpha)u)} \geq \frac{\bar{F}_1(v(u) + x - u)}{\bar{F}_1(v(u))}$$

for $0 \leq u \leq x$, and hence $I(0 \leq u \leq x)\left[\bar{F}_1(g_1(\alpha)x) - \frac{\bar{F}_1(v(u)-u+x)}{\bar{F}_1(v(u))}\bar{F}_1(g_1(\alpha)u)\right]$ is non-negative for any $u \geq 0$. Also it can be shown that, from the IFR property of X_1 and $g_1(\alpha)u \leq v(u)$,

$$\frac{d}{du}\frac{\bar{F}_1(g_1(\alpha)u)}{\bar{F}_1(v(u))} \geq 0.$$

Since $u - v(u)$ is increasing in $u \geq 0$, we have $\frac{\bar{F}_1(v(u)-u+x)}{\bar{F}_1(v(u))}\bar{F}_1(g_1(\alpha)u)$ to be increasing in $u \geq 0$, and hence

$$I(0 \leq u \leq x)\left[\bar{F}_1(g_1(\alpha)x) - \frac{\bar{F}_1(v(u)-u+x)}{\bar{F}_1(v(u))}\bar{F}_1(g_1(\alpha)u)\right]$$

is decreasing in $u \geq 0$. Therefore, from Lemma 6.2.1, the first term on the RHS of (6.13) is non-negative for $x \geq 0$.

Note that $X_2 \leq_{hr} X_3$ if and only if $[X_2 - t|X_2 > t] \leq_{st} [X_3 - t|X_3 > t]$ for any $t \geq 0$ (see Shaked and Shanthikumar (2007), and so it follows that

$$\frac{\bar{F}_2(w(u) + x - u)}{\bar{F}_2(w(u))} \leq \frac{\bar{F}_3(w(u) + x - u)}{\bar{F}_3(w(u))}$$

for any $0 \leq u \leq x$, and thus

$$\frac{\bar{F}_2(w(u) - u + x)}{\bar{F}_2(w(u))}\bar{F}_2(g_2(1-\alpha)u) \leq \frac{\bar{F}_3(w(u) - u + x)}{\bar{F}_3(w(u))}\bar{F}_3(g_2(1-\alpha)u).$$

Hence, the second term on the RHS of (6.13) is non-negative, and so the desired result follows directly.

As in the cumulative exposure model of Nelson (1990), we may suppose that, for any $u \geq 0$,

$$F_{\mathcal{P}_1}(u) = F_1(g_1(\alpha)u) \leq F_1(v(u)),$$

which means that the virtual age $v(u)$ would produce at least as much population cumulative fraction of component C_1 failure as the age u does under partial load condition. Then, we can obtain, for $u \geq 0$,

$$F_1^{-1}F_{\mathcal{P}_1}(u) = g_1(\alpha)u \leq v(u).$$

Thus, it is reasonable to assume that $g_1(\alpha)u \leq v(u)$ in the above theorem.

From the proof of Theorem 6.2.2, if we assume that the virtual age functions would produce the same population cumulative fraction of units failing as the age u does under partial load condition, that is, $g_1(\alpha)u = v(u)$ and $g_2(1-\alpha)u = w(u)$ for all $u \geq 0$, then the condition $X_2 \leq_{hr} X_3$ may be relaxed to the usual stochastic order, and the IFR property of X_1 may also be dropped, to obtain the following result.

Theorem 6.2.3 *Suppose* $g_1(\alpha)u = v(u)$ *and* $g_2(1-\alpha)u = w(u)$ *for all* $u \geq 0$, *and* $u - v(u)$ *is increasing in* $u \geq 0$. *If* $X_2 \leq_{st} X_3$, *then*

$$R_{1,2}(x) \leq R_{1,3}(x), \quad x \geq 0.$$

Following the lines of proof of Theorems 6.2.2 and 6.2.3, the following theorem can be established.

Theorem 6.2.4 *(a) Suppose* X_2 *is IFR,* $g_2(1-\alpha)u \leq w(u)$ *for all* $u \geq 0$, *and* $u - w(u)$ *is increasing in* $u \geq 0$. *If* $X_1 \leq_{hr} X_3$, *then* $R_{1,2}(x) \leq R_{3,2}(x)$ *for* $x \geq 0$;
(b) Suppose $g_1(\alpha)u = v(u)$ *and* $g_2(1-\alpha)u = w(u)$ *for all* $u \geq 0$, *and* $u - w(u)$ *is increasing in* $u \geq 0$. *If* $X_1 \leq_{st} X_3$, *then* $R_{1,2}(x) \leq R_{3,2}(x)$ *for* $x \geq 0$.

Let $R_X(x)$ be the reliability function of the lifetime of one load-sharing system with two independent components having lifetimes X_1 and X_2 under full load condition, and $R_Y(x)$ be the reliability function of the lifetime of another load-sharing system with two independent components having lifetimes Y_1 and Y_2 under full load condition. Suppose X_i $(i = 1, 2)$ has distribution function $F_i(t)$ and survival function $\bar{F}_i(t) = 1 - F_i(t)$ under full load condition, and Y_i $(i = 1, 2)$ has distribution function $G_i(t)$ and survival function $\bar{G}_i(t) = 1 - G_i(t)$ under full load condition. Assume that both X_1 and Y_1 have virtual age function $v(u)$ and assignment function $g_1(\alpha)$. Similarly, assume further that both X_2 and Y_2 have virtual age function $w(u)$ and assignment function $g_2(1-\alpha)$.

The following theorem shows that the stronger the components are in the sense of hazard rate order, the stronger the corresponding system is in the usual stochastic order.

Theorem 6.2.5 *Suppose X_i $(i = 1, 2)$ is IFR, $g_1(\alpha)u \le v(u)$, $g_2(1-\alpha)u \le w(u)$ for all $u \ge 0$, and $u - v(u)$ and $u - w(u)$ are increasing in $u \ge 0$. If $X_i \le_{hr} Y_i$ for $i = 1, 2$, then $R_X(x) \le R_Y(x)$, $x \ge 0$.*

Proof Note that, for any $x \ge 0$,

$$\begin{aligned}
&\bar{G}_1(g_1(\alpha)x)\bar{G}_2(g_2(1-\alpha)x) - \bar{F}_1(g_1(\alpha)x)\bar{F}_2(g_2(1-\alpha)x)\\
&= [\bar{G}_2(g_2(1-\alpha)x) - \bar{F}_2(g_2(1-\alpha)x)] \times [\bar{G}_1(g_1(\alpha)x) - \bar{F}_1(g_1(\alpha)x)]\\
&+\bar{F}_2(g_2(1-\alpha)x)[F_1(g_1(\alpha)x) - G_1(g_1(\alpha)x)] \ + \bar{F}_1(g_1(\alpha)x)[F_2(g_2(1-\alpha)x) - G_2(g_2(1-\alpha)x)].
\end{aligned}$$

Hence, after some algebraic computations, we find, for any $x \ge 0$,

$$\begin{aligned}
&R_Y(x) - R_X(x)\\
&= [\bar{G}_2(g_2(1-\alpha)x) - \bar{F}_2(g_2(1-\alpha)x)] \times [\bar{G}_1(g_1(\alpha)x) - \bar{F}_1(g_1(\alpha)x)]\\
&+ \int_0^x \left[\frac{\bar{G}_2(w(u)-u+x)}{\bar{G}_2(w(u))}\bar{G}_2(g_2(1-\alpha)u)s - \frac{\bar{F}_2(w(u)-u+x)}{\bar{F}_2(w(u))}\bar{F}_2(g_2(1-\alpha)u)\right] dG_1(g_1(\alpha)u)\\
&+ \int_0^x \left[\bar{F}_2(g_2(1-\alpha)x) - \frac{\bar{F}_2(w(u)-u+x)}{\bar{F}_2(w(u))}\bar{F}_2(g_2(1-\alpha)u)\right] d[F_1(g_1(\alpha)u) - G_1(g_1(\alpha)u)]\\
&+ \int_0^x \left[\frac{\bar{G}_1(v(u)-u+x)}{\bar{G}_1(v(u))}\bar{G}_1(g_1(\alpha)u) - \frac{\bar{F}_1(v(u)-u+x)}{\bar{F}_1(v(u))}\bar{F}_1(g_1(\alpha)u)\right] dG_2(g_2(1-\alpha)u)\\
&+ \int_0^x \left[\bar{F}_1(g_1(\alpha)x) - \frac{\bar{F}_1(v(u)-u+x)}{\bar{F}_1(v(u))}\bar{F}_1(g_1(\alpha)u)\right] d[F_2(g_2(1-\alpha)u) - G_2(g_2(1-\alpha)u)] \qquad (6.14)
\end{aligned}$$

It follows that, from $X_i \le_{hr} Y_i$ for $i = 1, 2$,

$$[\bar{G}_2(g_2(1-\alpha)x) - \bar{F}_2(g_2(1-\alpha)x)] \times [\bar{G}_1(g_1(\alpha)x) - \bar{F}_1(g_1(\alpha)x)] \ge 0$$

for any $x \ge 0$. Furthermore, as in the proof of Theorem 6.2.2, each of the last four terms in the RHS of (6.14) can be shown to be nonnegative for $x \ge 0$. This completes the proof of the theorem.

From the proof of Theorem 6.2.5, the following result can be obtained easily.

Theorem 6.2.6 *Suppose $g_1(\alpha)u = v(u)$, $g_2(1-\alpha)u = w(u)$ for all $u \ge 0$, and $u - v(u)$, $u - w(u)$ are increasing in $u \ge 0$. If $X_i \le_{st} Y_i$ for $i = 1, 2$, then $R_X(x) \le R_Y(x)$, $x \ge 0$.*

We now present an example showing the complexity of the reliability function for a load sharing system even when the component lifetimes are distributed as exponential.

Example 6.2.7 Suppose X_1 and X_2 are independent exponential variables with respective parameters λ and β. Then, according to (6.2), we obtain that, for any $x > 0$,

$$R(x) = e^{-\lambda g_1(\alpha)x} e^{-\beta g_2(1-\alpha)x} + \int_0^x \beta g_2(1-\alpha) e^{-[\lambda x - \lambda u + \lambda g_1(\alpha)u]} e^{-\beta g_2(1-\alpha)u} du$$
$$+ \int_0^x \lambda g_1(\alpha) e^{-[\beta x - \beta u + \beta g_2(1-\alpha)u]} e^{-\lambda g_1(\alpha)u} du. \tag{6.15}$$

After some algebraic computations, the reliability function in (6.15), for $\alpha \neq 0$ or $\alpha \neq 1$, can be simplified as

$$R(x) = \frac{\beta g_2(1-\alpha)}{A(\lambda,\beta,\alpha)-\lambda} e^{-\lambda x} + \frac{\lambda g_1(\alpha)}{A(\lambda,\beta,\alpha)-\beta} e^{-\beta x} + \left(1 - \frac{\beta g_2(1-\alpha)}{A(\lambda,\beta,\alpha)-\lambda} - \frac{\lambda g_1(\alpha)}{A(\lambda,\beta,\alpha)-\beta}\right) e^{-A(\lambda,\beta,\alpha)x},$$

where $A(\lambda, \beta, \alpha) = \lambda g_1(\alpha) + \beta g_2(1-\alpha)$. It is clear that this reliability function can be written as a generalized mixture of exponential reliability functions with corresponding parameters λ, β and $A(\lambda, \beta, \alpha)$, where the mixture weights may be negative.

Furthermore, some particular cases of the reliability function can be examined further.

First, let us consider the case when $\beta = \lambda$, and $g_1(\alpha) + g_2(1-\alpha) \neq 1$. In this case,

$$R(x) = \frac{1}{1 - (g_1(\alpha) + g_2(1-\alpha))} e^{-\lambda[g_1(\alpha)+g_2(1-\alpha)]x} - \frac{g_1(\alpha) + g_2(1-\alpha)}{1 - (g_1(\alpha) + g_2(1-\alpha))} e^{-\lambda x}.$$

Secondly, for the cases when $\alpha = 0$, or $\alpha = 1$, or $g_1(\alpha) = \alpha$, $g_2(1-\alpha) = 1-\alpha$ for any $\alpha \geq 0$, the reliability function reduces to

$$R(x) = \frac{\beta}{\beta - \lambda} e^{-\lambda x} - \frac{\lambda}{\beta - \lambda} e^{-\beta x}$$

for $\lambda \neq \beta$.

Finally, consider the case when $\beta = \lambda$ and $g_1(\alpha) + g_2(1-\alpha) = 1$. In this case,

$$R(x) = (1 + \lambda x)e^{-\lambda x},$$

which is just a gamma distribution with shape parameter 2 and scale parameter λ.

6.2.3 *On the Residual Life of Load-Sharing Systems*

In this section, we investigate some stochastic comparisons of the residual lifetime of the load-sharing parallel system given that the weaker one of the two components under partial load is still alive at some time $t \geq 0$; that is,

$$(T(X_{1,\mathcal{P}_1}, X_{2,\mathcal{P}_2}) - t \,|\, \min(X_{1,\mathcal{P}_1}, X_{2,\mathcal{P}_2}) \geq t).$$

Theorem 6.2.8 gives the reliability function of $(T(X_{1,\mathcal{P}_1}, X_{2,\mathcal{P}_2}) - t \,|\, \min(X_{1,\mathcal{P}_1}, X_{2,\mathcal{P}_2}) \geq t)$.

Theorem 6.2.8 *Let $R^t_{X_{1,\mathcal{P}_1}, X_{2,\mathcal{P}_2}}(\cdot)$ be the reliability function of the conditional random variable*

$$(T(X_{1,\mathcal{P}_1}, X_{2,\mathcal{P}_2}) - t \,|\, \min(X_{1,\mathcal{P}_1}, X_{2,\mathcal{P}_2}) \geq t).$$

Then, for any $x \geq 0, t \geq 0$,

$$\begin{aligned} R^t_{X_{1,\mathcal{P}_1}, X_{2,\mathcal{P}_2}}(x) = & \int_t^\infty \frac{\bar{F}_2(t + x - y + w(y))}{\bar{F}_2(w(y))} \bar{F}_2(g_2(1-\alpha)y) dF_1(g_1(\alpha)y) / [\bar{F}_1(g_1(\alpha)t)\bar{F}_2(g_2(1-\alpha)t)] \\ & + \int_t^\infty \frac{\bar{F}_1(t + x - y + v(y))}{\bar{F}_1(v(y))} \bar{F}_1(g_1(\alpha)y) dF_2(g_2(1-\alpha)y) / [\bar{F}_1(g_1(\alpha)t)\bar{F}_2(g_2(1-\alpha)t)]. \end{aligned}$$

Proof For any $x > 0, t > 0$, we have

$$\begin{aligned} R^t_{X_{1,\mathcal{P}_1}, X_{2,\mathcal{P}_2}}(x) &= P(T(X_{1,\mathcal{P}_1}, X_{2,\mathcal{P}_2}) - t > x \,|\, \min(X_{1,\mathcal{P}_1}, X_{2,\mathcal{P}_2}) \geq t) \\ &= \frac{P(T - t > x, \min(X_{1,\mathcal{P}_1}, X_{2,\mathcal{P}_2}) \geq t)}{P(\min(X_{1,\mathcal{P}_1}, X_{2,\mathcal{P}_2}) \geq t)} \\ &= \frac{P(T - t > x, t \leq X_{1,\mathcal{P}_1} \leq X_{2,\mathcal{P}_2})}{P(X_{1,\mathcal{P}_1} \geq t, X_{2,\mathcal{P}_2} \geq t)} \\ &\quad + \frac{P(T - t > x, t \leq X_{2,\mathcal{P}_2} \leq X_{1,\mathcal{P}_1})}{P(X_{1,\mathcal{P}_1} \geq t, X_{2,\mathcal{P}_2} \geq t)} \\ &= \frac{P(X_{1,\mathcal{P}_1} + X_2^* > t + x, t \leq X_{1,\mathcal{P}_1} \leq X_{2,\mathcal{P}_2})}{P(X_{1,\mathcal{P}_1} \geq t, X_{2,\mathcal{P}_2} \geq t)} \\ &\quad + \frac{P(X_{2,\mathcal{P}_2} + X_1^* > t + x, t \leq X_{2,\mathcal{P}_2} \leq X_{1,\mathcal{P}_1})}{P(X_{1,\mathcal{P}_1} \geq t, X_{2,\mathcal{P}_2} \geq t)}, \end{aligned}$$

where X_i^* represents the excess lifetime of unit i after the other unit fails, $i = 1, 2$. Note that

$$
\begin{aligned}
&P(X_{1,\mathcal{P}_1}+X_2^*>t+x, t\le X_{1,\mathcal{P}_1}\le X_{2,\mathcal{P}_2})\\
&=\int_0^{\infty} P(X_{1,\mathcal{P}_1}+X_2^*>t+x, t\le X_{1,\mathcal{P}_1}\le X_{2,\mathcal{P}_2}|X_{1,\mathcal{P}_1}=y)dF_{X_{1,\mathcal{P}_1}}(y)\\
&=\int_t^{\infty} P(X_2^*>t+x-y|X_{2,\mathcal{P}_2}\ge y, X_{1,\mathcal{P}_1}=y)P(X_{2,\mathcal{P}_2}\ge y|X_{1,\mathcal{P}_1}=y)dF_{X_{1,\mathcal{P}_1}}(y)\\
&=\int_t^{\infty}\frac{\bar{F}_2(t+x-y+w(y))}{\bar{F}_2(w(y))}\bar{F}_2(g_2(1-\alpha)y)dF_1(g_1(\alpha)y),
\end{aligned}
$$

where, of course $t+x-y\ge 0$ and $t+x-y+w(y)\ge 0$. In a similar way, we can obtain

$$
P(X_{2,\mathcal{P}_2}+X_1^*>t+x, t\le X_{2,\mathcal{P}_2}\le X_{1,\mathcal{P}_1})=\int_t^{\infty}\frac{\bar{F}_1(t+x-y+v(y))}{\bar{F}_1(v(y))}\bar{F}_1(g_1(\alpha)y)dF_2(g_2(1-\alpha)y).
$$

Hence, the result follows immediately.

To establish some of results, we introduce the following lemma (Barlow and Proschan (1975)).

Lemma 6.2.9 *Let W be a Lebesgue-Stieltjes measure, not necessarily positive. If $\int_t^{+\infty} dW(x)\ge 0$ for all t, then $\int_t^{+\infty} h(x)dW(x)\ge 0$ for any non-negative and increasing function $h(x)$.*

Theorem 6.2.10 *Suppose $g_1(\alpha)y=v(y)$, $g_2(1-\alpha)y=w(y)$ for all $y\ge 0$, and $y-w(y)$ increases in $y\ge 0$. If $X_1\le_{hr} Y_1$ and $X_1\le_{rh} Y_1$, then, for any $t\ge 0$,*

$$
(T(X_{1,\mathcal{P}_1}, X_{2,\mathcal{P}_2})-t|\min(X_{1,\mathcal{P}_1}, X_{2,\mathcal{P}_2})\ge t)\le_{st}(T(Y_{1,\mathcal{P}_1}, X_{2,\mathcal{P}_2})-t|\min(Y_{1,\mathcal{P}_1}, X_{2,\mathcal{P}_2})\ge t),
$$

where $T(Y_{1,\mathcal{P}_1}, X_{2,\mathcal{P}_2})$ is the lifetime of the load-sharing system with units C_1 and C_2 having respective lifetime Y_1 and X_2 under full load, and $Y_{1,\mathcal{P}_1}$ is the lifetime of component C_1 under partial load condition.

Proof Let $R^t_{Y_{1,\mathcal{P}_1},X_{2,\mathcal{P}_2}}(x)$ be the reliability function of the random variable $(T(Y_{1,\mathcal{P}_1}, X_{2,\mathcal{P}_2})-t|\min(Y_{1,\mathcal{P}_1}, X_{2,\mathcal{P}_2})\ge t)$. Then, from Theorem 6.2.8, for any $x\ge 0, t\ge 0$,

$$
R^t_{Y_{1,\mathcal{P}_1},X_{2,\mathcal{P}_2}}(x)=\frac{\int_t^{\infty}\bar{G}_1(t+x-y+v(y))dF_2(g_2(1-\alpha)y)}{\bar{G}_1(g_1(\alpha)t)\bar{F}_2(g_2(1-\alpha)t)}+\frac{\int_t^{\infty}\bar{F}_2(t+x-y+w(y))dG_1(g_1(\alpha)y)}{\bar{G}_1(g_1(\alpha)t)\bar{F}_2(g_2(1-\alpha)t)},
$$

where G_1 and $\bar{G}_1$ are the distribution and survival functions of Y_1, respectively. It is observed that

$$
\begin{aligned}
R^t_{Y_{1,\mathcal{P}_1},X_{2,\mathcal{P}_2}}(x)-R^t_{X_{1,\mathcal{P}_1},X_{2,\mathcal{P}_2}}(x)&=A\int_t^{\infty}\bar{G}_1(t+x-y+v(y))\bar{F}_1(g_1(\alpha)t)dF_2(g_2(1-\alpha)y)\\
&\quad-A\int_t^{\infty}\bar{F}_1(t+x-y+v(y))\bar{G}_1(g_1(\alpha)t)dF_2(g_2(1-\alpha)y)\\
&\quad+A\bar{F}_1(g_1(\alpha)t)\int_t^{\infty}\bar{F}_2(t+x-y+w(y))dG_1(g_1(\alpha)y)\\
&\quad-A\bar{G}_1(g_1(\alpha)t)\int_t^{\infty}\bar{F}_2(t+x-y+w(y))dF_1(g_1(\alpha)y)
\end{aligned}
$$

$$= A \int_t^{\infty} [\bar{F}_1(g_1(\alpha)t)\bar{G}_1(t + x - y + v(y))$$
$$- \bar{G}_1(g_1(\alpha)t)\bar{F}_1(t + x - y + v(y))]dF_2(g_2(1 - \alpha)y)$$
$$+ A \int_t^{\infty} \bar{F}_2(t + x - y + w(y))$$
$$d_y[\bar{F}_1(g_1(\alpha)t)G_1(g_1(\alpha)y) - F_1(g_1(\alpha)y)\bar{G}_1(g_1(\alpha)t)], \tag{6.16}$$

where

$$A = \frac{1}{\bar{F}_1(g_1(\alpha)t)\bar{G}_1(g_1(\alpha)t)\bar{F}_2(g_2(1 - \alpha)t)}.$$

Clearly, $x \geq y - t \geq (1 - g_1(\alpha))(y - t)$, and $t + x - y + v(y) - g_1(\alpha)t = x - (1 - g_1(\alpha))(y - t) \geq 0$. Hence, the condition $X_1 \leq_{hr} Y_1$ implies that

$$\frac{\bar{G}_1(t + x - y + v(y))}{\bar{F}_1(t + x - y + v(y))} \geq \frac{\bar{G}_1(g_1(\alpha)t)}{\bar{F}_1(g_1(\alpha)t)};$$

that is,

$$\bar{F}_1(g_1(\alpha)t)\bar{G}_1(t + x - y + v(y)) \geq \bar{G}_1(g_1(\alpha)t)\bar{F}_1(t + x - y + v(y)),$$

which indicates that the first term of (6.16) is non-negative.

On the other hand, for any $t \geq 0$,

$$\int_t^{\infty} \bar{F}_2(t + x - y + w(y))d_y[\bar{F}_1(g_1(\alpha)t)G_1(g_1(\alpha)y) - F_1(g_1(\alpha)y)\bar{G}_1(g_1(\alpha)t)]$$

$$= \int_0^{\infty} I_{\{t \leq y \leq \infty\}}\bar{F}_2(t + x - y + w(y))d_y[\bar{F}_1(g_1(\alpha)t)G_1(g_1(\alpha)y) - F_1(g_1(\alpha)y)\bar{G}_1(g_1(\alpha)t)].$$

The condition $X_1 \leq_{rh} Y_1$ implies that

$$F_1(g_1(\alpha)y)(\bar{F}_1(g_1(\alpha)t)\frac{G_1(g_1(\alpha)y)}{F_1(g_1(\alpha)y)} - \bar{G}_1(g_1(\alpha)t))$$

is increasing in $y \geq 0$. Thus, for any $x \geq 0$,

$$\int_x^{\infty} d_y[\bar{F}_1(g_1(\alpha)t)G_1(g_1(\alpha)y) - F_1(g_1(\alpha)y)\bar{G}_1(g_1(\alpha)t)]$$

$$= \int_x^\infty d_y[F_1(g_1(\alpha)y)(\bar{F}_1(g_1(\alpha)t)\frac{G_1(g_1(\alpha)y)}{F_1(g_1(\alpha)y)} - \bar{G}_1(g_1(\alpha)t))]$$
$$\geq 0.$$

Moreover, $I_{\{t \leq y \leq \infty\}} \bar{F}_2(t + x - y + w(y))$ increases in $y \geq 0$. Therefore, by Lemma 6.2.9, the second term of (6.16) is non-negative.

Combing the above facts, the desired result is obtained.

Proceeding in a analogous manner, we can establish the following theorem.

Theorem 6.2.11 *Suppose $g_1(\alpha)y = w(y)$, $g_2(1-\alpha)y = v(y)$ for $y \geq 0$, and $y - v(y)$ is increasing in $y \geq 0$. If $X_2 \leq_{hr} Y_2$ and $X_2 \leq_{rh} Y_2$, then, for any $t \geq 0$,*

$$(T(X_{1,\mathcal{P}_1}, X_{2,\mathcal{P}_2}) - t \mid \min(X_{1,\mathcal{P}_1}, X_{2,\mathcal{P}_2}) \geq t) \leq_{st} (T(X_{1,\mathcal{P}_1}, Y_{2,\mathcal{P}_2}) - t \mid \min(X_{1,\mathcal{P}_1}, Y_{2,\mathcal{P}_2}) \geq t),$$

where $T(X_{1,\mathcal{P}_1}, Y_{2,\mathcal{P}_2})$ is the lifetime of the load-sharing system with two components C_1 and C_2 having respective lifetime X_1 and Y_2 under full load condition, and $Y_{2,\mathcal{P}_2}$ is the lifetime of component C_2 under partial load condition.

By combining Theorems 6.2.10 and 6.2.11, we can easily obtain the following theorem.

Theorem 6.2.12 *Suppose $g_1(\alpha)y = w(y)$, $g_2(1-\alpha)y = v(y)$ for all $y \geq 0$, and both $y - w(y)$ and $y - v(y)$ are increasing in $y \geq 0$. If $X_i \leq_{hr} Y_i$ and $X_i \leq_{rh} Y_i$, for $i = 1, 2$, then, for any $t \geq 0$,*

$$(T(X_{1,\mathcal{P}_1}, X_{2,\mathcal{P}_2}) - t \mid \min(X_{1,\mathcal{P}_1}, X_{2,\mathcal{P}_2}) \geq t) \leq_{st} (T(Y_{1,\mathcal{P}_1}, Y_{2,\mathcal{P}_2}) - t \mid \min(Y_{1,\mathcal{P}_1}, Y_{2,\mathcal{P}_2}) \geq t),$$

where $T(Y_{1,\mathcal{P}_1}, Y_{2,\mathcal{P}_2})$ is the lifetime of the load-sharing system with two components C_1 and C_2 having respective lifetime Y_1 and Y_2 under full load condition.

6.2.4 Optimal Allocation of a Load Standby in a Series System

In this section, we consider the problem of allocating a load standby component in a two-component series system by stochastically optimizing the lifetime of the system.

We denote the lifetime of a series system having two independent components C_1 and C_3 with corresponding lifetimes X_1 and X_3 under full load condition by $\min\{X_1, X_3\}$. In order to decrease the load of components C_1, C_3 or improve the system performance, we can allocate another component C_2 (independent of C_1 and C_3) with lifetime X_2 to components C_1 or C_3, and make component C_2 take partial load from components C_1 or C_3. Assume that X_1 and X_3 both have virtual age function $v(u)$ and have assignment function $g_1(\alpha)$. Also, assume that X_2 has virtual age function $w(u)$ and assignment function $g_2(1-\alpha)$. Denote by T_1 and T_2

the lifetimes of the systems when component C_2 is allocated to components C_1 and C_3, respectively.

Theorem 6.2.13 *Suppose* $g_1(\alpha)u \leq v(u)$ *for all* $u \geq 0$. *If* $X_1 \leq_{hr} X_3$, *then* $T_1 \geq_{st} T_2$.

Proof Let $R_1(x)$ and $R_2(x)$ be the reliability functions of T_1 and T_2, respectively. Then from (6.9), we have for any $x \geq 0$,

$$\begin{aligned} R_1(x) = &\ \bar{F}_2(g_2(1-\alpha)x)\bar{F}_1(g_1(\alpha)x)\bar{F}_3(x) \\ &+ \bar{F}_3(x)\int_0^x \frac{\bar{F}_1(v(u)-u+x)}{\bar{F}_1(v(u))}\bar{F}_1(g_1(\alpha)u)dF_2(g_2(1-\alpha)u) \\ &+ \bar{F}_3(x)\int_0^x \frac{\bar{F}_2(w(u)-u+x)}{\bar{F}_2(w(u))}\bar{F}_2(g_2(1-\alpha)u)dF_1(g_1(\alpha)u), \end{aligned}$$

$$\begin{aligned} R_2(x) = &\ \bar{F}_2(g_2(1-\alpha)x)\bar{F}_3(g_1(\alpha)x)\bar{F}_1(x) \\ &+ \bar{F}_1(x)\int_0^x \frac{\bar{F}_3(v(u)-u+x)}{\bar{F}_3(v(u))}\bar{F}_3(g_1(\alpha)u)dF_2(g_2(1-\alpha)u) \\ &+ \bar{F}_1(x)\int_0^x \frac{\bar{F}_2(w(u)-u+x)}{\bar{F}_2(w(u))}\bar{F}_2(g_2(1-\alpha)u)dF_3(g_1(\alpha)u). \end{aligned}$$

Hence,

$$\begin{aligned} & R_1(x) - R_2(x) \\ = &\ \bar{F}_2(g_2(1-\alpha)x) \times [\bar{F}_1(g_1(\alpha)x)\bar{F}_3(x) - \bar{F}_3(g_1(\alpha)x)\bar{F}_1(x)] \\ + &\int_0^x \left[\bar{F}_3(x)\frac{\bar{F}_1(v(u)-u+x)}{\bar{F}_1(v(u))}\bar{F}_1(g_1(\alpha)u) - \bar{F}_1(x)\frac{\bar{F}_3(v(u)-u+x)}{\bar{F}_3(v(u))}\bar{F}_3(g_1(\alpha)u)\right] dF_2(g_2(1-\alpha)u) \\ + &\int_0^x \frac{\bar{F}_2(w(u)-u+x)}{\bar{F}_2(w(u))}\bar{F}_2(g_2(1-\alpha)u)d[\bar{F}_3(x)F_1(g_1(\alpha)u) - \bar{F}_1(x)F_3(g_1(\alpha)u)]. \end{aligned} \tag{6.17}$$

The condition $X_1 \leq_{hr} X_3$ implies that the first term on the RHS of (6.17) is non-negative. Also, the second term on the RHS of (6.17) is non-negative due the fact that

$$\begin{aligned} \frac{\bar{F}_3(x)}{\bar{F}_1(x)} &\geq \frac{\bar{F}_3(v(u)-u+x)}{\bar{F}_1(v(u)-u+x)} \\ &\geq \frac{\bar{F}_3(v(u)-u+x)}{\bar{F}_1(v(u)-u+x)}\frac{\bar{F}_3(g_1(\alpha)u)\bar{F}_1(v(u))}{\bar{F}_1(g_1(\alpha)u)\bar{F}_3(v(u))}, \end{aligned}$$

since $X_1 \leq_{hr} X_3$ and $g_1(\alpha)u \leq v(u)$ implies that

$$\frac{\bar{F}_3(g_1(\alpha)u)\bar{F}_1(v(u))}{\bar{F}_1(g_1(\alpha)u)\bar{F}_3(v(u))} \leq 1.$$

Moreover, $X_1 \leq_{hr} X_3$ implies that $\bar{F}_3(g_1(\alpha)u_2)/\bar{F}_3(g_1(\alpha)u_1) \geq \bar{F}_1(g_1(\alpha)u_2)/\bar{F}_1(g_1(\alpha)u_1)$ for $u_2 \geq u_1 \geq 0$, and hence

$$\frac{\bar{F}_3(g_1(\alpha)u_2) - \bar{F}_3(g_1(\alpha)u_1)}{\bar{F}_3(g_1(\alpha)u_1)} \geq \frac{\bar{F}_1(g_1(\alpha)u_2) - \bar{F}_1(g_1(\alpha)u_1)}{\bar{F}_1(g_1(\alpha)u_1)},$$

which is equivalent to

$$\frac{\bar{F}_3(g_1(\alpha)u_2) - \bar{F}_3(g_1(\alpha)u_1)}{\bar{F}_1(g_1(\alpha)u_2) - \bar{F}_1(g_1(\alpha)u_1)} \leq \frac{\bar{F}_3(g_1(\alpha)u_1)}{\bar{F}_1(g_1(\alpha)u_1)}.$$

Note that for $0 \leq u_1 \leq x$ and $0 \leq g_1(\alpha) \leq 1$,

$$\frac{\bar{F}_3(g_1(\alpha)u_1)}{\bar{F}_1(g_1(\alpha)u_1)} \leq \frac{\bar{F}_3(x)}{\bar{F}_1(x)},$$

and so we have

$$\frac{\bar{F}_3(g_1(\alpha)u_2) - \bar{F}_3(g_1(\alpha)u_1)}{\bar{F}_1(g_1(\alpha)u_2) - \bar{F}_1(g_1(\alpha)u_1)} \leq \frac{\bar{F}_3(x)}{\bar{F}_1(x)};$$

that is,

$$\frac{F_3(g_1(\alpha)u_1) - F_3(g_1(\alpha)u_2)}{F_1(g_1(\alpha)u_1) - F_1(g_1(\alpha)u_2)} \leq \frac{\bar{F}_3(x)}{\bar{F}_1(x)}.$$

Therefore, for $x \geq u_2 \geq u_1 \geq 0$ and $0 \leq g_1(\alpha) \leq 1$,

$$\bar{F}_3(x)F_1(g_1(\alpha)u_2) - \bar{F}_1(x)F_3(g_1(\alpha)u_2) \geq \bar{F}_3(x)F_1(g_1(\alpha)u_1) - \bar{F}_1(x)F_3(g_1(\alpha)u_1);$$

that is, $\bar{F}_3(x)F_1(g_1(\alpha)u) - \bar{F}_1(x)F_3(g_1(\alpha)u)$ is increasing in $0 \leq u \leq x$. This shows that the third term on the RHS of (6.17) is non-negative. This completes the proof of the theorem.

The above theorem displays that for a series system with two independent components under full load condition, it is optimal to allocate one independent load standby to the stochastically weaker component.

6.3 Stochastic Comparisons of Standby Systems

In this section, we study the performance of the load-sharing parallel system and discuss some stochastic comparisons among various standby systems with two common components.

It is very easy to obtain that the cold standby system is better than the hot standby system in the sense of usual stochastic order, that is, $\max\{X_1, X_2\} \leq_{st} X_1 + X_2$.

In order to characterize the lifetime of the components through two regimes, the initial age of a component in the new regime should be defined when it is switched from the partial load to the full load. According to the description of the cumulative exposure model of Nelson (1990), Finkelstein (2007), we may suppose that, for any $u \geq 0$,

$$F_{1,\mathcal{P}_1}(u) = F_1(g_1(\alpha)u) = F_1(v(u)).$$

This assumption indicates that the virtual age $v(u)$ would produce as much population cumulative fraction of the component C_1 failure as the age u does under partial load. The above equality implies

$$F_1^{-1} F_{1,\mathcal{P}_1}(u) = g_1(\alpha)u = v(u).$$

In a similar way, the virtual age $w(u)$ for another component X_2 can be defined as well.

Theorem 6.3.1 *If* $g_1(\alpha)u \leq v(u)$ *for any* $u \geq 0$, *then* $T(X_{1,\mathcal{P}_1}, X_2) \geq_{st} \max\{X_1, X_2\}$.

Proof According to Eqs. (6.10) and (6.11) and by some combinations, for any $t \geq 0$,

$$R_{X_{1,\mathcal{P}_1},X_2}(t) - R_{\max\{X_1,X_2\}}(t) = \int_0^t [\frac{\bar{F}_1(t-u+v(u))}{\bar{F}_1(v(u))}\bar{F}_1(g_1(\alpha)u) - \bar{F}_1(t)]dF_2(u).$$

Noting $\bar{F}_1(t-u+v(u)) \geq \bar{F}_1(t)$ for $t \geq u \geq 0$, and $\bar{F}_1(g_1(\alpha)u)/\bar{F}_1(v(u)) \geq 1$ for the case of $g_1(\alpha)u \leq v(u)$. Also, $F_2(u)$ is increasing in $u \geq 0$. Therefore, $R_{X_{1,\mathcal{P}_1},X_2}(t) - R_{\max\{X_1,X_2\}}(t) \geq 0$.

The above result shows that the warm standby system is better than the hot standby system in the sense of usual stochastic order under the condition that, the virtual age $v(u)$ would produce at least population cumulative fraction of the component C_1 failure as the age u does under partial load condition.

The next example indicates that the condition $g_1(\alpha)u \leq v(u)$ of Theorem 6.3.1 is not deleted.

Example 6.3.2 Assume that both X_1 and X_2 are distributed as uniform distribution $U(0, 1)$, and $g_1(\alpha)u = 0.5u$, $v(u) = 0.2u$. Then, for any $t \geq 0$, it can be easily obtained

$$R_{X_{1,\mathcal{P}_1},X_2}(t) - R_{\max\{X_1,X_2\}}(t) = \frac{5}{8}\int_0^t [0.5t + 0.4u - 0.5]du = t(0.7t - 0.5),$$

which is negative for $t \in (0, \frac{5}{7})$. This means the result in Theorem 6.1 does not hold.

Recall that X is said to be NBU if $(X - t|X > t) \leq_{st} X$ for any $t \geq 0$. That is, X is NBU if and only if $\bar{F}(x+t) \leq \bar{F}(t)\bar{F}(x)$ for $x \geq 0$ and $t \geq 0$; See Barlow and Proschan (1975).

Theorem 6.3.3 *(a) If $g_1(\alpha)u \geq v(u)$ for any $u \geq 0$, then $T(X_{1,\mathcal{P}_1}, X_2) \leq_{st} X_1 + X_2$; (b) If X_1 is NBU, then $T(X_{1,\mathcal{P}_1}, X_2) \leq_{st} X_1 + X_2$.*

Proof According to Eqs. (6.1) and (6.12), for any $t \geq 0$,

$$R_{X_1+X_2}(t) - R_{X_{1,\mathcal{P}_1},X_2}(t) = \int_0^t [\bar{F}_1(t-u) - \frac{\bar{F}_1(t-u+v(u))}{\bar{F}_1(v(u))}\bar{F}_1(g_1(\alpha)u)]dF_2(u).$$

No matter which of conditions (a) and (b) holds, we can easily prove that the term in integral is nonnegative. Therefore, the results follow immediately.

The above results indicate that the cold standby system is better than the warm standby system under some specific conditions.

Example 6.3.4 below shows that the condition of Theorem 6.3.3 must be satisfied.

Example 6.3.4 Assume that the survival function of X_1 is $\bar{F}_1(t) = 1/(1+t)$ for $t \geq 0$. In (a) of the Theorem 6.3.3, let $g_1(\alpha)u = u/5$, $v(u) = u/2$ for $u \geq 0$. Obviously, the condition $g_1(\alpha)u \geq v(u)$ does not hold. It is noted,

$$R_{X_1+X_2}(t) - R_{X_{1,\mathcal{P}_1},X_2}(t) = \int_0^t [\frac{1}{1+t-u} - \frac{1}{1+t-\frac{1}{2}u} \cdot \frac{1+\frac{1}{2}u}{1+\frac{1}{5}u}]d\bar{F}_2(u) \leq 0,$$

which means $T(X_{1,\mathcal{P}_1}, X_2) \nleq_{st} X_1 + X_2$.

In (b) of Theorem 6.3.3, let $\bar{F}_1(t) = 1/(1+t)$. Obviously, X_1 is not NBU. And

$$R_{X_1+X_2}(t) - R_{X_{1,\mathcal{P}_1},X_2}(t) = \int_0^t [\frac{1}{1+t-u} - \frac{1}{1+t-u+v(u)} \cdot \frac{1+v(u)}{1+g_1(\alpha)u}]dF_2(u).$$

If let $v(u) = 0.8u$, $g_1(\alpha)u = 0.1u$, then

$$\begin{aligned} \frac{1}{1+t-u} - \frac{1}{1+t-u+v(u)} \cdot \frac{1+v(u)}{1+g_1(\alpha)u} &= \frac{1}{1+t-u} - \frac{1}{1+t-0.2u} \cdot \frac{1+0.8u}{1+0.1u} \\ &= \frac{0.78u^2 - 0.7ut - 0.8u}{(1+t-u)(1+t-0.2u)(1+0.1u)} \\ &\leq \frac{0.78u^2 - 0.7u^2 - 0.8u}{(1+t-u)(1+t-0.2u)(1+0.1u)} \\ &= \frac{0.08u^2 - 0.8u}{(1+t-u)(1+t-0.2u)(1+0.1u)}. \end{aligned}$$

It is clear, $R_{X_1+X_2}(t) - R_{X_{1,\mathcal{P}_1},X_2}(t) \leq 0$, for $u \in [0, 10]$. It implies $T(X_{1,\mathcal{P}_1}, X_2) \nleq_{st} X_1 + X_2$.

The following theorem proves that the load standby system is better than the warm standby system in the sense of usual stochastic order.

Theorem 6.3.5 *Suppose $u - v(u)$ increases with respect to u, and $g_1(\alpha)u \leq v(u)$, $g_2(1-\alpha)u \leq w(u)$. Then, $T(X_{1,\mathcal{P}_1}, X_{2,\mathcal{P}_2}) \geq_{st} T(X_{1,\mathcal{P}_1}, X_2)$.*

Proof It is noted, for $t \geq 0$,

$$R_{X_{1,\mathcal{P}_1},X_2}(t) = \bar{F}_2(t)\bar{F}_1(g_1(\alpha)t) + \int_0^t \bar{F}_2(t)dF_1(g_1(\alpha)u) + \int_0^t \frac{\bar{F}_1(t-u+v(u))}{\bar{F}_1(g_1(\alpha)u)}\bar{F}_1(g_1(\alpha)u)\,dF_2(u).$$

According to Eqs. (6.9) and (6.10), for any $t \geq 0$,

$$\begin{aligned} R_{X_{1,\mathcal{P}_1},X_{2,\mathcal{P}_2}}(t) - R_{X_{1,\mathcal{P}_1},X_2}(t) &= [\bar{F}_1(g_1(\alpha)t)\bar{F}_2(g_2(1-\alpha)t) - \bar{F}_1(g_1(\alpha)t)\bar{F}_2(t)] \\ &+ \int_0^t \frac{\bar{F}_1(v(u)-u+t)}{\bar{F}_1(v(u))}\bar{F}_1(g_1(\alpha)u)d[F_2(g_2(1-\alpha)u) - F_2(u)] \\ &+ \int_0^t [\frac{\bar{F}_2(w(u)-u+t)}{\bar{F}_2(w(u))}\bar{F}_2(g_2(1-\alpha)u) - \bar{F}_2(t)]dF_1(g_1(\alpha)u) \\ &= \int_0^t [F_2(u) - F_2(g_2(1-\alpha)u)]d\frac{\bar{F}_1(v(u)-u+t)}{\bar{F}_1(v(u))}\bar{F}_1(g_1(\alpha)u) \\ &+ \int_0^t [\frac{\bar{F}_2(w(u)-u+t)}{\bar{F}_2(w(u))}\bar{F}_2(g_2(1-\alpha)u) - \bar{F}_2(t)]dF_1(g_1(\alpha)u). \end{aligned}$$

The first term in the last equality above is nonnegative due to the fact that $\bar{F}_1(v(u)-u+t)\bar{F}_1(g_1(\alpha)u)/\bar{F}_1(v(u))$ is increasing in $u \geq 0$ under the condition that, $u - v(u)$ is increasing in u and $g_1(\alpha)u \leq v(u)$. Also, $g_2(1-\alpha)u \leq w(u)$ implies that $\bar{F}_2(g_2(1-\alpha)u) \geq \bar{F}_2(w(u))$, and hence the integraled function in the second integral is nonnegative. Thus the second term of the above equality is nonnegative. Combining above facts, the result follows immediately.

The following example shows that the conditions $g_1(\alpha)u \leq v(u)$ and $g_2(1-\alpha)u \leq w(u)$ in Theorem 6.3.5, are not deleted.

Example 6.3.6 Assume that X_1, X_2 are exponential distributions with parameters $\lambda_1, \lambda_2 = 1$, respectively. Let $g_1(\alpha)u = 2u/5 > v(u) = 3u/10$ and $g_2(1-\alpha)u = u/2 \geq w(u) = u/5$, then

$$R_{X_{1,\mathcal{P}_1},X_{2,\mathcal{P}_2}}(t) - R_{X_{1,\mathcal{P}_1},X_2}(t) = \frac{19}{10}e^{-\frac{9}{10}t} - \frac{9}{10}e^{-t} - \frac{7}{2}e^{-t} + \frac{5}{2}e^{-\frac{7}{5}t}.$$

As we can see from the Fig. 6.2, $R_{X_{1,\mathcal{P}_1},X_{2,\mathcal{P}_2}}(t) - R_{X_{1,\mathcal{P}_1},X_2}(t) \leq 0$, for $0 \leq t \leq 8$, while $R_{X_{1,\mathcal{P}_1},X_{2,\mathcal{P}_2}}(t) - R_{X_{1,\mathcal{P}_1},X_2}(t) \geq 0$, for $8 < t < \infty$.

Theorem 6.3.7 *Assume that the lifetimes X_1 and X_2 of the components C_1 and C_2 of load-sharing system are distributed as exponential with failure rates λ_1 and λ_2, respectively. If $g_1(\alpha) + g_2(1-\alpha) = 1$ for any $0 \leq \alpha \leq 1$, then,*

$$T(X_{1,\mathcal{P}_1}, X_{2,\mathcal{P}_2}) \geq_{st} \max\{X_1, X_2\}.$$

Proof First, we prove the result for the case of $\lambda_1 \neq \lambda_2$. For any $x \geq 0$,

$$R_{\max\{X_1,X_2\}}(x) = e^{-\lambda_1 x} + e^{-\lambda_2 x} - e^{-(\lambda_1+\lambda_2)x}.$$

It can be computed from (6.2) that

$$\begin{aligned}
R_{X_{1,\mathcal{P}_1},X_{2,\mathcal{P}_2}}(x) &= e^{-[\lambda_1 g_1(\alpha)+\lambda_2 g_2(1-\alpha)]x} + \frac{\lambda_2}{\lambda_1-\lambda_2} \times [e^{[\lambda_1 g_2(1-\alpha)-\lambda_2 g_2(1-\alpha)]x} - 1]e^{-\lambda_1 x} \\
&\quad + \frac{\lambda_1}{\lambda_2-\lambda_1} \times [e^{[\lambda_2 g_1(\alpha)-\lambda_1 g_1(\alpha)]x} - 1]e^{-\lambda_2 x} \\
&= \frac{\lambda_1}{\lambda_1-\lambda_2} e^{-\lambda_2 x} - \frac{\lambda_2}{\lambda_1-\lambda_2} e^{-\lambda_1 x}.
\end{aligned}$$

Therefore, in this case,

$$\begin{aligned}
R_{X_{1,\mathcal{P}_1},X_{2,\mathcal{P}_2}}(x) - R_{\max\{X_1,X_2\}}(x) &= \frac{\lambda_2}{\lambda_1-\lambda_2} e^{-\lambda_2 x} - \frac{\lambda_1}{\lambda_1-\lambda_2} e^{-\lambda_1 x} + e^{-(\lambda_1+\lambda_2)x} \\
&= e^{-(\lambda_1+\lambda_2)x} \times [\frac{\lambda_2}{\lambda_1-\lambda_2} e^{\lambda_1 x} - \frac{\lambda_1}{\lambda_1-\lambda_2} e^{\lambda_2 x} + 1] \\
&= e^{-(\lambda_1+\lambda_2)x} h(x),
\end{aligned}$$

where $h(x) = \frac{\lambda_2}{\lambda_1-\lambda_2} e^{\lambda_1 x} - \frac{\lambda_1}{\lambda_1-\lambda_2} e^{\lambda_2 x} + 1$. It is noted that $h'(x) = \frac{\lambda_1\lambda_2}{\lambda_1-\lambda_2}[e^{\lambda_1 x} - e^{\lambda_2 x}]$ is always nonnegative for any $x \geq 0$. Hence, $h(x)$ is increasing in $x \geq 0$. Moreover, $h(0) = 0$, and so $h(x) \geq 0$ for any $x \geq 0$. That is, $R_{X_{1,\mathcal{P}_1},X_{2,\mathcal{P}_2}}(x) - R_{\max\{X_1,X_2\}}(x) \geq 0$ for any $x \geq 0$.

On the other hand, for the case when $\lambda_1 = \lambda_2$, $R_{X_{1,\mathcal{P}_1},X_{2,\mathcal{P}_2}}(x) = (\lambda_1 x + 1)e^{-\lambda_1 x}$ and $R_{\max\{X_1,X_2\}}(x) = (2 - e^{-\lambda_1 x})e^{-\lambda_1 x}$. It can be easily shown that $R_{X_{1,\mathcal{P}_1},X_{2,\mathcal{P}_2}}(x) \geq R_{\max\{X_1,X_2\}}(x)$, for any $x \geq 0$. Thus, the result follows.

The above theorem shows the load-sharing parallel system is better than the usual parallel system in the sense of usual stochastic order given that the sum of allocation proportion function of load is one.

One may ask whether the cold standby system is better than the load-standby system or not? The following example gives a negative answer.

Example 6.3.8 Assume X_1 and X_2 are distributed uniformly in intervals $[0, b]$ $(0 < b)$ and $[0, d]$ $(0 < d)$, respectively. That is, $\bar{F}_1(t) = (b-t)/b$ when $t \in (0, b)$, and $\bar{F}_2(t) = (d-t)/d$ when $t \in (0, d)$. Suppose $g_1(\alpha)u = v(u)$ and $g_2(1-\alpha)u = w(u)$ for any $u \geq 0$. Then, by Eqs. (6.2) and (6.4), for any $t \geq 0$,

$$R_{X_{1,\mathcal{P}_1},X_{2,\mathcal{P}_2}}(t) = \frac{1}{bd}[bd - \int_0^t (w(u) + v(u))du],$$

and

$$R_{X_1+X_2}(t) = \frac{1}{bd}[bd - \frac{t^2}{2}].$$

If $w(u) + v(u) = u$, then $\int_0^t (w(u) + v(u))du = \frac{t^2}{2}$, which in turn implies $R_{X_{1,\mathcal{P}_1}, X_{2,\mathcal{P}_2}}(t) = R_{X_1+X_2}(t)$ for any $t \geq 0$. If $w(u) + v(u) > u$, then $\int_0^t (w(u) + v(u))du > \frac{t^2}{2}$, which means that $R_{X_{\mathcal{P}_1}, X_{\mathcal{P}_2}}(t) < R_{X_1+X_2}(t)$. If $w(u) + v(u) < u$, then $\int_0^t (w(u) + v(u))du < \frac{t^2}{2}$, which implies that $R_{X_{\mathcal{P}_1}, X_{\mathcal{P}_2}}(t) > R_{X_1+X_2}(t)$.

Theorem 6.3.9 *(a) If $g_1(\alpha)u \leq v(u)$ for any $u \geq 0$, then $T(X_{1,\mathcal{P}_1}, X_{2,\mathcal{P}_2}) \geq_{st} X_1$;*
(b) If $g_2(1-\alpha)u \leq w(u)$ for any $u \geq 0$, then $T(X_{1,\mathcal{P}_1}, X_{2,\mathcal{P}_2}) \geq_{st} X_2$.

Proof (a) Noting that $\bar{F}_1(x) = \bar{F}_1(x)\bar{F}_2(g_2(1-\alpha)x) + \bar{F}_1(x)F_2(g_2(1-\alpha)x)$, then, for any $x \geq 0$,

$$\begin{aligned} R_{X_{1,\mathcal{P}_1}, X_{2,\mathcal{P}_2}}(x) - \bar{F}_1(x) &= \bar{F}_1(g_1(\alpha)x)\bar{F}_2(g_2(1-\alpha)x) - \bar{F}_1(x)\bar{F}_2(g_2(1-\alpha)x) \\ &+ \int_0^x \frac{\bar{F}_1(v(u)-u+x)}{\bar{F}_1(v(u))}\bar{F}_1(g_1(\alpha)u)dF_2(g_2(1-\alpha)u) - \bar{F}_1(x)F_2(g_2(1-\alpha)x) \\ &+ \int_0^x \frac{\bar{F}_2(w(u)-u+x)}{\bar{F}_2(w(u))}\bar{F}_2(g_2(1-\alpha)u)dF_1(g_1(\alpha)u) \\ &= \bar{F}_1(g_1(\alpha)x)\bar{F}_2(g_2(1-\alpha)x) - \bar{F}_1(x)\bar{F}_2(g_2(1-\alpha)x) \\ &+ \int_0^x [\frac{\bar{F}_1(v(u)-u+x)}{\bar{F}_1(v(u))}\bar{F}_1(g_1(\alpha)u) - \bar{F}_1(x)]dF_2(g_2(1-\alpha)u) \\ &+ \int_0^x \frac{\bar{F}_2(w(u)-u+x)}{\bar{F}_2(w(u))}\bar{F}_2(g_2(1-\alpha)u)dF_1(g_1(\alpha)u). \end{aligned} \tag{6.18}$$

Clearly, $\bar{F}_1(g_1(\alpha)x)\bar{F}_2(g_2(1-\alpha)x) \geq \bar{F}_1(x)\bar{F}_2(g_2(1-\alpha)x)$. So we only show that the second term of (6.18) is nonnegative. Since $g_1(\alpha)u \leq v(u)$,

$$\bar{F}_1(v(u)-u+x)\frac{\bar{F}_1(g_1(\alpha)u)}{\bar{F}_1(v(u))} \geq \bar{F}_1(v(u)-u+x) \geq \bar{F}_1(x),$$

for any $0 \leq u \leq x$. Furthermore, $F_2(g_2(1-\alpha)u)$ is increasing in $u \geq 0$. Thus the second term is nonnegative and hence the result of (a) follows immediately.

The proof of (b) is very similar to that of (a) and hence is omitted for the sake of brevity.

For the (a) of Theorem 6.3.9, if let $g_1(\alpha)u = 2u/5 > v(u) = 3u/10$ in Example 6.3.6, then the conclusion does not hold. So the condition $g_1(\alpha)u \leq v(u)$ must be needed.

Remark 6.3.10 From Theorem 6.3.9, in the case when the age u produces at most population cumulative fraction of units failing under partial load as the virtual age functions does, that is, $v(u) \geq g_1(\alpha)u$ and $w(u) \geq g_2(1-\alpha)u$ for all $u \geq 0$, then the load-sharing system with two units would be better than any one of two components in the usual stochastic order.

6.4 Parameter Estimation Under Load-Sharing Systems

In the literature, most of the work has focused on the characterization of system reliability and by assuming that the parameters of the lifetime distribution are known. This assumption is unrealistic since, in practice, the parameters of the lifetime distribution of components are usually unknown. So, it is important to consider the estimation of the model parameters.

6.4.1 Introduction and Preliminaries

In practice, redundancy techniques are usually used to improve system reliability. In most cases, when analyzing redundancy, independence of the components within the system is assumed. However, many systems are load-sharing, so the assumption of independence is't valid any more. In a load-sharing system, if a component fails, the same workload has to be shared by the remaining components of the system, resulting in an increased load shared by each surviving component. Originally, Daniels (1945) and Rosen (1964) used the load-sharing system models to study the reliability of composite materials. They regarded a bundle of fibers as a parallel system subject to a constant load. After a fiber breaks yarn bundles tend to spread the strength load on the remaining operating fibers. Gross et al. (1971) observed that two organ subsystems in human body typically show this pattern.

Load-sharing models have a wide range of engineering applications. They are extensively used in the textile engineering, material science and testing (fatigue and crack growth) by Carlson and Kardomateas (1996), mechanical engineering, and civil and structural engineering (welded joint on large support structures) disciplines by Kvam and Pena (2005). Further, load-sharing models have interesting applications in nuclear reactor safety, software reliability by Jelinski and Moranda (1972), distributed computing by Wang and Morris (1985), combat modeling by Kvam and Day (2001), modeling the incubation period for the human immunodeficiency virus (HIV) by Jewell and Kalbfleisch (1996), and condition-based maintenance by Amari et al. (2006), etc.

Estimating load-sharing parameters is becoming an important topic in recent years. Kim and Kvam (2004) considered a k-component parallel system. They found the ML (maximum likelihood) estimators of the failure rate parameters of component lifetimes. Also, a likelihood ratio test for testing equality of failure rates against the alternative that they are monotone was derived there. Singh et al. (2006) further modified the model by using a constant hazard at the start and then a linearly increasing hazard rate after failures of a certain number of components. Kvam and Pena (2005) discussed load sharing models which are special cases of dynamic models where performance of the system changes as components fail or their performance deteriorates. Also, they obtained the estimator of the component baseline cumulative hazard function and discussed its asymptotic distribution. McCool (2006) modelled the time to failure as a two-parameters Weibull distribution and tested the hypothesis that the failure of the first component in a parallel system shortens the life of the remaining components of the same system. Singh et al. (2009) have also researched the k-components load-sharing parallel system model with discrete set up. Park (2010) investigated the parameter estimation for such systems when the lifetime distributions of components are exponential or Weibull, and the closed-form ML and best unbiased (BU) estimators were obtained under equal load-sharing rule in the case of exponential distribution of component lifetimes. Singh and Gupta (2012) dealed with the classical and Bayesian estimates of the parameters of a k-components load-sharing parallel system model in which each component lifetime follows Lindley distribution.

6.4.2 Parameter Estimation Under Exponential Distribution

For convenience, we assume that the component lifetimes for n load-sharing systems are observable, and $g_1(\alpha) = g_2(1-\alpha) = g(\alpha)$ for any $\alpha \geq 0$. Suppose the random variables X_{i1} and X_{i2} represent the lifetimes of the two components under full load condition in the ith ($i = 1, \ldots, n$) parallel system, and that the spacings T_{ij} are the times between the jth and $(j-1)$th failures for the ith system, for $j = 1, 2$.

First, assume that the two components are identically distributed as exponential with common parameter λ. It can then be shown that the likelihood function for the ith system is given by

$$L_i(\alpha, \lambda | t_{i1}, t_{i2}) = 2\lambda^2 g(\alpha) \exp\{-\lambda(2g(\alpha)t_{i1} + t_{i2})\},$$

and so the likelihood function for the sample of n systems is

$$L(\alpha, \lambda | \mathbf{T}) = 2^n \lambda^{2n} g^n(\alpha) \exp\{-\lambda \sum_{i=1}^{n} (2g(\alpha)t_{i1} + t_{i2})\},$$

where $\mathbf{T} = \{t_{ij};\ 1 \leq i \leq n,\ j = 1, 2\}$. Then, we have the log-likelihood function to be

$$\ln L(\alpha, \lambda | \mathbf{T}) = n \ln 2 + 2n \ln \lambda + n \ln g(\alpha) - \lambda \sum_{i=1}^{n} (2g(\alpha)t_{i1} + t_{i2}).$$

The corresponding likelihood equations are

$$\frac{\partial lnL}{\partial \lambda} = \frac{2n}{\lambda} - \sum_{i=1}^{n} (2g(\alpha)t_{i1} + t_{i2})$$

and

$$\frac{\partial lnL}{\partial\alpha}=\frac{ng'(\alpha)}{g(\alpha)}-\lambda\sum_{i=1}^{n}2g'(\alpha)t_{i1},$$

provided the function g is differentiable. Solving the above equations, we obtain the maximum likelihood estimates of the parameters λ and α as

$$\hat{\lambda}=\frac{n}{\sum_{i=1}^{n}t_{i2}},$$

$$\hat{\alpha}=g^{-1}\left(\frac{\sum_{i=1}^{n}t_{i2}}{2\sum_{i=1}^{n}t_{i1}}\right).$$

If we let $\bar{T}_2=\frac{1}{n}\sum_{i=1}^{n}t_{i2}$ and $\bar{T}_1=\frac{1}{n}\sum_{i=1}^{n}t_{i1}$, then $\hat{\lambda}=\frac{1}{\bar{T}_2}$ and $\hat{\alpha}=g^{-1}(\frac{\bar{T}_2}{2\bar{T}_1})$.

Furthermore, the corresponding Fisher information matrix is

$$\mathbf{I}=\frac{n}{\lambda}\begin{pmatrix}\frac{2}{\lambda} & \frac{g'(\alpha)}{g(\alpha)}\\ \frac{g'(\alpha)}{g(\alpha)} & \lambda\left(\frac{g'(\alpha)}{g(\alpha)}\right)^2\end{pmatrix}.$$

Thus, the variance-covariance matrix is

$$\Sigma=I^{-1}=\begin{pmatrix}\frac{\lambda^2}{n} & -\frac{\lambda g(\alpha)}{ng'(\alpha)}\\ -\frac{\lambda g(\alpha)}{ng'(\alpha)} & \frac{2}{n}\left(\frac{g(\alpha)}{g'(\alpha)}\right)^2\end{pmatrix}.$$

Let $\Sigma=\frac{1}{n}\Sigma^*$. Then, asymptotically, we have

$$\sqrt{n}(\widehat{\lambda}-\lambda,\widehat{\alpha}-\alpha)\longrightarrow N(\mathbf{0},\Sigma^*).$$

Next, assume that the two components are identically distributed as two-parameter Weibull with scale parameter $\lambda>0$ and shape parameter $k>0$, that is, for $x>0$, with density function

$$f(x;\lambda,k)=\frac{k}{\lambda}\left(\frac{x}{\lambda}\right)^{k-1}e^{-(\frac{x}{\lambda})^k}.$$

After some algebraic computations, we have the likelihood function for the sample of n systems as

$$L(g(\alpha),\lambda,k|\mathbf{T})=2^nk^{2n}g^{nk}(a)\lambda^{-2nk}\prod_{i=1}^{n}(t_{i1}t_{i2})^{k-1}\times e^{-\lambda^{-k}[2g^k(a)\sum_{i=1}^{n}t_{i1}^k+\sum_{i=1}^{n}t_{i2}^k]},$$

where $\mathbf{T}=\{t_{ij};1\le i\le n,j=1,2\}$. Then, we have the log-likelihood function to be

$$\ln L(\alpha, \lambda, k|\mathbf{T}) = n\ln 2 + 2n\ln k + nk\ln(g(a)) - 2nk\ln\lambda + (k-1)\sum_{i=1}^{n}\ln(t_{i1}t_{i2})$$

$$-2\lambda^{-k}g^{k}(\alpha)\sum_{i=1}^{n}t_{i1}^{k} - \lambda^{-k}\sum_{i=1}^{n}t_{i2}^{k}.$$

The corresponding likelihood equations are

$$\frac{\partial lnL}{\partial\lambda} = \frac{-2nk}{\lambda} + 2k\lambda^{-k-1}g^{k}(\alpha)\sum_{i=1}^{n}t_{i1}^{k} + k\lambda^{-k-1}\sum_{i=1}^{n}t_{i2}^{k} = 0,$$

$$\frac{\partial lnL}{\partial\alpha} = \frac{kng'(\alpha)}{g(\alpha)} - 2k\lambda^{-k}g^{-k-1}(\alpha)g'(\alpha)\sum_{i=1}^{n}t_{i1}^{k} = 0,$$

$$\frac{\partial lnL}{\partial k} = \frac{2n}{k} + n\ln g(\alpha) - 2n\ln\lambda + \sum_{i=1}^{n}\ln(t_{i1}t_{i2}) + 2\lambda^{-k}g^{k}(\alpha)\ln(\lambda)\sum_{i=1}^{n}t_{i1}^{k} + \lambda^{-k}\ln(\lambda)\sum_{i=1}^{n}t_{i2}^{k}$$

$$-2\lambda^{-k}g^{k}(\alpha)\ln(g(\alpha))\sum_{i=1}^{n}t_{i1}^{k} - \sum_{i=1}^{n}(t_{i2}^{k}\ln t_{i2}) - 2\lambda^{-k}g^{k}(\alpha)\sum_{i=1}^{n}(t_{i1}^{k}\ln t_{i1}) = 0.$$

Then, we obtain

$$\lambda^{k} = \frac{1}{n}\sum_{i=1}^{n}t_{i2}^{k},$$

which yields

$$\alpha = g^{-1}\left(\sqrt[k]{\frac{\sum_{i=1}^{n}t_{i2}^{k}}{2\sum_{i=1}^{n}t_{i1}^{k}}}\right).$$

Therefore, we obtain

$$\frac{\sum_{i=1}^{n}(t_{i1}^{k}\ln t_{i1})}{\sum_{i=1}^{n}t_{i1}^{k}} + \frac{\sum_{i=1}^{n}(t_{i2}^{k}\ln t_{i2})}{\sum_{i=1}^{n}t_{i2}^{k}} = \frac{2}{k} + \frac{1}{k}\ln\left(\frac{n}{2\sum_{i=1}^{n}t_{i2}^{k}}\right) + \frac{\ln(t_{i1}t_{i2})}{n} + \ln\left(\frac{2\sum_{i=1}^{n}t_{i1}^{k}}{n}\right).$$

The above equation is a nonlinear equation in parameter k. By using some iterative methods such as the Gauss-Seidel method (see Ortega and Rheinboldt 1970), the MLE of k can be obtained. Then, the MLEs of the parameters $g(\alpha)$ and λ can be obtained.

It is quite complicated to obtain the Fisher information in this case, and so we just present the second derivatives of the log-likelihood function with respect to the parameters here.

$$\frac{\partial^2 lnL}{\partial\lambda^2} = \frac{2nk}{\lambda^2} - 2k(k+1)\lambda^{-k-2}g^k(\alpha)\sum_{i=1}^{n} t_{i1}^k - k(k+1)\lambda^{-k-2}\sum_{i=1}^{n} t_{i2}^k$$

$$\frac{\partial^2 lnL}{\partial\alpha^2} = \frac{kn(g''(\alpha)g(\alpha) - (g'(\alpha))^2)}{g^2(\alpha)} + 2k(k+1)\,\lambda^{-k}g^{-k-2}(\alpha)(g'(\alpha))^2\sum_{i=1}^{n} t_{i1}^k$$
$$- 2k\lambda^{-k}g^{-k-1}(\alpha)g''(\alpha)\sum_{i=1}^{n} t_{i1}^k,$$

$$\frac{\partial^2 lnL}{\partial k^2} = \frac{2n}{k^2} + 4\lambda^{-k}g^k(\alpha)\ln(\lambda)\sum_{i=1}^{n}(t_{i1}^k \ln t_{i1}) + 4\lambda^{-k}g^k(\alpha)\ln(g\alpha)\sum_{i=1}^{n}(t_{i1}^k \ln t_{i1}),$$
$$+ 4\lambda^{-k}g^k(\alpha)\ln(g(\alpha))\ln(\lambda)\sum_{i=1}^{n} t_{i1}^k - 2\ln^2(\lambda)\lambda^{-k}g^k(\alpha)\sum_{i=1}^{n} t_{i1}^k$$
$$- 2\ln^2(g(\alpha))\lambda^{-k}g^k(\alpha)\sum_{i=1}^{n}(t_{i1}^k - 2\lambda^{-k}g^k(\alpha)\sum_{i=1}^{n}(t_{i1}^k \ln^2 t_{i1})$$
$$- \lambda^{-k}\sum_{i=1}^{n}(t_{i2}^k \ln^2 t_{i2}) + 2\sum_{i=1}^{n}(t_{i2}^k \ln t_{i2}) - \lambda^{-k}\ln^2(\lambda)\sum_{i=1}^{n} t_{i2}^k,$$

$$\frac{\partial^2 lnL}{\partial\lambda\partial\alpha} = -2k^2\lambda^{-k-1}g^{k-1}(\alpha)g'(\alpha)\sum_{i=1}^{n} t_{i1}^k,$$

$$\frac{\partial^2 lnL}{\partial k\partial\alpha} = \frac{ng'(\alpha)}{g(\alpha)} - 2\lambda^{-k}g^{-k-1}(\alpha)g'(\alpha)\sum_{i=1}^{n} t_{i1}^k + 2\lambda^{-k}g^{-k-1}(\alpha)g'(\alpha)\ln(\lambda)\sum_{i=1}^{n} t_{i1}^k$$
$$- 2k\lambda^{-k}g^{-k-1}(\alpha)g'(\alpha)\ln(g\alpha)\sum_{i=1}^{n} t_{i1}^k - 2k\lambda^{-k}g^{-k-1}(\alpha)g'(\alpha)\sum_{i=1}^{n}(t_{i1}^k \ln t_{i1}),$$

$$\frac{\partial^2 lnL}{\partial k\partial\lambda} = 2\lambda^{-k-1}g^k(\alpha)\sum_{i=1}^{n} t_{i1}^k - 2k\lambda^{-k-1}g^k(\alpha)\ln(\lambda)\sum_{i=1}^{n} t_{i1}^k + 2k\lambda^{-k-1}g^k(\alpha)\ln(g\alpha)\sum_{i=1}^{n} t_{i1}^k$$
$$+ k\lambda^{-k-1}\sum_{i=1}^{n}(t_{i2}^k \ln t_{i2}) + \frac{-2n}{\lambda} + 2k\lambda^{-k-1}g^k(\alpha)\ln(\lambda)\sum_{i=1}^{n} t_{i1}^k + 2k\lambda^{-k-1}g^k(\alpha)\sum_{i=1}^{n}(t_{i1}^k \ln t_{i1})$$
$$+ \lambda^{-k-1}\sum_{i=1}^{n} t_{i2}^k - k\lambda^{-k-1}\ln(\lambda)\sum_{i=1}^{n} t_{i1}^k.$$

6.4.3 Parameter Estimation Under Exponential Pareto Distribution

The section deals with the parameter estimation for the load-sharing parallel system when the underlying distribution of component lifetime is EP (Exponential Pareto). The appropriate iterative method is used to get the parameter estimators. And the asymptotic distribution of the proposed ML (maximum likelihood) estimator is presented. For the model simplified, the ML and BAU (best asymptotic unbiased) estimators of parameters are obtained. Moreover, the confidence intervals for the parameters are constructed by using parameter Bootstrap technique. In the end, some appropriate simulation methods and an example are used to substantiate the proposed methods.

6.4.3.1 Introduction and Preliminaries

Gupta et al. (1998) introduced the EP (exponential Pareto) distribution as a lifetime model, with the reliability function

$$\bar{F}(x) = 1 - [1 - (1+x)^{-\alpha}]^{\theta}, x > 0, \alpha > 0, \theta > 0,$$

and probability density function

$$f(x) = \alpha\theta[1 - (1+x)^{-\alpha}]^{\theta-1}(1+x)^{-(\alpha+1)}, x > 0, \alpha > 0, \theta > 0,$$

where α and θ are two positive shape parameters. When $\theta = 1$, it was called as standard Pareto distribution of the second kind (see Johnson et al. 1994).

Here, we give the application of EP distribution. On the one hand, if we construct a parallel system with n IID (independent and identically distributed) components whose lifetime distributions are EP$(\alpha, 1)$, then the system lifetime is an EP(α, n). That is, the family of EP distribution is closed under maxima. If variables $X_1, \ldots, X_n$ are independent and $X_i \sim$ EP(α, θ_i) $(i = 1, 2, \ldots, n)$, then $\max(X_1, \ldots, X_n)$ is distributed as EP$(\alpha, \Sigma_{i=1}^{n}\theta_i)$. Therefore, the EP distribution is closed under parallel structure with respect to shape parameter θ. On the other hand, suppose that $X_1, \ldots, X_n$ are independent, and $X_1, \ldots, X_p$ $(1 \le p < n)$ are standard Pareto distribution, i.e., EP$(\alpha_1, 1)$, and $X_{p+1}, \ldots, X_n$ are distributed as EP$(\alpha_2, 1)$. Let $Y_1 = \max\{X_1, \ldots, X_p\}$, $Y_2 = \max\{X_{p+1}, \ldots, X_n\}$, and hence $Y_1 \sim$ EP(α_1, p), $Y_2 \sim$ EP$(\alpha_2, n-p)$. As $\max\{Y_1, Y_2\} = \max\{X_1, \ldots, X_n\}$, the model can be described by multiple-outlier Pareto model. That is, a parallel system with the multiple-outlier Pareto components can be considered as a parallel system with two heterogeneous components.

EP distribution, depending on both the exponential and Pareto distribution, is very useful in reliability and risk theory. Al-Kadim and Boshi (2013) investigated its moments and corresponding parameter estimation. Oanceaa et al. (2017) used it to model the distribution of the gross personal income and income inequality in Romania. The Bayes and classical estimators have been obtained by Afify (2010) for a two-parameters EP distribution when sample are available from type I and type II censoring schemes.

Though researchers have focused on estimating load-sharing parameters as described in the literature, the case of EP distribution of component lifetime has been overlooked. Therefore, we are interested in developing a procedure of parameter estimation for this distribution.

The rest of this section is arranged as follows. At first, the model is constructed for the load-sharing parallel system when the underlying distribution of component lifetime is EP. Then, the ML and BAU (best asymptotic unbiased) estimators of unknown parameters are researched. Moreover, the CI (confidence interval) and the asymptotic distribution for the parameters are constructed. In the end, some simulation methods and an example are used to substantiate the proposed methods.

6.4.3.2 Model Description

For developing a load-sharing parallel system model and estimating its parameters, it is assumed that:

(i) Each load-sharing parallel system has k components, bearing a common constant load which is equally distributed among its working components.
(ii) And n such IID systems are put on test and their failure times are recorded. Let t_{ij} $(i = 1, 2, \ldots, n, j = 1, 2, \ldots, k)$ be the time spacing between $(j - 1)$th and jth component failure in the ith parallel system.
(iii) The failure of a component within the system equally increases the load on the remaining surviving components through the quantity increase of the shape parameters.
(iv) To begin with, the component lifetimes are the EP distribution having two positive shape parameters α_1, θ_1. Upon the first failure of the component, the initial shape parameters change to α_2, θ_2 and so on. After the $(k - 1)$-th failure of the component, the shape parameters of the last component would be α_k, θ_k.

6.4.3.3 Parameter Estimation

Let $\mathcal{A} = (\alpha_1, \alpha_2, \ldots, \alpha_k)^{'}, \Theta = (\theta_1, \theta_2, \ldots, \theta_k)^{'}$ in the space $(0, \infty)^k$, with $\alpha_j > 0$, $\theta_j > 0$ for $j = 1, \ldots, k$. It can be shown that the likelihood function for the ith system is

$$
\begin{aligned}
&L_i(\mathcal{A}, \Theta | t_{i1}, t_{i2}, \ldots, t_{ik}) \\
&= k! \prod_{j=1}^{k} \alpha_j \theta_j (1 + t_{ij})^{-(\alpha_j+1)} [1 - (1 + t_{ij})^{-\alpha_j}]^{\theta_j - 1} \{1 - [1 - (1 + t_{ij})^{-\alpha_j}]^{\theta_j}\}^{k-j},
\end{aligned}
$$

and the likelihood function based on n samples is

$$L(\mathcal{A},\Theta|\mathbf{T})=(k!)^n\prod_{i=1}^{n}\prod_{j=1}^{k}\alpha_j\theta_j(1+t_{ij})^{-(\alpha_j+1)}[1-(1+t_{ij})^{-\alpha_j}]^{\theta_j-1}\{1-[1-(1+t_{ij})^{-\alpha_j}]^{\theta_j}\}^{k-j},$$

where $\mathbf{T}=\{t_{ij}, i=1,2,\ldots,n;\ j=1,2,\ldots,k\}$. Therefore, the log-likelihood function is

$$\begin{aligned}l(\mathcal{A},\Theta|\mathbf{T})&=n\ln(k!)+\sum_{i=1}^{n}\sum_{j=1}^{k}(k-j)\ln\{1-[1-(1+t_{ij})^{-\alpha_j}]^{\theta_j}\}\\&+\sum_{i=1}^{n}\sum_{j=1}^{k}\big[(\theta_j-1)\ln[1-(1+t_{ij})^{-\alpha_j}]-(\alpha_j+1)\ln(1+t_{ij})+\ln(\alpha_j\theta_j)\big].\end{aligned}$$

In order to obtain the ML estimator $(\hat{\mathcal{A}},\hat{\Theta})$ of $(\mathcal{A},\Theta)$, we have the following likelihood equations

$$\begin{aligned}\frac{\partial l(\mathcal{A},\Theta|\mathbf{T})}{\partial\alpha_j}&=\frac{n}{\alpha_j}\\&+\sum_{i=1}^{n}\frac{\ln(1+t_{ij})}{(1+t_{ij})^{\alpha_j}}\left[\frac{(\theta_j-1)\bar{F}(t_{ij};\alpha_j,\theta_j)-(k-j)\theta_jF(t_{ij};\alpha_j,\theta_j)}{\bar{F}(t_{ij};\alpha_j,\theta_j)\sqrt[\theta_j]{F(t_{ij};\alpha_j,\theta_j)}}-(1+t_{ij})^{\alpha_j}\right]=0,\end{aligned}\tag{6.19}$$

and

$$\frac{\partial l(\mathcal{A},\Theta|\mathbf{T})}{\partial\theta_j}=\frac{n}{\theta_j}+\sum_{i=1}^{n}\frac{1}{\theta_j}\ln F(t_{ij};\alpha_j,\theta_j)\times\left[1-(k-j)\frac{F(t_{ij};\alpha_j,\theta_j)}{\bar{F}(t_{ij};\alpha_j,\theta_j)}\right]=0,\tag{6.20}$$

for $j=1,\ldots,k$, where $F(t_{ij};\alpha_j,\theta_j)=[1-(1+t_{ij})^{-\alpha_j}]^{\theta_j}$, $\bar{F}(t_{ij};\alpha_j,\theta_j)=1-F(t_{ij};\alpha_j,\theta_j)$.

It is clear that it is impossible to get a closed form for the parameter $(\mathcal{A},\Theta)$. So the equations in (6.19) and (6.20) can only be solved for 2k-parameters using suitable iterative techniques. As described in Yun and Cha (2010), the Gauss-Seidel method (see Birnbaum and Saunders, 1958), is especially well suited for the log-likelihood equations in this problem, which can solve the following nonlinear equations

$$Q_j=\frac{n}{\alpha_j}+\sum_{i=1}^{n}\frac{\ln(1+t_{ij})}{(1+t_{ij})^{\alpha_j}}\left[\frac{(\theta_j-1)\bar{F}(t_{ij};\alpha_j,\theta_j)-(k-j)\theta_jF(t_{ij};\alpha_j,\theta_j)}{\bar{F}(t_{ij};\alpha_j,\theta_j)\sqrt[\theta_j]{F(t_{ij};\alpha_j,\theta_j)}}-(1+t_{ij})^{\alpha_j}\right]=0,\tag{6.21}$$

and

$$D_j = \frac{n}{\theta_j} + \sum_{i=1}^{n} \frac{1}{\theta_j} \ln F(t_{ij}; \alpha_j, \theta_j) \times \left[1 - (k - j)\frac{F(t_{ij}; \alpha_j, \theta_j)}{\bar{F}(t_{ij}; \alpha_j, \theta_j)}\right] = 0, \quad (6.22)$$

for $j = 1, \ldots, k$.

The numerical method to get the parameter estimators are given as following steps:

(a) For $j = 1, \ldots, k$, choose initial solution $\alpha_j^{(0)}$, solve $D_j = 0$ in (6.22) for $\theta_j^{(0)}$. Denote the solution as $\theta_j^{(1)}$.
(b) For $j = 1, \ldots, k$, substitute $\theta_j^{(1)}$ for θ_j in $Q_j = 0$ in (6.21) for $\alpha_j^{(1)}$. Denote the solution as $\alpha_j^{(1)}$.
(c) Repeat steps (a) and (b) in the same manner continuously until $\hat{\alpha}_j$ and $\hat{\theta}_j$ $(j = 1, \ldots, k)$ converge to what is needed under certain rule.

In the following, we give the asymptotic distribution of the proposed ML estimator. We must differentiate the representations again in (6.19) and (6.20) with respect to α_j and θ_j, denoted by $\frac{\partial^2 l(\mathcal{A},\Theta|\mathbf{T})}{\partial \alpha_j^2}$, $\frac{\partial^2 l(\mathcal{A},\Theta|\mathbf{T})}{\partial \theta_j^2}$ and $\frac{\partial^2 l(\mathcal{A},\Theta|\mathbf{T})}{\partial \alpha_j \partial \theta_j}$ for $j = 1, \ldots, k$. Unfortunately, they are very difficult to obtain, so we just describe the process here. The Fisher's information matrix of the parameter vector $(\mathcal{A}, \Theta)$ is the following partitioned matrix form

$$\mathcal{F}(\mathcal{A}, \Theta) = \begin{pmatrix} \mathcal{F}^{(1)}(\mathcal{A}, \Theta) & 0 & \cdots & 0 \\ 0 & \mathcal{F}^{(2)}(\mathcal{A}, \Theta) & \cdots & 0 \\ \vdots & \vdots & \ddots & \vdots \\ 0 & 0 & \cdots & \mathcal{F}^{(k)}(\mathcal{A}, \Theta) \end{pmatrix},$$

where $\mathcal{F}^{(m)}(\mathcal{A}, \Theta) = (F_{ij}^{(m)})$ is a 2×2 matrix for $m = 1, \ldots, k$, whose elements are defined by $F_{11}^{(m)} = -\mathrm{E}[\frac{@^2 \mathrm{l}(\mathcal{A},\Theta|\mathbf{T})}{@^2_{\mathrm{m}}}]$, $F_{12}^{(m)} = -\mathrm{E}[\frac{\partial^2 l(\mathcal{A},\Theta|\mathbf{T})}{\partial \alpha_m \partial \theta_m}] = F_{21}^{(m)}$, and $F_{22}^{(m)} = -\mathrm{E}[\frac{@^2 \mathrm{l}(\mathcal{A},\Theta|\mathbf{T})}{@`^2_{\mathrm{m}}}]$. The asymptotic distribution of $(\hat{\alpha}_1, \hat{\theta}_1, \ldots, \hat{\alpha}_k, \hat{\theta}_k)' - (\alpha_1, \theta_1, \ldots, \alpha_k, \theta_k)'$ is $N_{2k}(\mathbf{0}, \mathcal{F}^{-1}(\mathcal{A}, \Theta))$, where $\mathcal{F}(\mathcal{A}, \Theta)$ can be replaced by observed Fisher information matrix with corresponding elements: $F_{11}^{(m)} = -\frac{\partial^2 l(\mathcal{A},\Theta|\mathbf{T})}{\partial \alpha_m^2} |_{\alpha_m = \hat{\alpha}_m, \theta_m = \hat{\theta}_m}$, $F_{12}^{(m)} = -\frac{\partial^2 l(\mathcal{A},\Theta|\mathbf{T})}{\partial \alpha_m \partial \theta_m} |_{\alpha_m = \hat{\alpha}_m, \theta_m = \hat{\theta}_m} = F_{21}^{(m)}$, and $F_{22}^{(m)} = -\frac{\partial^2 l(\mathcal{A},\Theta|\mathbf{T})}{\partial \theta_m^2} |_{\alpha_m = \hat{\alpha}_m, \theta_m = \hat{\theta}_m}$.

Next, for simplicity, we investigate the ML estimator for the EP distribution when $\theta = 1$. In this case, the reliability function with one parameter α is

$$\bar{F}(x) = (1 + x)^{-\alpha}, x > 0, \alpha > 0.$$

The corresponding probability density function is

$$f(x) = \alpha(1 + x)^{-(\alpha+1)}, x > 0, \alpha > 0.$$

And the probability density function for the jth ($j = 1, 2, \ldots, k$) failure time of the ith ($i = 1, 2, \ldots, n$) system is

$$f(t_{ij}|\alpha_j) = (k - j + 1)\alpha_j(1 + t_{ij})^{-\alpha_j(k-j+1)-1}. \tag{6.23}$$

So the likelihood function for the ith ($i = 1, 2, \ldots, n$) system is

$$L_i(\alpha_1, \alpha_2, \ldots, \alpha_k|t_{i1}, t_{i1}, \ldots, t_{ik}) = k!\prod_{j=1}^{k}\alpha_j(1 + t_{ij})^{-\alpha_j(k-j+1)-1},$$

and the likelihood function based on n samples is

$$L(\alpha_1, \alpha_2, \ldots, \alpha_k|\mathbf{T}) = (k!)^n(\prod_{j=1}^{k}\alpha_j)^n\prod_{i=1}^{n}\prod_{j=1}^{k}(1 + t_{ij})^{-\alpha_j(k-j+1)-1}. \tag{6.24}$$

The log-likelihood function is

$$l(\alpha_1, \alpha_2, \ldots, \alpha_k|\mathbf{T}) = n\ln k! + n\sum_{j=1}^{k}\ln\alpha_j - \sum_{i=1}^{n}\sum_{j=1}^{k}[\alpha_j(k - j + 1) - 1]\ln(1 + t_{ij}).$$

By differentiating the above log-likelihood function, we have

$$\frac{\partial l}{\partial\alpha_j} = n\frac{1}{\alpha_j} - (k - j + 1)\sum_{i=1}^{n}\ln(1 + t_{ij}),\ j = 1, \ldots, k.$$

Thus, it is easy to obtain the ML estimator

$$\hat{\alpha}_j = \frac{1}{(k - j + 1)} \times \frac{n}{\sum_{i=1}^{n}\ln(1 + t_{ij})},\ j = 1, \ldots, k. \tag{6.25}$$

It can be easily proven that $\ln(1 + t_{ij})$ is distributed as exponential with parameter $(k - j + 1)\alpha_j$ for $j = 1, 2, \ldots, k$, and hence $\sum_{i=1}^{n}\ln(1 + t_{ij})$ follows Gamma distribution with shape parameter n and scale parameter $(k - j + 1)\alpha_j$. Therefore, $E[\hat{\alpha}_j] = \frac{n}{n-1}\alpha_j$, that is, $\hat{\alpha}_j$ is a biased estimator of α_j for $j = 1, 2, \ldots, k$. While $\tilde{\alpha}_j = \frac{n-1}{n}\hat{\alpha}_j$ is an unbiased estimator and it is also a BAU one. Furthermore, $MSE(\hat{\alpha}_j) = \frac{n+2}{(n-1)(n-2)}\alpha_j^2$, and $MSE(\tilde{\alpha}_j) = \frac{1}{n-2}\alpha_j^2$, where MSE is the abbreviation of mean square error. So their RE (relative efficiency) under MSE is

$$\mathrm{RE} = \frac{\mathrm{MSE}(\hat{\alpha}_j)}{\mathrm{MSE}(\tilde{\alpha}_j)} = \frac{n + 2}{n - 1}.$$

By the RE, the unbiased estimator outperforms the ML one. Moreover,

$$\frac{\partial^2 l}{\partial \alpha_j^2} = n\frac{-1}{\alpha_j^2}, j = 1, \ldots, k, \frac{\partial^2 l}{\partial \alpha_i \partial \alpha_j} = \frac{\partial^2 l}{\partial \alpha_j \partial \alpha_i} = 0, i \neq j, i, j = 1, \ldots, k.$$

It is clear that the $k \times k$ Hessian matrix $\{\partial^2 l / \partial \alpha_i \partial \alpha_j\}$ is negative definite, so there exists a vector $\mathcal{A}$ in this subspace that yields a global maximum for (6.24).

Let $\mathcal{A} = (\alpha_1, \alpha_2, \ldots, \alpha_k)'$ in the space $(0, \infty)^k$, with $\alpha_j > 0$ for $j = 1, \ldots, k$. Then the Fisher's information matrix of the parameter vector $\mathcal{A}$ contained in the sample is

$$\mathcal{F}(\mathcal{A}) = n \begin{pmatrix} \frac{1}{\alpha_1^2} & 0 & \cdots & 0 \\ 0 & \frac{1}{\alpha_2^2} & \cdots & 0 \\ \vdots & \vdots & \ddots & \vdots \\ 0 & 0 & \cdots & \frac{1}{\alpha_k^2} \end{pmatrix}.$$

Thus, from the property of ML estimator, $\sqrt{n}(\hat{\mathcal{A}} - \mathcal{A}) \to N(\mathbf{0}, \Sigma(\mathcal{A}))$, where

$$\Sigma(\mathcal{A}) = \begin{pmatrix} \alpha_1^2 & 0 & \cdots & 0 \\ 0 & \alpha_2^2 & \cdots & 0 \\ \vdots & \vdots & \ddots & \vdots \\ 0 & 0 & \cdots & \alpha_k^2 \end{pmatrix},$$

is a variance-covariance matrix. And clearly, the variance of the ML estimator does not attain Cramer-Rao Lower Bound, but it does when n approaches infinity. Thus it is a BAU estimator. Usually, $\Sigma(\mathcal{A})$ can be estimated by $\Sigma(\hat{\mathcal{A}})$ in practice.

Joint confidence statements, based on the likelihood ratio, can be constructed for parameters. The inverse of the observed Fisher information matrix $\mathcal{F}$ provides an estimator of the covariance in the large sample normal distribution. For large samples, the approximate $1 - \alpha$ $(0 < \alpha < 1)$ confidence ellipsoid for $\mathcal{A} \in (0, \infty)^k$ is

$$(\hat{\mathcal{A}} - \mathcal{A})' \Sigma^{-1}(\mathcal{A})(\hat{\mathcal{A}} - \mathcal{A}) \leq \chi^2_{k,\alpha}$$

centered at the ML estimator $\hat{\mathcal{A}}$. Here, $\chi^2_{k,\alpha}$ is the upper αth quantile of the χ^2 distribution with k degrees of freedom.

6.4.3.4 Simulation Study

In this section, we do some simulation studies by using our estimation methods, given that the underlying distribution of the component lifetimes is EP with one unknown parameter.

Sample data for twenty identical load-sharing systems with three components ($k = 3$) are generated by letting $\alpha_1 = 1/3, \alpha_2 = 2/3, \alpha_3 = 8$ (three suitable parameters

Table 6.1 Failure time samples

n	t_{i1}	t_{i2}	t_{i3}
1	0.23688814	0.17286444	0.02693109
2	1.64626412	1.07479356	0.129351543
3	0.77849784	0.54006839	0.07462433
4	3.65893310	2.17113591	0.21209263
5	3.54495477	2.11277102	0.20834569
6	0.37474544	0.26959942	0.04058554
7	0.38210172	0.27469126	0.04127994
8	4.04724671	2.36737048	0.22428299
9	22.8887855	9.80551705	0.48687431
10	7.06013731	3.78362173	0.29805414
11	0.5.718918	0.39397837	0.05692135
12	1.00398844	0.6843076	0.09077933
13	2.15918769	1.36963563	0.15464088
14	2.48 070677	1.54830014	0.16871449
15	0.5879336 2	0.4145694	0.05950752
16	1.27892276	0. 8548005	0.10845024
17	1.01240918	0.68961 29	0.09135121
18	0.01396067	0.01045233	0.00173452
19	3.40192424	2.03900855	0.20352557
20	7.52087718	3.987274	0.30710519

are selected here to make the confidence ellipse display more sensible), see Table 6.1. By (6.23), their ML estimators $\hat{\alpha}_1 = 0.3102$, $\hat{\alpha}_2 = 0.6203$, $\hat{\alpha}_3 = 7.4436$. Also the contour lines of confidence ellipses for (α_1, α_2), (α_1, α_3) and (α_2, α_3) at confidence regions $1 - \alpha = \{80, 90, 95, 99\%\}$ are plotted in Figs. 6.3, 6.4 and 6.5, respectively, which convey the strong positive correlation between $\hat{\alpha}_1$, $\hat{\alpha}_2$ and $\hat{\alpha}_3$. In the figures, the small blank points represent the generated sample points, and the blue solid ones represent the center of the ellipse.

The ML and BAU results are summarized in Table 6.2.

The individual approximate $100(1-\beta)\%$ $(0 < \beta < 1)$ CI for α_i is $(\hat{\alpha}_i - z_{\beta/2}\hat{\sigma}_i/\sqrt{n}, \hat{\alpha}_i + z_{\beta/2}\hat{\sigma}_i/\sqrt{n})$, $i = 1, 2, \ldots, k$, where $z_{\beta/2}$ is the upper $\beta/2$-th quantile of a standard normal distribution, and $\hat{\alpha}_i$ is the ML estimator of the parameter α_i.

In the following, two kinds of parametric confidence intervals based on bootstrapping technique such as percentile bootstrap (boot-p) and bootstrap-t (boot-t) methods are constructed, respectively; see Efron (1982) and Hall (1988) for more details.

Algorithm for bootstrap-p method

[1] Generate sample $T_{ij} (i = 1, 2, \ldots, n; j = 1, 2, \ldots, k)$ of size n using (6.23).

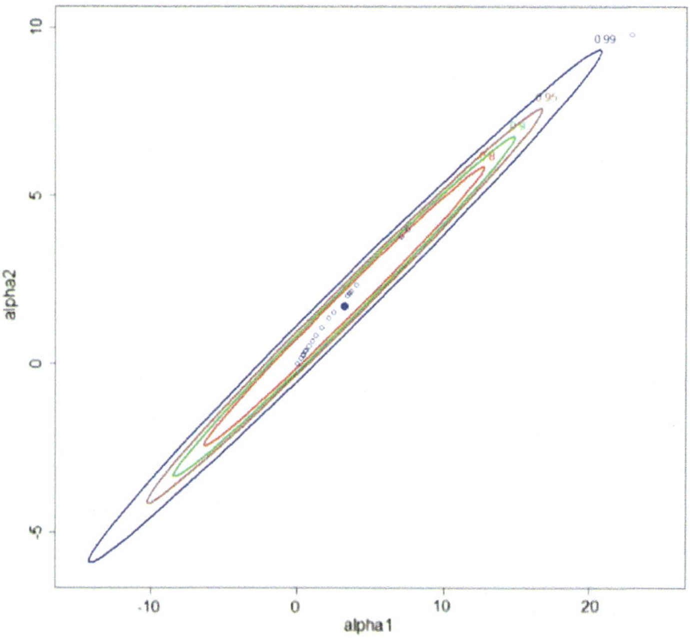

Fig. 6.3 Confidence regions (80, 90, 95, 99%) for (α_1, α_2)

[2] Generate a bootstrap sample T_{ij}^* of size n from the sample T_{ij}. Further, use the Bootstrap sample to obtain Bootstrap point estimator for the parameter $\mathcal{A} = (\alpha_1, \alpha_2, \ldots, \alpha_k)'$, denoted by $\widehat{\mathcal{A}}^* = (\hat{\alpha}_1^*, \hat{\alpha}_2^*, \ldots, \hat{\alpha}_k^*)'$.
[3] Repeat step [2] B-times.
[4] Let $\widehat{\mathcal{A}}_{(1)}^*, \widehat{\mathcal{A}}_{(2)}^*, \ldots, \widehat{\mathcal{A}}_{(B)}^*$ be the ordered statistics of Bootstrap estimate $\widehat{\mathcal{A}}_1^*, \widehat{\mathcal{A}}_2^*, \ldots, \widehat{\mathcal{A}}_B^*$. Then $1-\gamma$ $(0 < \gamma < 1)$ **CI** for $\mathcal{A}^*$ is $[\widehat{\mathcal{A}}_{(\gamma B/2)}^*, \widehat{\mathcal{A}}_{((1-\gamma/2)B)}^*]$.

Algorithm for bootstrap-t method

[1] Generate sample $T_{ij} (i = 1, 2, \ldots, n; j = 1, 2, \ldots, k)$ of size n using (6.23).
[2] Generate a bootstrap sample T_{ij}^* of size n from the sample T_{ij}. Then obtain the Bootstrap point estimator of the parameter $\mathcal{A} = (\alpha_1, \alpha_2, \ldots, \alpha_k)'$, denoted by $\widehat{\mathcal{A}}^* = (\hat{\alpha}_1^*, \hat{\alpha}_2^*, \ldots, \hat{\alpha}_k^*)'$.
[3] Compute bootstrap pivot as $T_j^* = (\hat{\alpha}_j^* - \hat{\alpha}_j)/\hat{\sigma}_j^*$ for $j = 1, 2, \ldots, k$, where $\hat{\sigma}_j^*$ is the Bootstrap estimate of the j-th standard variance.
[4] Repeat steps [2]–[3] B-times.
[5] Let $T_{(1)}^*, T_{(2)}^*, \ldots, T_{(B)}^*$ be the ordered values of $T_1^*, T_2^*, \ldots, T_B^*$.
[6] The $1-\gamma$ $(0 < \gamma < 1)$ Bootstrap-t confidence region for α_j $(j = 1, 2, \ldots, k)$ is written as $[\hat{\alpha}_j - \hat{\sigma}_j^* T_{(B(1-\gamma/2))}^*, \hat{\alpha}_j - \hat{\sigma}_j^* T_{(B\gamma/2)}^*]$.

By the help of R software, we can calculate different estimates with their standard errors (SE) and 95% Bootstrap CI with their widths according to different sample numbers n, under the condition that $k = 3, \alpha_1 = 1/3, \alpha_2 = 2/3, \alpha_3 = 8$, see

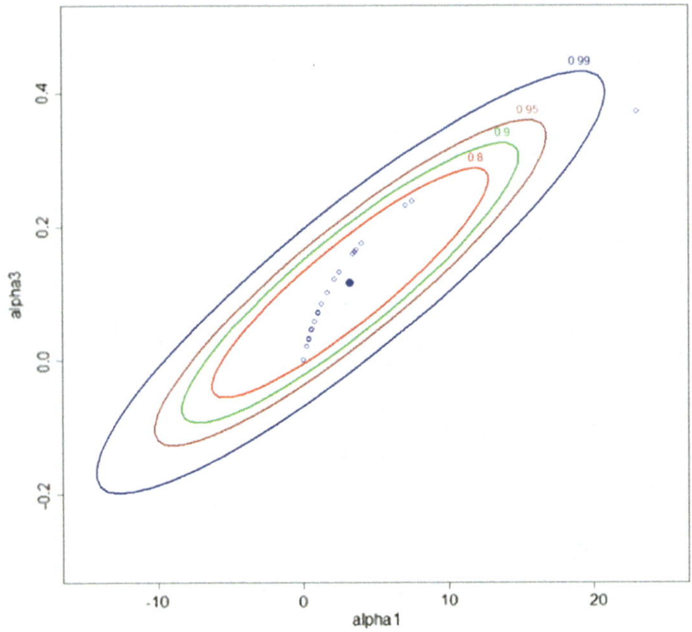

Fig. 6.4 Confidence regions (80, 90, 95, 99%) for (α_1, α_3)

Table 6.3. Here we have adopted 5000 bootstrap programs ($B = 5000$). And after combining Table 6.2 with Table 6.3, we can see that Bootstrap estimates are better than ML ones, slightly inferior to BAU ones. Furthermore, we can see, from Table 6.3, the Bootstrap point estimators for α approach the real value and the parametric CIs are more accurate as the sample size increases.

6.4.3.5 Illustrative Example

To illustrate the use of the estimation methods proposed in this part, the following example is discussed.

Nelson (1982) presented data on the time to breakdown of an insulating fluid in an accelerated test conducted at various test voltages and used the exponential Pareto distribution to analyze this data set. Mahmoud et al. (2016) estimated the parameters for the exponentiated Pareto distribution based on progressively type-II right censored data (log data of Table 6.4 from the $n = 19$ observations recorded at 34 kV in Nelson's table (see P. 228 of Nelson 1982)). For the purposes of illustrating the methods discussed in this article, we select this set of log data in Table 6.4 as Mahmoud et al. (2016).

Using the formula of Eq. (6.23), we obtain that the BAU and ML of the parameter α_1 are 0.5070387 and 0.4803524, respectively. Moreover, we calculate different

Table 6.2 ML estimator vs. BAU one

	α_1	α_2	α_3
ML estimator	0.310151902448029	0.620303804792817	7.4436456349709
BAU estimator	0.294644307325628	0.589288614553176	7.0714633532224

Table 6.3 95% bootstrap confidence intervals (CI)

Estimates	n			
	20	50	75	100
	$\hat{\alpha}_1 = 0.3181332$	$\hat{\alpha}_1 = 0.3127915$	$\hat{\alpha}_1 = 0.3126375$	$\hat{\alpha}_1 = 0.3119628$
	[0.0007411142]	[0.0004549371]	[0.0003746743]	[0.0003198372]
Bootstrap	$\hat{\alpha}_2 = 0.6395655$	$\hat{\alpha}_2 = 0.6273043$	$\hat{\alpha}_2 = 0.6246412$	$\hat{\alpha}_2 = 0.6241701$
estimates	[0.001502244]	[0.0009151824]	[0.0007451514]	[0.0006353488]
$[SE]$	$\hat{\alpha}_3 = 7.652486$	$\hat{\alpha}_3 = 7.514289$	$\hat{\alpha}_3 = 7.504781$	$\hat{\alpha}_3 = 7.475801$
	[0.01834039]	[0.01086252]	[0.008823432]	[0.007598901]
	$\alpha_1 \in$	$\alpha_1 \in$	$\alpha_1 \in$	$\alpha_1 \in$
	(0.2324167,	(0.2564988,	(0.2665643,	(0.2713061,
	0.4386673)	0.3844500)	0.3689980)	0.3580020)
	{0.2062506}	{0.1279512}	{0.1024337}	{0.0866959}
Bootstrap-p	$\alpha_2 \in$	$\alpha_2 \in$	$\alpha_2 \in$	$\alpha_2 \in$
CI	(0.4691763,	(0.5148109,	(0.5333431,	(0.5406065,
$\{widths\}$	0.8847640)	0.7649915)	0.7413725)	0.7177437)
	{0.4155877}	{0.2501806}	{0.2080294}	{0.1771372}
	$\alpha_3 \in$	$\alpha_3 \in$	$\alpha_3 \in$	$\alpha_3 \in$
	(5.58279,	(6.178553,	(6.343337,	(6.502809,
	10.54648)	9.174695)	8.786609)	8.596271)
	{4.96369}	{2.996142}	{2.443272}	{2.093462}
	$\alpha_1 \in$	$\alpha_1 \in$	$\alpha_1 \in$	$\alpha_1 \in$
	(−0.9696164,	(−0.008189074,	(0.1659010,	(0.2077289,
	0.4254538)	0.356209438)	0.3455138)	0.3418191)
	{1.3950702}	{0.364398512}	{0.1796128}	{0.1340902}
Bootstrap-t	$\alpha_2 \in$	$\alpha_2 \in$	$\alpha_2 \in$	$\alpha_2 \in$
CI	(−0.8990139,	(0.1831282,	(0.3777831,	(0.4289629,
$\{widths\}$	0.7812601)	0.7078775)	0.6964646)	0.6837629)
	{1.680274}	{0.5247493}	{0.3186815}	{0.2548}
	$\alpha_3 \in$	$\alpha_3 \in$	$\alpha_3 \in$	$\alpha_3 \in$
	(−0.2069366,	(4.525305,	(5.402058,	(5.822500,
	9.0530891)	8.512641)	8.360744)	8.304901)
	{9.2600257}	{3.987336}	{2.958686}	{2.482401}

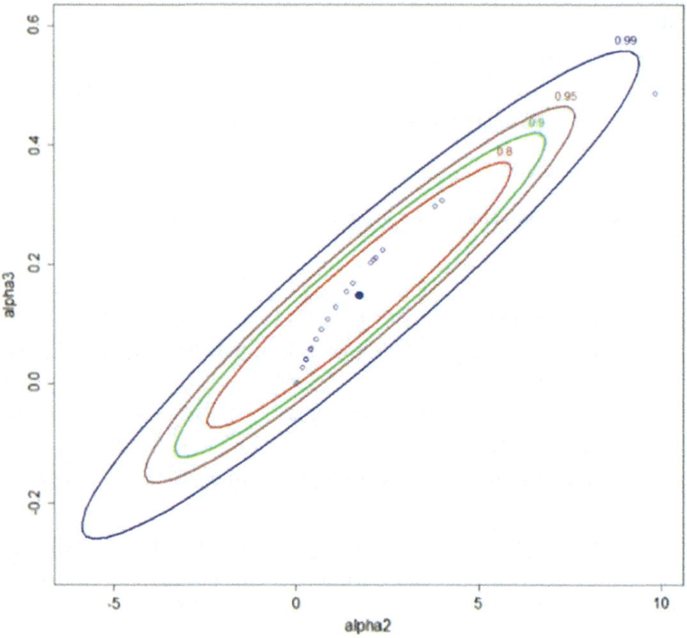

Fig. 6.5 Confidence regions (80, 90, 95, 99%) for (α_2, α_3)

Table 6.4 Sample observations

−1.6608	−0.2485	−0.0409	0.2700	1.0224	1.1505	
1.14231	1.5411	1.5789	1.8718	1.9947	2.0806	
2.1126	2.4898	3.4578	3.4818	3.5237	3.6030	4.2889

estimates with their standard errors (SE) and 95% Bootstrap CI with their widths according to different sample numbers n, given that $k = 1$, see Table 6.5. Here we have adopted 5000 bootstrap programs ($B = 5000$).

Table 6.5 95% Bootstrap CI

Bootstrap	n			
	19	50	75	100
Estimates	$\hat{\alpha}_1 = 0.4880658$	$\hat{\alpha}_1 = 0.4836804$	$\hat{\alpha}_1 = 0.4824427$	$\hat{\alpha}_1 = 0.4819396$
$[SE]$	[0.0009071813]	[0.0005393407]	[0.0004432456]	[0.0003736753]
	$\alpha_1 \in$	$\alpha_1 \in$	$\alpha_1 \in$	$\alpha_1 \in$
Boot-p	(0.3851791,	(0.4174839,	(0.4262332,	(0.4342543,
CI	0.6338479)	0.5660315)	0.5481114)	0.5371907)
$\{widths\}$	{0.2486688}	{0.1485476}	{0.1218782}	{0.1029364}
	$\alpha_1 \in$	$\alpha_1 \in$	$\alpha_1 \in$	$\alpha_1 \in$
Boot-t	(−0.07900374,	(0.3251621,	(0.3677690,	(0.3918383,
CI	0.57590171)	0.5367134)	0.5282657)	0.5235150)
$\{widths\}$	{0.65490545}	{0.2115513}	{0.1604967}	{0.1316767}

6.4.3.6 Conclusion

In the section we investigate the parameter estimation for the load-sharing parallel system when the underlying distribution of component lifetime is exponential Pareto. We use the appropriate iterative method to discuss the parameter estimators for the distribution with two unknown shape parameters. Also, the asymptotic distribution of estimated statistics is constructed. For the model simplified, the closed-form ML estimator and the confidence interval of parameters are provided. Also some appropriate simulations and an illustrative example are used to substantiate the proposed methods.

Chapter 7
The Conditional Spacings and Their Properties

This chapter is composed of four parts. The first part is mainly for the conditional spacings and normalized conditional ones based on a series system. In the second part, the case of the (normalized) conditional spacings based on unfailed components is investigated based on the fact that an $(n-k+1)$-out-of-$n$system or a general coherent system with certain signature just fails at certain time t. The third part is about the conditional spacings and their properties based on a series system with n heterogeneous Exponential component lifetimes. In the last part, the generalized conditional spacings and their properties are investigated.

7.1 Conditional Spacings Under Series System with IID Components

In this section, the concepts of conditional spacings and normalized conditional ones are introduced when a series system with independent and identically distributed components fails at time t. First, the survival function and the joint distribution of conditional spacings are obtained. Based on this, some stochastic comparisons among different conditional spacings are conducted. Also the case of conditional spacings when the component lifetimes are Exponential is investigated. In the end, some statistical properties of conditional spacings are given, and the MTP2 property of the joint density for the conditional spacings is investigated too.

7.1.1 Introductions and Preliminaries

Suppose $X_1, X_2, \ldots, X_n$ are independent nonnegative random variables with a common distribution function F, and $X_{i,n}$ $(i \in \{1, 2, \ldots, n\})$ is the ith smallest order

Z. Zhang, *Conditional Order Statistics and Conditional Spacings*,
https://doi.org/10.1007/978-981-96-7151-9_7

statistic among $X_1, X_2, \ldots, X_n$. It is well-known that the differences between consecutive observations, namely, for $i \in \{1, \ldots, n-1\}$, $D_{i,n} = (X_{i+1,n} - X_{i,n})$, and $D^*_{i,n} = (n-i)\left(X_{i+1,n} - X_{i,n}\right)$, are called spacing and normalized spacing, respectively. Spacings have been studied for distribution-free goodness-of-fit tests, chi-square, as well as Kolmogorov-Smirnov etc., by Cochran (1952), and Darling (1957). In a goodness-of-fit problem, the statistician wishes to decide whether or not a sample of n random variables could be independent and identically distributed random variables with a certain specified distribution function. In many of the more "non-parametric" situations, this problem may be reduced to that of testing whether or not the given observations have come from a Uniform population. In the survival context spacings correspond to times elapsed between successive failures of components in a system. In stochastic auction theory, $D_{n-1,n-1}$ and $D_{2,n}$ are of particular interest, which represent auction rents in buyer s auction and reverse auction in the second-price business auction, respectively, (see Xu and Li 2008).

Various stochastic comparisons on magnitudes and variability between $D_{i,n}$ ($D^*_{i,n}$) and $D_{i+1,n}$ ($D^*_{i+1,n}$), also between $D_{i,n}$ ($D^*_{i,n}$) and $D_{i,n+1}$ ($D^*_{i,n+1}$), based on a set of independent random variables or based on two different sets of random variables, are researched in the past. For example, Barlow and Proschan (1966) gained several stochastic orderings between the successive normalized spacings from decreasing failure rate distributions, that is, $D^*_{i,n} \leq_{st} D^*_{i+1,n}$ and $D^*_{i,n+1} \leq_{st} D^*_{i,n}$ for $i \in \{1, \ldots, n-1\}$, given that the component lifetimes have decreasing failure rate. Similar results hold for the increasing failure rate case with the inequalities reversed in above results. Later Pledger and Proschan (1971) partially extended this result to the case when the random variables are independent with proportional decreasing failure rates. Kochar and Kirmani (1995) strengthened the above results from stochastic ordering to hazard rate ordering. For more details about spacings based on independent and identically distributed random variables, readers may refer to papers by Hu and Zhuang (2005, 2006), and Kochar and Korwar (1996) etc. Some interesting partial ordering results on order statistics and spacings from independent but non-identically random variables have been obtained by Kochar and Rojo (1996), Kochar and Xu (2011) etc. Also, on the topic of spacings, readers may refer to the books by Arnold et al. (1992), Balakrishnan (2007), Balakrishnan and Rao (1998), and David and Nagaraja (2003).

It is well known that the lifetime of an $(n-k+1)$-out-of-n system are just the order statistics $X_{k,n}$ of random variables $X_1, X_2, \ldots, X_n$. In particular, the lifetimes of series and parallel systems correspond to the first and the last order statistics, respectively. When a series system with n components fails at some time, there is still $n-1$ components that remain alive. The remaining unfailed components can be used for life tests and other purposes. Based on this, many literatures have been mainly for the aging and stochastic behaviors of the residual life of these surviving components. As a matter of fact, the spacings that are constructed by the residual lifetimes from these unfailed components are more important in life tests etc. But unfortunately, this topic was ignored in the past. Hence in this paper, we naturally consider it and do some researches on this.

Given that a series system with n independent and identically distributed components failed at time t, for $i \in \{2, \dots, n-1\}$ and fixed $t > 0$, let

$$D^t_{i,n-1} = \left(X_{i+1,n} - X_{i,n} \mid X_{1,n} = t\right), D^{t,*}_{i,n-1} = (n-i)\left(X_{i+1,n} - X_{i,n} \mid X_{1,n} = t\right)$$

be the corresponding conditional spacing and normalized conditional spacing based on $n-1$ unfailed components, which may be regarded as the conditional differences between consecutive observations.

7.1.2 The Stochastic and Statistical Properties of Conditional Spacings

In this section, at first, we mainly focus on the marginal and joint survival functions of conditional spacings and normalized conditional ones, and some stochastic comparisons among a set of order statistics or two different sets of order statistics. Then, we research the case when the component lifetimes are independent and identically distributed as exponential. Naturally some statistical properties of conditional spacings are given. Finally, the dependence of conditional spacings is investigated.

Let $X_1, \dots, X_n$ be independent nonnegative random variables with a common distribution function F, survival function $\bar{F}$ and density function f, and $X_{i,n}$ be the ith smallest order statistic among $X_1, \dots, X_n$. Let $Y_{i,n-1}$ be the ith order statistics of independent random variables $Y_1, \dots, Y_{n-1}$, which have the same distribution as $(X_1 - t|X_1 > t)$. Denote the distribution function, density one and survival one of random variable Y_1 by $F_Y(\cdot)$, $f_Y(\cdot)$ and $\bar{F}_Y(\cdot)$, respectively.

As the first result, we give the survival function of conditional spacings.

Theorem 7.1.1 *For $i \in \{2, \dots, n-1\}$, fixed $t > 0$ and any $x > 0$, the survival function of $D^t_{i,n-1}$ is*

$$P(D^t_{i,n-1} > x) = \binom{n-1}{i-1} E[\bar{F}^{n-i}_Y(Y_{i-1,i-1} + x)],$$

where $Y_{i-1,i-1} = \max\{Y_1, \dots, Y_{i-1}\}$ and $E(\cdot)$ is the mean value of a random variable.

Proof For fixed $t > 0$ and any $x > 0$,

$$\begin{aligned} P(D^t_{i,n-1} > x) &= P\left(X_{i+1,n} - X_{i,n} > x \mid X_{1,n} = t\right) \\ &= P\left(Y_{i,n-1} - Y_{i-1,n-1} > x\right) \\ &= \int_0^{+\infty} P\left(Y_{i,n-1} - y > x | Y_{i-1,n-1} = y\right) f_{Y_{i-1,n-1}}(y) dy \\ &= (n-1)\binom{n-2}{i-2} \int_0^{+\infty} \bar{F}^{n-i}_Y(y+x) F^{i-2}_Y(y) f_Y(y) dy \end{aligned}$$

$$
\begin{aligned}
&= \binom{n-1}{i-1} \int_0^{+\infty} \bar{F}_Y^{n-i}(y+x)(i-1)F_Y^{i-2}(y) f_Y(y)dy \\
&= \binom{n-1}{i-1} \int_0^{+\infty} \bar{F}_Y^{n-i}(y+x)dF_{Y_{i-1,i-1}}(y) \\
&= \binom{n-1}{i-1} E[\bar{F}_Y^{n-i}(Y_{i-1,i-1}+x)],
\end{aligned}
$$

where $f_{Y_{i-1,n-1}}$ is the density of $Y_{i-1,n-1}$. That is, the result holds.

Remark As shown in the above theorem, we can obtain that, for $i \in \{2, \ldots, n-1\}$, fixed $t > 0$ and any $x > 0$, the survival function of normalized conditional spacing $D_{i,n-1}^{t,*}$ is

$$
\begin{aligned}
P(D_{i,n-1}^{t,*} > x) = &\ P\left((n-i)(X_{i+1,n} - X_{i,n} \mid X_{1,n} = t) > x\right) \\
= &\binom{n-1}{i-1} E[\bar{F}_Y^{n-i}(Y_{i-1,i-1} + \frac{x}{n-i})].
\end{aligned}
$$

Since $X_1, X_2, \ldots, X_n$ are independent with a common distribution function F, then for fixed $t > 0$, the joint distribution of $Y_{1,n-1}, Y_{2,n-1}, \ldots, Y_{n-1,n-1}$ is given as follows

$$
\begin{aligned}
& f_{Y_{1,n-1},Y_{2,n-1},\ldots,Y_{n-1,n-1}}(y_1, y_2, \ldots, y_{n-1}) \\
&= \begin{cases} (n-1)! f_{Y_1,Y_2,\ldots,Y_{n-1}}(y_1, y_2, \ldots, y_{n-1}), & 0 < y_1 < y_2 < \ldots < y_{n-1} \\ 0, & \text{Otherwise} \end{cases} \\
&= \begin{cases} \frac{(n-1)!}{\bar{F}^{n-1}(t)} f(t+y_1) f(t+y_2) \ldots f(t+y_{n-1}), & 0 < y_1 < y_2 < \ldots < y_{n-1} \\ 0, & \text{Otherwise,} \end{cases}
\end{aligned}
$$

where $f_{Y_1,Y_2,\ldots,Y_{n-1}}(y_1, y_2, \ldots, y_{n-1})$ is the joint density function of $Y_1, Y_2, \ldots, Y_{n-1}$. Therefore, by some computations, we obtain that, for any $d_2 > 0, \ldots, d_{n-1} > 0$, the joint density of conditional spacings $D_{2,n-1}^t, \ldots, D_{n-1,n-1}^t$ is

$$
\begin{aligned}
& f_{D_{2,n-1}^t,\ldots,D_{n-1,n-1}^t}(d_2, \ldots, d_{n-1}) \\
&= \begin{cases} (n-1)! \int_0^{+\infty} f_{Y_2}(x) f_{Y_3}(d_2+x) \ldots f_{Y_n}(d_2 + \cdots + d_{n-1} + x)dx, & d_i \geq 0 \\ 0, & \text{Otherwise} \end{cases} \\
&= \begin{cases} \frac{(n-1)!}{\bar{F}^{n-1}(t)} \int_0^{+\infty} f(x+t) f(d_2+x+t) \ldots f(d_2 + \cdots + d_{n-1} + x + t)dx, & d_i \geq 0 \\ 0, & \text{Otherwise.} \end{cases}
\end{aligned}
$$

Obviously, $D_{2,n-1}^t, \ldots, D_{n-1,n-1}^t$ are not independent and so aren't $D_{2,n-1}^{t,*}, \ldots, D_{n-1,n-1}^{t,*}$.

In what follows, we construct some stochastic comparisons among conditional spacings. To get the main results on this, we introduce some useful lemmas due to Kochar and Kirmani (1995).

Lemma 7.1.2 *Let $\psi_1(x, y)$ and $\psi_2(x, y)$ be positive real-valued functions such that*

(i) $\psi_2(x, y)$ is TP2, that is, for $y_1 \le y_2$, $\psi_2(x, y_2)/\psi_2(x, y_2)$ is nondecreasing in x;
(ii) For $y_1 \le y_2$, $\psi_1(x, y_2)/\psi_2(x, y_1)$ is nondecreasing in x;
(iii) For each fixed x, $\psi_1(x, y)/\psi_2(x, y)$ is nondecreasing in y.

Then, for $x_1 \le x_2$, $y_1 \le y_2$, it holds

$$0 \le \psi_1(x_2, y_2)\psi_2(x_1, y_1) - \psi_1(x_1, y_1)\psi_2(x_2, y_2) \ge \psi_1(x_1, y_2)\psi_2(x_2, y_1) - \psi_1(x_2, y_1)\psi_2(x_1, y_2).$$

Lemma 7.1.3 *Suppose F is DFR, and for $2 \le i \le n-1$, $x > 0$, $y > 0$, $\psi_1(x, y) = \bar{F}^{n-i}[x/(n-i) + y]$, $\psi_2(x, y) = \bar{F}^{n-i+1}[x/(n-i+1) + y]$. Then ψ_1 and ψ_2 satisfy the conditions of Lemma 7.1.2.*

Lemma 7.1.4 *Let ψ_1 and ψ_2 satisfy the conditions of Lemma 7.1.2, and (Y_1, Y_2) be a bivariate random vector with joint density $f(y_1, y_2)$ satisfying $f(y_1, y_2) \ge f(y_2, y_1)$ for $y_1 \le y_2$. Then, for $x_1 \le x_2$, $E[\psi_1(x_1, Y_2)]/E[\psi_2(x_1, Y_1)] \le E[\psi_1(x_2, Y_2)]/E[\psi_2(x_2, Y_1)]$.*

Theorem 7.1.5 *Let $X_1, \dots, X_n$ be independent and identically distributed component lifetimes of a series system with X_1 being a DFR distribution. Then, for fixed $t > 0$,*

(1) $D^{t,*}_{2,n-1} \le_{hr} D^{t,*}_{3,n-1} \le_{hr} \cdots \le_{hr} D^{t,*}_{n-1,n-1}$;
(2) $D^{t,*}_{i,n-1} \le_{hr} D^{t,*}_{i,n-2}$, for fixed $i \in \{2, \dots, n-2\}$.

Proof Noting X_1 is DFR if and only if $Y_1 = [X_1 - t|X_1 > t]$ is DFR for fixed $t > 0$.

(1) By the above remark, for $i \in \{2, \dots, n-1\}$, fixed $t > 0$ and any $x > 0$, the survival function of $D^{t,*}_{i,n-1}$ is

$$P(D^{t,*}_{i,n-1} > x) = \binom{n-1}{i-1} E[\bar{F}_Y^{n-i}(Y_{i-1,i-1} + \frac{x}{n-i})],$$

and

$$P(D^{t,*}_{i-1,n-1} > x) = \binom{n-1}{i-2} E[\bar{F}_Y^{n-i+1}(Y_{i-2,i-2} + \frac{x}{n-i+1})],$$

where $Y_{i-2,i-2} = \max\{Y_1, \dots, Y_{i-2}\}$. For $x_1 \leqslant x_2$ and $i \in \{2, \dots, n-1\}$, it needs to prove

$$\frac{P\left[D^{t,*}_{i,n-1} > x_2\right]}{P\left[D^{t,*}_{i-1,n-1} > x_2\right]} \geqslant \frac{P\left[D^{t,*}_{i,n-1} > x_1\right]}{P\left[D^{t,*}_{i-1,n-1} > x_1\right]},$$

which is equivalent to

$$\frac{E\left[\bar{F}_Y^{n-i}(x_2/(n-i)+Y_{i-1,i-1})\right]}{E\left[\bar{F}_Y^{n-i+1}(x_2/(n-i+1)+Y_{i-2,i-2}\right]} \geq \frac{E\left[\bar{F}_Y^{n-i}(x_1/(n-i)+Y_{i-1,i-1})\right]}{E\left[\bar{F}_Y^{n-i+1}(x_1/(n-i+1)+Y_{i-2,i-2})\right]}.$$

For $x > 0$, $y > 0$, let

$$\psi_1(x,y) = \bar{F}_Y^{n-i}\left(\frac{x}{n-i}+y\right), \quad \psi_2(x,y) = \bar{F}_Y^{n-i+1}\left(\frac{x}{n-i+1}+y\right).$$

Then the above inequality about expectations is equivalent to

$$\frac{E\left[\psi_1\left(x_2, Y_{i-1,i-1}\right)\right]}{E\left[\psi_2\left(x_2, Y_{i-2,i-2}\right)\right]} \geqslant \frac{E\left[\psi_1\left(x_1, Y_{i-1,i-1}\right)\right]}{E\left[\psi_2\left(x_1, Y_{i-2,i-2}\right)\right]}.$$

It can be seen that the result follows from Lemmas 7.1.3 and 7.1.4 immediately.
(2) It is sufficient to prove that, for $x_1 \leqslant x_2$ and $i \in \{2, \ldots, n-1\}$,

$$\frac{P\left[D_{i,n-2}^{t,*} > x_2\right]}{P\left[D_{i,n-1}^{t,*} > x_2\right]} \geqslant \frac{P\left[D_{i,n-2}^{t,*} > x_1\right]}{P\left[D_{i,n-1}^{t,*} > x_1\right]}.$$

From the above remark, we have

$$P\left[D_{i,n-2}^{t,*} > x\right] = \binom{n-2}{i-1} E\left[\psi_1\left(x, Y_{i-1,i-1}\right)\right],$$

and

$$P\left[D_{i,n-1}^{t,*} > x\right] = \binom{n-1}{i-1} E\left[\psi_2\left(x, Y_{i-1,i-1}\right)\right].$$

Proving the above inequality is equivalent to showing

$$\frac{E\left[\psi_1\left(x_2, Y_{i-1,i-1}\right)\right]}{E\left[\psi_2\left(x_2, Y_{i-1,i-1}\right)\right]} \geqslant \frac{E\left[\psi_1\left(x_1, Y_{i-1,i-1}\right)\right]}{E\left[\psi_1\left(x_1, Y_{i-1,i-1}\right)\right]}.$$

Then the above inequality follows from Lemmas 7.1.3 and 7.1.4 immediately and hence the result holds.

Remark The results of this theorem are of course valid for the $D_{i,n-1}^t$, that is, $D_{2,n-1} \leq_{hr} D_{3,n-1}^t \leq_{hr} \cdots \leq_{hr} D_{n-1,n-1}^t$, and $D_{i,n-1}^t \leq_{hr} D_{i,n-2}^t$, $n-2 \geq i$ for fixed i.

Since, X_1 is DFR if and only if $Y_1 = [X_1 - t|X_1 > t]$ is DFR, the following results are easily obtained according to Kochar and Kirmani (1995). Hence their proofs are omitted.

Theorem 7.1.6 *Let $X_1, \ldots, X_n$ be independent and identically distributed component lifetimes of a series system with X_1 being a DFR distribution. Then*

(1) $D^t_{i,n-1} \leq_{disp} D^t_{i+1,n-1}$ *for* $i \in \{2, \ldots, n-2\}$;
(2) $Var(D^t_{i,n-1}) \leq Var(D^t_{i+1,n-1})$ *for* $i \in \{2, \ldots, n-2\}$;
(3) $D^t_{i,n-1} \leq_{disp} D^t_{i,n-2}$ *for* $i \in \{2, \ldots, n-2\}$;
(4) $Var(D^t_{i,n-1}) \leq Var(D^t_{i+1,n-2})$ *for* $i \in \{2, \ldots, n-3\}$,

where $Var(\cdot)$ *is the variance of a random variable.*

Recall that $\boldsymbol{x} \overset{\star}{\ll} \boldsymbol{y}$ if $\sum_{i=1}^{j} y_i \leqslant \sum_{i=1}^{j} x_i$ for $j \in \{1, \ldots, n-1\}$ and $\sum_{i=1}^{n} y_i = \sum_{i=1}^{n} x_i$, where $\boldsymbol{x} = (x_1, x_2, \ldots, x_n)$ and $\boldsymbol{y} = (y_1, y_2, \ldots, y_n)$. Let $\{x_{(1)} \leqslant \ldots \leqslant x_{(n)}\}$ and $\{y_{(1)} \leqslant \ldots \leqslant y_{(n)}\}$ denote the increasing arrangement of the components of the vector $\boldsymbol{x} = (x_1, x_2, \ldots, x_n)$ and $\boldsymbol{y} = (y_1, y_2, \ldots, y_n)$, respectively. The vector $\boldsymbol{y}$ is said to majorize the vector $\boldsymbol{x}$ (written as $\boldsymbol{x} \overset{m}{\leqslant} \boldsymbol{y}$) if $\sum_{i=1}^{j} y_{(i)} \leqslant \sum_{i=1}^{j} x_{(i)}$ for $j \in \{1, \ldots, n-1\}$, and $\sum_{i=1}^{n} y_{(i)} = \sum_{i=1}^{n} x_{(i)}$. A real-valued function ϕ defined on a set $\mathcal{A} \subset \mathbb{R}^n$ is said to be (a) ISO* if $\phi(\boldsymbol{x}) \leqslant \phi(\boldsymbol{y}), \forall \boldsymbol{x} \overset{\star}{\ll} \boldsymbol{y}$; (b) Schur-convex (Schur-concave) on $\mathcal{A}$ if $\boldsymbol{x} \overset{m}{\leqslant} \boldsymbol{y} \Rightarrow \phi(\boldsymbol{x}) \leqslant (\geqslant)\phi(\boldsymbol{y})$. For more details about the above definitions, please see Robertson and Wright (1982).

Below, we shall prove that the joint density of the normalized spacings is ISO* when the joint density of the parent observations is convex. To get this result, we introduce the following lemma (see Kochar and Kirmani 1995).

Lemma 7.1.7 *Let* $d_i \geqslant 0, d_i' \geqslant 0, i \in \{1, \ldots, n-1\}$, *be real numbers and let* $\boldsymbol{d}' = (d_1', \ldots, d_{n-1}')$, $\boldsymbol{d} = (d_1, \ldots, d_{n-1})$. *Assume that*

$$u_j = \sum_{i=1}^{j} \frac{d_i}{n-i+1}, \quad v_j = \sum_{i=1}^{j} \frac{d_i'}{n-i+1}, \quad j = 1, \ldots, n-1.$$

Then

$$\boldsymbol{d}' \overset{*}{\ll} \boldsymbol{d} \Longrightarrow \boldsymbol{v} \overset{m}{\leqslant} \boldsymbol{u},$$

where $\boldsymbol{v} = (v_1, \ldots, v_{n-1})$ *and* $\boldsymbol{u} = (u_1, \ldots, u_{n-1})$.

Theorem 7.1.8 *Assume* $D^{t,*}_{1,n-1} = (n-1)(X_{2,n} - X_{1,n} \mid X_{1,n} = t)$ *for convenience. If let the joint density* $f_X(x_1, \ldots, x_n)$ *of* $X = (X_1, \ldots, X_n)$ *be convex. Then, for fixed* $t > 0$, *the joint density of* $\boldsymbol{D} = \left(D^{t,*}_{1,n-1}, \ldots, D^{t,*}_{n-1,n-1}\right)$ *is ISO*.*

Proof Let $Y_1, \ldots, Y_{n-1}$ be independent random variables, which have the same distribution as $(X_1 - t | X_1 > t)$. It is noted that the joint density of $Y' = \left(Y_{1,n-1}, \ldots, Y_{n-1,n-1}\right)$ is

$$f_{Y'}(y_1, y_2, \ldots, y_{n-1}) = \begin{cases} \sum_P f_{Y_1,Y_2,\ldots,Y_{n-1}}\left(y_{j_1}, y_{j_2}, \ldots, y_{j_{n-1}}\right) & \text{if } 0 < y_{j_1} \leqslant y_{j_2} \leqslant \ldots \leqslant y_{j_{n-1}} \\ 0 & \text{otherwise,} \end{cases}$$

where $f_{Y_1,Y_2,\ldots,Y_{n-1}}$ is the joint density function of $Y = (Y_1, Y_2, \ldots, Y_{n-1})$ and $\sum_P$ is summation over all permutations $(j_1, j_2, \ldots, j_{n-1})$ of $n-1$ integers $\{1, 2, \ldots, n-$

1}. Therefore, by some computations, we can obtain the joint density of the normalized conditional spacings $\boldsymbol{D}^* = \left(D^{t,*}_{1,n-1}, \ldots, D^{t,*}_{n-1,n-1}\right)$ is given as

$$
\begin{aligned}
& f_{\boldsymbol{D}}\left(d_1, d_2, \ldots, d_{n-1}\right) \\
= & \sum_P f_{Y_1, Y_2, \ldots, Y_{n-1}}\left(\frac{d_1}{n-1}, \frac{d_1}{n-1}+\frac{d_2}{n-2}, \ldots, \frac{d_1}{n-1}+\frac{d_2}{n-2}+\cdots+\frac{d_{n-2}}{2}+d_{n-1}\right),
\end{aligned}
$$

where $d_i \geq 0$ for $i = 1, \ldots, n-1$. Also, that $f_X\left(x_1, \ldots, x_{n-1}, x_n\right)$ is convex implies that $f_Y\left(x_1, \ldots, x_{n-1}\right)$ is convex. Then, according to Marshall and Olkin (2010),

$$
\sum_P f_{Y_1, Y_2, \ldots, Y_{n-1}}\left(\frac{d_1}{n-1}, \frac{d_1}{n-1}+\frac{d_2}{n-2}, \ldots, \frac{d_1}{n-1}+\frac{d_2}{n-2}+\cdots+\frac{d_{n-2}}{2}+d_{n-1}\right)
$$

is a Schur-convex function and hence the result follows from Lemma 7.1.7 immediately.

Example 7.1.9 Suppose that $X_1, \ldots, X_n$ are independent Exponential distributions with identical hazard rate λ, then for $i \in \{2, \ldots, n-1\}$, fixed $t > 0$ and any $x > 0$, by the procedure of the proof of Theorem 7.1.1, we have

$$
\begin{aligned}
P(D^t_{i,n-1} > x) = & \ P\left(X_{i+1,n} - X_{i,n} > x \mid X_{1,n} = t\right) \\
= & \ (n-1)\binom{n-2}{i-2} \int_0^{+\infty} e^{-(n-i)\lambda(y+x)}\left(1-e^{-\lambda y}\right)^{i-2} \lambda e^{-\lambda y} dy \\
= & \ e^{-(n-i)\lambda x} \int_0^{+\infty} f_{X_{i,n-1}}(y) dy \\
= & \ e^{-(n-i)\lambda x},
\end{aligned}
$$

and $P(D^{t,*}_{i,n-1} > x) = e^{-\lambda x}$. That is, $D^t_{i,n-1}$ is Exponential distribution and depend on the subscript i of spacing, while $D^{t,*}_{i,n-1}$ is Exponential distribution and does not depend on the subscript i of spacing, with the same hazard rate λ as X_1.

Obviously, under the Exponential assumption of parent distribution, for fixed $t > 0$, we have: (1) the joint density function of $D^t_{2,n-1}, \ldots, D^t_{n-1,n-1}$ is the product of the marginal Exponential densities, and hence $D^t_{2,n-1}, \ldots, D^t_{n-1,n-1}$ are independent Exponential random variable with corresponding parameters $(n-2)\lambda, \ldots, \lambda$; (2) $D^{t,*}_{2,n-1}, \ldots, D^{t,*}_{n-1,n-1}$ are independent and identically distributed Exponential random variable with the same rate λ as X_1; (3) $(n-i)\lambda D^t_{i,n}$ is independent and distributed as Exponential with parameter 1, hence $\sum_{i=2}^{n-1}(n-i)\lambda D^t_{i,n}$ is Gamma distribution with parameters $n-2$ and 1. Moreover, $2\sum_{i=2}^{n-1}(n-i)\lambda D^t_{i,n}$ is Chi'-square distribution with degree $n-2$.

In the following, let's look at the stochastic comparisons among conditional spacings when the distributions of component lifetimes are Exponential. According to Example 7.1.9, it is not difficult to get the following results and hence their proofs are omitted here.

Theorem 7.1.10 *Let $X_1, \ldots, X_n$ be independent and identically distributed component lifetimes of a series system, and X_1 be an Exponential distribution. Then, for fixed $t > 0$, $D^t_{2,n-1} \le_{lr} D^t_{3,n-1} \le_{lr} \cdots \le_{lr} D^t_{n-1,n-1}$, and $D^t_{n-1,n-1} \le_{lr} D^t_{n-2,n-1} \le_{lr} \cdots \le_{lr} D^t_{2,n-1}$.*

Theorem 7.1.11 *Suppose $X_1, \ldots, X_n$ are independent and identically distributed component lifetimes of a series system, and $Z_1, \ldots, Z_n$ are independent and identically distributed component lifetimes of another series system, and $X_{i,n}$, $Z_{i,n}$ are corresponding order statistics. Assume that $D^{X,t}_{i,n-1} = \left(X_{i+1,n} - X_{i,n} \mid X_{1,n} = t\right)$, $D^{Z,t}_{i,n-1} = \left(Z_{i+1,n} - Z_{i,n} \mid Z_{1,n} = t\right)$. If X_1 and Z_1 are both Exponential distributions, such that $X_1 \le_{lr} Z_1$, then, $D^{X,t}_{i,n-1} \le_{lr} D^{Y,t}_{i,n-1}$, $i \in \{2, \ldots, n-1\}$.*

Obviously, the results of Theorems 16, and 17 are of course valid for the case of conditional normalized spacings. That is, $D^{t,*}_{2,n-1} \le_{lr} D^{t,*}_{3,n-1} \le_{lr} \cdots \le_{lr} D^{t,*}_{n-1,n-1}$, $D^{t,*}_{n-1,n-1} \le_{lr} D^{t,*}_{n-2,n-1} \le_{lr} \cdots \le_{lr} D^{t,*}_{2,n-1}$, and $D^{X,t,*}_{i,n-1} \le_{lr} D^{Y,t,*}_{i,n-1}$, $i \in \{2, \ldots, n-1\}$.

In the end, let us investigate the dependence of conditional spacings. A set of random variables $X_1, X_2, \ldots, X_n$ are said to be MTP_2 dependent if their joint density function is MTP_2. It is shown by Karlin and Rinott (1980) that the spacings of a random sample from a distribution with log-convex density $f(x)$ are MTP_2 dependent. It is not difficult to find that $f(x)$ is log-convex implies $f(x+t)/\bar{F}(t)$ is log-convex in x, for fixed $t > 0$, hence we have the following theorem.

Theorem 7.1.12 *Let $X_1, \ldots, X_n$ be independent and identically distributed component lifetimes of a series system, with X_1 having absolutely continuous density function $f(x)$. If $f(x)$ is log-convex, then, for fixed $t > 0$, $D^t_{1,n-1}, \ldots, D^t_{n-1,n-1}$ are MTP_2 dependent.*

Khaledi and Kochar (2000) proved that if the joint probability density function of $X_1, \ldots, X_n$ is permutation symmetric, TP_2 in pairs, and log-convex in each argument, then their spacings $D_{1,n}, \ldots, D_{n,n}$ are MTP_2 dependent. That is, they extended to the case when random variables are dependent. For the conditional spacings, the similar result also holds as stated in the following theorem.

Theorem 7.1.13 *Suppose $X_1, \ldots, X_n$ are exchangeable random variables with absolutely continuous joint density function $f_X(x_1, \ldots, x_n)$, satisfying the following conditions: (1) $f_X(x_1, \ldots, x_n)$ is TP_2 in pairs; (2) $f_X(x_1, \ldots, x_n)$ is log-convex in each argument when remaining arguments are fixed; (3) The first partial derivative of $f_X(x_1, \ldots, x_n)$ with respect to x_i exists for $i = 1, \ldots, n$. Then, for fixed $t > 0$, $D^t_{1,n-1}, \ldots, D^t_{n-1,n-1}$ are MTP_2 dependent.*

Proof By the proof procedure of Theorem 7.1.8, we can obtain the joint density of the conditional spacings $\boldsymbol{D} = \left(D^t_{1,n-1}, \ldots, D^t_{n-1,n-1}\right)$ is

$$f_{\boldsymbol{D}}\left(d_1, d_2, \ldots, d_{n-1}\right) = (n-1)! f_{Y_1, Y_2, \ldots, Y_{n-1}}\left(d_1, d_1 + d_2, \ldots, d_1 + d_2 + \cdots + d_{n-2} + d_{n-1}\right),$$

for $d_i \geq 0, i = 1, \ldots, n-1$, where $Y_1, Y_2, \ldots, Y_{n-1}$ are exchangeable random variables having the same distribution as $(X_1 - t|X_1 > t)$. It can be seen that the joint density $f_{Y_1,Y_2,\ldots,Y_{n-1}}$ satisfies all the conditions of the theorem. Noting the remaining proof is similar to that of Theorem 2.1 of Khaledi and Kochar (1999) and hence it is omitted.

Remark In the above theorem, if the conditions are changed to: (1) $f_X(x_1, \ldots, x_n)$ is RR_2 in pairs; (2) $f_X(x_1, \ldots, x_n)$ is log-convave in each argument, then for fixed $t > 0$, the joint density of $D^t_{1,n-1}, \ldots, D^t_{n-1,n-1}$ is RR_2 in pairs.

7.1.3 Conclusions

When one of components from a series system with n independent and identically distributed components fails at time $t(> 0)$, the other components still remain alive. These unfailed components may be used for life tests or other purposes. Motivated by this, this paper introduces the concepts of conditional spacings and normalized conditional ones given the component lifetimes are independent and identically distributed. First, the marginal survival function and their joint survival one of conditional spacings are obtained. Then, the hazard rate order: $D^{t,*}_{i,n-1} \preceq_{hr} D^{t,*}_{i+1,n-1}$ and $D^{t,*}_{i,n-1} \preceq_{hr} D^{t,*}_{i,n-2}$ are conducted. We conjecture the likelihood ratio order between them may hold, which is an open problem. Also, other stochastic behaviors of the conditional spacings are gained when the parent distribution of component lifetimes is DFR. Moreover, it is obtained that $D^{t,*}_{i,n-1} \preceq_{lr} D^{t,*}_{i+1,n-1}$ and $D^{t,*}_{i,n-1} \preceq_{lr} D^{t,*}_{i,n-2}$ when the component lifetimes are independent and identically distributed as Exponential, and also $D^t_{2,n-1}, \ldots, D^t_{n-1,n-1}$ are independent spacings. And for two different series system having two different sets of components, we get the likelihood ratio order between corresponding spacings given that there exists the likelihood ratio order between their component lifetimes. And some statistical properties and the dependence of conditional spacings are investigated in the end.

7.2 Conditional Spacings Under Coherent Systems with IID Components

In this section, we mainly focus on the (normalized) conditional spacings based on unfailed components when an $(n-k+1)$-out-of-$n$system or a coherent system with signature in (7.2) just fails at time t. Firstly, the section is mainly for stochastic behaviors of conditional spacings based on an $(n-k+1)$-out-of-n system. The (joint) reliability functions of conditional spacings are obtained and some comparisons among conditional spacings are conducted. Then, some stochastic behaviors of conditional spacings based on a coherent system with signature form in (7.2) are investigated. It is obtained there that the reliability function of conditional spacings based on a coherent system can be represented as a mixture of those of the conditional spacings based on an $(n-k+1)$-out-of-n system with conditional signature as mixture coefficients.

7.2.1 Introductions and Preliminaries

It is well known that the lifetime of an $(n-k+1)$-out-of-n system is just the order statistics $X_{k,n}$ of random variables $X_1, X_2, \ldots, X_n$, and when the $(n-k+1)$-out-of-n system just fails at some time, there is still $n-k$ components that remain alive. Also, it is noted, for a coherent system of size n having its signature vector to be

$$s = (s_1, s_2, \ldots, s_r, 0, \ldots, 0), r = 1, \ldots, n-1, s_r > 0, \tag{7.1}$$

$\sum_{k=1}^{r} s_k = 1$, this system with signature in (7.1) has the property that the components with lifetimes $X_{j,n}$ $(j = r+1, \ldots, n)$ would never fail at the time of failure of the system.

Given that an $(n-k+1)$-out-of-n system with independent and identically distributed components just fails at time t, for $i \in \{k+1, \ldots, n-1\}$ and fixed $t > 0$, let

$$D^t_{i-k,n-k-1} = \left(X_{i+1,n} - X_{i,n} \mid X_{k,n} = t\right), D^{t,*}_{i-k,n-k-1} = (n-i)\left(X_{i+1,n} - X_{i,n} \mid X_{k,n} = t\right)$$

be the corresponding conditional spacing and normalized conditional spacing based on $n-k$ unfailed components. Given that a coherent system having independent and identically distributed components, with signature form

$$s = (s_1, s_2, \ldots, s_r, 0, \ldots, 0), r = 1, \ldots, n-3, s_r > 0, \tag{7.2}$$

just fails at time t, let

$$S^t_{i-r,n-r-1} = \left(X_{i+1,n} - X_{i,n} \mid T = t\right), S^{t,*}_{i-r,n-r-1} = (n-i)\left(X_{i+1,n} - X_{i,n} \mid T = t\right)$$

be the corresponding conditional spacing and normalized conditional spacing based on $n-r$ unfailed components for $i = r+1, \ldots, n-1$. These conditional spacings may be regarded as the conditional differences between consecutive observations.

7.2.2 On Conditional Spacings Based on an $(n-k+1)$-Out-of-n System

Let $X_1, \ldots, X_n$ be independent nonnegative random variables with a common distribution function F, reliability function $\bar{F}$ and density function f, and $X_{i,n}$ be the ith smallest order statistic among $X_1, \ldots, X_n$. Let $Y_{i,n-k}$ be the ith order statistics of independent random variables $Y_1, \ldots, Y_{n-k}$, which have the same distribution as $(X_1 - t|X_1 > t)$. Denote the distribution function, density one and reliability one of random variable Y_1 by $F_Y(\cdot)$, $f_Y(\cdot)$ and $\bar{F}_Y(\cdot)$, respectively.

As the first result, we give the reliability function of (normalized) conditional spacings.

Theorem 7.2.1 *For $i \in \{k+1, \ldots, n-1\}$, fixed $t > 0$ and any $x > 0$, the reliability functions of $D^t_{i-k,n-k-1}$ and $D^{t,*}_{i-k,n-k-1}$ are*

$$P(D^t_{i-k,n-k-1} > x) = \binom{n-k}{i-k} E[\bar{F}_t^{n-i}(Y_{i-k,i-k} + x)],$$

$$P(D^{t,*}_{i-k,n-k-1} > x) = \binom{n-k}{i-k} E[\bar{F}_t^{n-i}(Y_{i-k,i-k} + \frac{x}{n-i})],$$

respectively, where $Y_{i-k,i-k} = \max\{Y_1, \ldots, Y_{i-k}\}$ and $E(\cdot)$ is the mean value of a random variable.

Proof For fixed $t > 0$, any $x > 0$, and $i = k+1, \ldots, n-1$,

$$\begin{aligned}
P(D^t_{i-k,n-k-1} > x) &= P(X_{i+1,n} - X_{i,n} > x \mid X_{k,n} = t) \\
&= \int_t^{\infty} P(X_{i+1,n} - X_{i,n} > x \mid X_{i,n} = y) f_{X_{i,n}|X_{k,n}=t}(y|t)dy \\
&= (i-k)\binom{n-k}{i-k} \int_t^{\infty} [\frac{\bar{F}(x+y)}{\bar{F}(t)}]^{n-i}[1 - \frac{\bar{F}(y)}{\bar{F}(t)}]^{i-k-1} \frac{f(y)}{\bar{F}(t)} dy \\
&= (i-k)\binom{n-k}{i-k} \int_0^{\infty} \bar{F}_t^{n-i}(x+y)[1 - \bar{F}_t^{i-k-1}(y)] f_t(y)dy \\
&= \binom{n-k}{i-k} \int_0^{+\infty} \bar{F}_t^{n-i}(y+x) dF_{Y_{i-k,i-k}}(y) \\
&= \binom{n-k}{i-k} E[\bar{F}_Y^{n-i}(Y_{i-k,i-k} + x)],
\end{aligned}$$

where $\bar{F}_t(x+u) = \bar{F}(t+x+u)/\bar{F}(t)$ and $f_t(u) = f(t+u)/\bar{F}(t)$ are the reliability function and density one of $(X_1 - t|X_1 > t)$ (or Y_1), respectively, and $f_{X_{i,n}|X_{k,n}=t}(y|t)$ is the conditional density of $X_{i,n}$ given $X_{k,n} = t$. And hence, it is easy to obtain

$$P(D^{t,*}_{i-k,n-k-1} > x) = \binom{n-k}{i-k} E[\bar{F}_Y^{n-i}(Y_{i-k,i-k} + \frac{x}{n-i})].$$

The following Theorems 7.2.2 and 7.2.5 give some comparisons between $D^t_{i,n-k-1}$ $(D^{t,*}_{i,n-k-1})$ and $D^t_{i+1,n-k-1}(D^{t,*}_{i+1,n-k-1})$, also $D^t_{i,n-k-1}$ $(D^{t,*}_{i,n-k-1})$ and $D^t_{i,n-k-2}(D^{t,*}_{i,n-k-2})$. The proof of Theorem 7.2.2 is similar to that of Theorem 7.1.5 of Zhang et al. (2022), and the proof of Theorem 7.2.5 is similar to that of Theorem 2 of Kochar and Kirmani (1995), and hence are omitted.

Theorem 7.2.2 *Let $X_1, \ldots, X_n$ be independent and identically distributed component lifetimes of an $(n-k+1)$-out-of-n system, and X_1 has a DFR distribution. Then, for fixed $t > 0$,*

(a) $D^{t,*}_{1,n-k-1} \le_{hr} D^{t,*}_{2,n-k-1} \le_{hr} \cdots \le_{hr} D^{t,*}_{n-k-1,n-k-1}$;
(b) $D^{t,*}_{j,n-k-1} \le_{hr} D^{t,*}_{j,n-k-2}$ *for each fixed* $j \in \{1, \ldots, n-k-2\}$.

The following example gives an illustration of Theorem 7.2.2.

Example 7.2.3 Suppose a series system has four components whose lifetimes are IID following the first kind of Pareto distribution, that is, for $x \ge 1$, the reliability function and density function of the component lifetime are $\bar{F}(x) = x^{-1}$ and $f(x) = x^{-2}$, respectively. It is clear that the component lifetime is DFR. Now, given this system has just failed at time t $(t > 0)$, we have

$$D^{t,*}_{1,2} = 2\left(X_{3,4} - X_{2,4} \mid X_{1,4} = t\right),\ D^{t,*}_{2,2} = \left(X_{4,4} - X_{3,4} \mid X_{1,4} = t\right).$$

Assume that the system just fails at time $t = 2$. Then, after some computations, we find

$$P(D^{2,*}_{1,2} > x) = 24\int_2^\infty \frac{1}{y^2(y+x/2)^2}dy = \frac{48[x^2 + 8x + 8(x+4)\ln\frac{4}{x+4}]}{x^3(x+4)},$$

$$P(D^{2,*}_{2,2} > x) = 48\int_2^\infty \frac{1}{y^2(y+x)} \times \left(\frac{1}{2} - \frac{1}{y}\right)dy = \frac{6(x+4)}{x^2} + \frac{24(x+2)}{x^3} \times \ln\frac{2}{x+2},$$

for $x > 0$. It can be seen, from Fig. 7.1, that the ratio of $P(D^{2,*}_{2,2} > x)$ to $P(D^{2,*}_{1,2} > x)$ is increasing in $x \in (0, 100)$, that is, $D^{2,*}_{1,2} \le_{hr} D^{2,*}_{2,2}$.

One may now ask whether the condition DFR can be dropped in Theorem 7.2.2. The following example gives a negative answer.

Example 7.2.4 Suppose a series system has four components whose lifetimes are IID following Weibull distribution with shape parameter 2 and scale parameter 1, that is, the reliability function of the component lifetime is $\bar{F}(x) = e^{-x^2}$, for $x > 0$. It is well-known that the component lifetimes belong to the IFR class for shape parameter 2. For simplicity, let $t = 0$. Then, after performing some computations, we find

$$P(D^{0,*}_{1,2} > x) = e^{-\frac{1}{2}x^2}\left[1 - \sqrt{\frac{\pi}{3}}xe^{\frac{1}{3}x^2}P\left(Z > \frac{1}{3}x^2\right)\right],$$

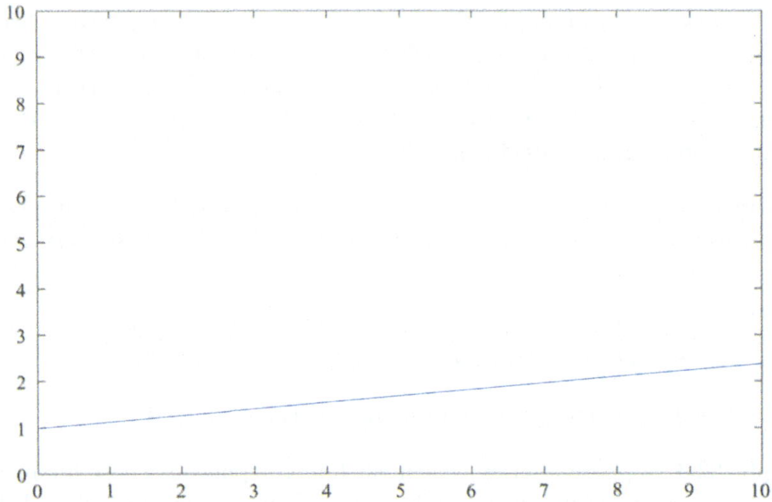

Fig. 7.1 Curve of $P(D_{2,2}^{2,*} > x)/P(D_{1,2}^{2,*} > x)$

$$P(D_{2,2}^{0,*} > x) = e^{-x^2}\left[1 - 3\sqrt{\frac{\pi}{2}}xe^{\frac{1}{2}x^2}P\left(Z > \frac{1}{2}x^2\right) + 2\sqrt{\frac{\pi}{3}}xe^{\frac{1}{3}x^2}P\left(Z > \frac{1}{3}x^2\right)\right],$$

where Z is a random variable that has Gamma distribution with shape parameter $1/2$ and scale parameter 1. From Fig. 7.2, we observe that the curve of $P(D_{2,2}^{0,*} > x)/P(D_{1,2}^{0,*} > x)$ is not increasing in $x \in (0, 10)$, that is, $D_{1,2}^{0,*} \nleq_{hr} D_{2,2}^{0,*}$.

Moreover, given that the system has just failed at time $t = 1$, then, after some computations, for $x > 0$, we find

Fig. 7.2 Curve of $P(D_{2,2}^{0,*} > x)/P(D_{1,2}^{0,*} > x)$

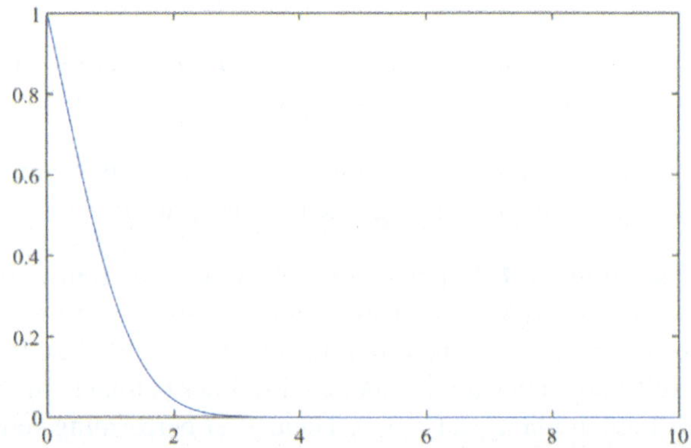

$$P(D_{1,2}^{1,*} > x) = e^{3-\frac{1}{6}x^2} \times \left[e^{-3(1+\frac{1}{3}x)^2} - \frac{\sqrt{3\pi}}{3} x P\left(Z > 3(1+\frac{x}{3})^2\right) \right],$$

$$P(D_{1,1}^{1,*} > x) = e^{2-\frac{x^2}{2}} \times \left[e^{-2(1+\frac{x}{2})^2} - \frac{\sqrt{2\pi}}{2} x P\left(Z > 2(1+\frac{x}{2})^2\right) \right].$$

According to Fig. 7.3, the curve of $P(D_{1,2}^{1,*} > x)/P(D_{1,1}^{1,*} > x)$ is increasing in $x \in (0, 2)$, that is, $D_{1,2}^{1,*} \not\leq_{hr} D_{1,1}^{1,*}$.

Theorem 7.2.5 *Let $X_1, \ldots, X_n$ be independent and identically distributed component lifetimes of an $(n-k+1)$-out-of-n system with X_1 being a DFR distribution. Then, for fixed $t > 0$, and $i \in \{k+1, \ldots, n-2\}$*

(a) $D_{i-k,n-k-1}^t \leq_{disp} D_{i+1-k,n-k-1}^t$;
(b) $Var(D_{i-k,n-k-1}^t) \leq Var(D_{i+1-k,n-k-1}^t)$;
(c) $D_{i-k,n-k-1}^t \leq_{disp} D_{i-k,n-k-2}^t$;
(d) $Var(D_{i-k,n-k-1}^t) \leq Var(D_{i-k,n-k-2}^t)$.

The following theorems are similar to those of conditional spacing based on a series system, and hence their proofs are omitted here.

Theorem 7.2.6 *Let $X_1, \ldots, X_n$ be independent and identically distributed component lifetimes of an $(n-k+1)$-out-of-n system with X_1 being a DFR distribution. For convenience, denote $D_{0,n-k-1}^{t,*} = (n-k)(X_{k+1,n} - X_{k,n} \mid X_{k,n} = t)$. If the joint density $f_X(x_1 \ldots, x_n)$ of $X = (X_1, \ldots, X_n)$ is convex, then, the joint density of $\boldsymbol{D} = \left(D_{0,n-k-1}^{t,*}, D_{1,n-k-1}^{t,*}, \ldots, D_{n-k-1,n-k-1}^{t,*}\right)$ is ISO* for fixed $t > 0$.*

Theorem 7.2.7 *Let $X_1, \ldots, X_n$ be independent and identically distributed component lifetimes of an $(n-k+1)$-out-of-n system, with X_1 having absolutely continuous density function $f(x)$. If $f(x)$ is log-convex, then, for fixed $t > 0$, $D_{1,n-k-1}^t, \ldots, D_{n-k-1,n-k-1}^t$ are MTP_2 dependent.*

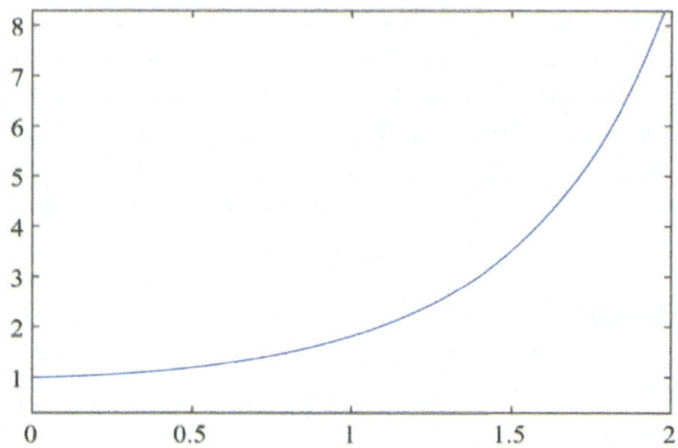

Fig. 7.3 Curve of $P(D_{1,2}^{1,*} > x)/P(D_{1,1}^{1,*} > x)$

Theorem 7.2.8 *Suppose $X_1, \ldots, X_n$ are exchangeable random variables with absolutely continuous joint density function $f_X(x_1, \ldots, x_n)$, satisfying the following conditions: (1) $f_X(x_1, \ldots, x_n)$ is TP_2 in pairs; (2) $f_X(x_1, \ldots, x_n)$ is log-convex in each argument when remaining arguments are fixed; (3) The first partial derivative of $f_X(x_1, \ldots, x_n)$ with respect to x_i exists for $i = 1, \ldots, n$. Then, for fixed $t > 0$, $D^t_{1,n-k-1}, \ldots, D^t_{n-k-1,n-k-1}$ are MTP_2 dependent.*

Remark In the above theorem, if the conditions are changed to: (1) $f_X(x_1, \ldots, x_n)$ is RR_2 in pairs; (2) $f_X(x_1, \ldots, x_n)$ is log-convave in each argument, then for fixed $t > 0$, the joint density of $D^t_{1,n-k-1}, \ldots, D^t_{n-k-1,n-k-1}$ is RR_2 in pairs.

In what follows, we consider the case when $X_1, \ldots, X_n$ are independent Exponential variables with identical hazard rate λ. According to Theorem 7.2.1, for $i \in \{k+1, \ldots, n-1\}$, fixed $t > 0$ and any $x > 0$, we have

$$P(D^t_{i-k,n-k-1} > x) = P\left(X_{i+1,n} - X_{i,n} > x \mid X_{k,n} = t\right) = e^{-(n-i)\lambda x},$$

and $P(D^{t,*}_{i-k,n-k-1} > x) = e^{-\lambda x}$. That is, $D^t_{i-k,n-k-1}$ is distributed as Exponential with hazard rate $(n-i)\lambda$, and $D^{t,*}_{i,n-1}$ is also Exponential distribution having the same hazard rate λ as X_1.

Theorem 7.2.9 *Suppose $X_1, \ldots, X_n$ are independent and identically distributed as Exponential with hazard rate λ. Then, for fixed $t > 0$, (1) the joint density of $D^t_{1,n-k-1}, \ldots, D^t_{n-k-1,n-k-1}$ is the product of the marginal Exponential densities, and hence $D^t_{1,n-k-1}, \ldots, D^t_{n-k-1,n-k-1}$ are independent Exponential random variable with corresponding parameters $(n-k-1)\lambda, \ldots, \lambda$; (2) $D^{t,*}_{1,n-k-1}, \ldots, D^{t,*}_{n-k-1,n-k-1}$ are independent and identically distributed Exponential random variable with the rate λ as X_1.*

Theorem 7.2.10 *Let $X_1, \ldots, X_n$ be independent and identically distributed component lifetimes of an $(n-k+1)$-out-of-n system, and X_1 be an Exponential distribution. Then, (1) $D^t_{1,n-k-1} \leq_{lr} D^t_{2,n-k-1} \leq_{lr} \cdots \leq_{lr} D^t_{n-k-1,n-k-1}$; (2) $D^t_{n-k-1,n-k-1} \leq_{lr} D^t_{n-k-2,n-k-1} \leq_{lr} \cdots \leq_{lr} D^t_{1,n-k-1}$.*

Theorem 7.2.11 *Suppose $X_1, \ldots, X_n$ are independent and identically distributed component lifetimes of an $(n-k+1)$-out-of-n system, and $Z_1, \ldots, Z_n$ are independent and identically distributed component lifetimes of another $(n-k+1)$-out-of-n system, and $X_{i,n}$, $Z_{i,n}$ are corresponding order statistics. Assume that $D^{X,t}_{i-k,n-k-1} = \left(X_{i+1,n} - X_{i,n} \mid X_{k,n} = t\right)$, $D^{Z,t}_{i-k,n-k-1} = \left(Z_{i+1,n} - Z_{i,n} \mid Z_{k,n} = t\right)$. If X_1 and Z_1 are both Exponential distributions, such that $X_1 \leq_{lr} Z_1$, then, $D^{X,t}_{i-k,n-k-1} \leq_{lr} D^{Z,t}_{i-k,n-k-1}$, $i \in \{k+1, \ldots, n-k-1\}$.*

Remark For $i \in \{k+1, \ldots, n-k-1\}$, fixed $t > 0$, let $D^{X,t,*}_{i-k,n-k-1} = (n-i)\left(X_{i+1,n} - X_{i,n} \mid X_{k,n} = t\right)$ and $D^{Z,t,*}_{i-k,n-k-1} = (n-i)\left(Z_{i+1,n} - Z_{i,n} \mid Z_{k,n} = t\right)$. Obviously, the results of Theorems 7.2.9, 7.2.10 and 7.2.11 are of course valid for the case of conditional normalized spacings. That is, $D^{t,*}_{1,n-k-1} \leq_{lr} D^{t,*}_{2,n-k-1} \leq_{lr}$

$\cdots \leq_{lr} D^{t,*}_{n-k-1,n-k-1}$, $D^{t,*}_{n-k-1,n-k-1} \leq_{lr} D^{t,*}_{n-k-2,n-k-1} \leq_{lr} \cdots \leq_{lr} D^{t,*}_{1,n-k-1}$, and $D^{X,t,*}_{i-k,n-k-1} \leq_{lr} D^{Z,t,*}_{i-k,n-k-1}$, $i \in \{k+1, \dots, n-k-1\}$.

In the end of this part, we give an application of conditional spacings in Type-II censoring. Given an $n-k+1$-out-of-n system fails at time t, there are $n-k$ components still alive. Let $Z_1, \dots, Z_{n-k}$ be the residual lifetimes of these $n-k$ surviving components, and $Z_{1,n-k}, \dots, Z_{n-k,n-k}$ be the corresponding order statistics. If we put these $n-k$ old items on a life-test and the test is terminated at the time of m–th ($m \leq n-k$) failure, then, the total time on test is $T_{m,n-k} = \Sigma_{i=1}^{m}(n-i)\left(Y_{i,n-k} - Y_{i-1,n-k}\right)$, where $Z_{0,n-k} = 0$. If these component lifetimes are independent and identically distributed as exponential with mean α ($\alpha > 0$) and hazard rate λ ($\alpha = 1/\lambda$), then, from Epstein (1960), $\hat{\alpha} = T_{m,n-k}/r$ is the uniformly minimum-variance unbiased estimator of α and $2T_{m,n-k}/m$ converges to χ^2_{2m} in distribution, where χ^2_{2m} denotes chi-square distribution with $2m$ degrees of freedom. It is well known that, based on this, many statistical estimation and testing procedures have been carried out. Thus, the results established provide some additional insight into these inferential aspects.

7.2.3 *On Conditional Spacings Based on a General Coherent System*

In this section, we consider the conditional spacings for a coherent system of size n having its signature vector as $s = (s_1, s_2, \dots, s_r, 0, \dots, 0)$, $r = 1, \dots, n-3$, $s_r > 0$.

The first theorem gives the reliability function of conditional spacing $S^t_{i-r,n-r-1}$.

Theorem 7.2.12 *For $i \in \{r+1, \dots, n-1\}$, fixed $t > 0$ and any $x > 0$, the reliability function of $S^t_{i-r,n-1}$ is*

$$P(S^t_{i-r,n-r-1} > x) = \sum_{k=1}^{r} s_k(t) P(D^t_{i-k,n-k-1} > x),$$

where $s_k(t) = P(T = X_{k,n} | T = t)$, a kind of conditional signature, is a probability of the j-th failure times of components given the system just fails at time t, such that $s_k(t) > 0$ and $\sum_{k=1}^{r} s_k(t) = 1$.

Proof For fixed $t > 0$, any $x > 0$, and $i = r+1, \dots, n-1$,

$$\begin{aligned}
P(S^t_{i-r,n-r-1} > x) &= P\left(X_{i+1,n} - X_{i,n} > x \mid T = t\right) \\
&= \sum_{k=1}^{r} P(X_{i+1,n} - X_{i,n} > x, T = X_{k,n} \mid T = t) \\
&= \sum_{k=1}^{r} P(X_{i+1,n} - X_{i,n} > x, \mid T = X_{k,n} = t) P(T = X_{k,n} \mid T = t)
\end{aligned}$$

$$= \sum_{k=1}^{r} P(X_{i+1,n} - X_{i,n} > x, \mid X_{k,n} = t) P(T = X_{k,n} \mid T = t)$$

$$= \sum_{k=1}^{r} s_k(t) P(X_{i+1,n} - X_{i,n} > x \mid X_{k,n} = t)$$

$$= \sum_{k=1}^{r} s_k(t) P(D^t_{i-k,n-k-1} > x).$$

That is, the result holds.

Remark By the above result, we can easily obtain, for fixed $t > 0$, any $x > 0$,

$$P(S^{t,*}_{i-r,n-r-1} > x) = \sum_{k=1}^{r} s_k(t) P(X_{i+1,n} - X_{i,n} > \frac{x}{n-i} \mid X_{k,n} = t),$$

for $i = r + 1, \ldots, n - 1$; Also, by the result, it can be seen that the reliability function of the conditional spacings based on a coherent system with signature as (7.2) can be represented as a mixture of those of the conditional spacings based on an $(n - k + 1)$-out-of-n system with conditional signature as mixture coefficients. Finally, we can easily obtain that $s_1(t)$ is decreasing in time t, while $s_r(t)$ is increasing in time t.

Remark If let $X_1, \ldots, X_n$ be independent Exponential variables with identical hazard rate λ. According to Theorem 7.2.12, for $i \in \{r + 1, \ldots, n - 1\}$, fixed $t > 0$ and any $x > 0$, we have

$$P(S^t_{i-r,n-r-1} > x) = e^{-(n-i)\lambda x},$$

and $P(D^{t,*}_{i-k,n-k-1} > x) = e^{-\lambda x}$. That is, $D^t_{i-k,n-k-1}$ is distributed as Exponential with hazard rate $(n - i)\lambda$, and $D^{t,*}_{i,n-1}$ is also Exponential distribution having the same hazard rate λ as X_1.

Theorems 7.2.13, 7.2.14 and 7.2.15 below are easily get when component lifetimes are Exponential.

Theorem 7.2.13 *Suppose $X_1, \ldots, X_n$ are independent and identically distributed as Exponential with hazard rate λ. Then, for fixed $t > 0$, (1) the joint density of $D^t_{1,n-k-1}, \ldots, D^t_{n-k-1,n-k-1}$ is the product of the marginal Exponential densities, and hence $D^t_{1,n-k-1}, \ldots, D^t_{n-k-1,n-k-1}$ are independent Exponential random variable with corresponding parameters $(n - k - 1)\lambda, \ldots, \lambda$; (2) $D^{t,*}_{1,n-k-1}, \ldots, D^{t,*}_{n-k-1,n-k-1}$ are independent and identically distributed Exponential random variable with the rate λ as X_1.*

Theorem 7.2.14 *Let $X_1, \ldots, X_n$ be independent and identically distributed component lifetimes of an $(n - k + 1)$-out-of-n system, and X_1 be an Exponential distribution. Then, (1) $D^t_{1,n-k-1} \leq_{lr} D^t_{2,n-k-1} \leq_{lr} \cdots \leq_{lr} D^t_{n-k-1,n-k-1}$; (2) $D^t_{n-k-1,n-k-1} \leq_{lr} D^t_{n-k-2,n-k-1} \leq_{lr} \cdots \leq_{lr} D^t_{1,n-k-1}$.*

Theorem 7.2.15 *Suppose $X_1, \ldots, X_n$ are independent and identically distributed component lifetimes of an $(n-k+1)$-out-of-n system, and $Z_1, \ldots, Z_n$ are independent and identically distributed component lifetimes of another $(n-k+1)$-out-of-n system, and $X_{i,n}$, $Z_{i,n}$ are corresponding order statistics. Assume that $D^{X,t}_{i-k,n-k-1} = \left(X_{i+1,n} - X_{i,n} \mid X_{k,n} = t\right)$, $D^{Z,t}_{i-k,n-k-1} = \left(Z_{i+1,n} - Z_{i,n} \mid Z_{k,n} = t\right)$. If X_1 and Z_1 are both Exponential distributions, such that $X_1 \le_{lr} Z_1$, then, $D^{X,t}_{i-k,n-k-1} \le_{lr} D^{Z,t}_{i-k,n-k-1}$, $i \in \{k+1, \ldots, n-k-1\}$.*

Remark For $i \in \{k+1, \ldots, n-k-1\}$, fixed $t > 0$, let $D^{X,t,*}_{i-k,n-k-1} = (n-i)\left(X_{i+1,n} - X_{i,n} \mid X_{k,n} = t\right)$ and $D^{Z,t,*}_{i-k,n-k-1} = (n-i)\left(Z_{i+1,n} - Z_{i,n} \mid Z_{k,n} = t\right)$. Obviously, the results of Theorems 7.2.13, 7.2.14 and 7.2.15 are of course valid for the case of conditional normalized spacings. That is, $D^{t,*}_{1,n-k-1} \le_{lr} D^{t,*}_{2,n-k-1} \le_{lr} \cdots \le_{lr} D^{t,*}_{n-k-1,n-k-1}$, $D^{t,*}_{n-k-1,n-k-1} \le_{lr} D^{t,*}_{n-k-2,n-k-1} \le_{lr} \cdots \le_{lr} D^{t,*}_{1,n-k-1}$, and $D^{X,t,*}_{i-k,n-k-1} \le_{lr} D^{Z,t,*}_{i-k,n-k-1}$, $i \in \{k+1, \ldots, n-k-1\}$.

Example 7.2.16 Consider a coherent system with its lifetime to be $T = \min\{X_1, X_2, X_3, \max\{X_4, X_5\}\}$, whose signature is $s = (3/5, 2/5, 0, 0, 0)$. After some computations, we have

$$s_1(t) = \frac{3\bar{F}(t)}{3\bar{F}(t) + 8F(t)},\ s_2(t) = \frac{8F(t)}{3\bar{F}(t) + 8F(t)},$$

for fixed $t > 0$. Therefore, for fixed $t > 0$, from Theorem 7.2.12, it follows

$$P(S^t_{1,2} > x) = s_1(t)P(X_{4,5} - X_{3,5} > x \mid X_{1,5} = t) + s_2(t)P(X_{4,5} - X_{3,5} > x \mid X_{2,5} = t),$$
$$P(S^t_{2,2} > x) = s_1(t)P(X_{5,5} - X_{4,5} > x \mid X_{1,5} = t) + s_2(t)P(X_{5,5} - X_{4,5} > x \mid X_{2,5} = t).$$

According to above Theorems 7.2.2, 7.2.5, the following theorem can be easily obtained.

Theorem 7.2.17 *Let $X_1, \ldots, X_n$ be independent and identically distributed component lifetimes of a coherent system with signature form as (7.2). Suppose X_1 is DFR. Then, for fixed $t > 0$, $S^{t,*}_{i-r,n-r-1} \le_{st} S^{t,*}_{i+1-r,n-r-1}$, and $Var(S^{t,*}_{i-r,n-r-1}) \le Var(S^{t,*}_{i+1-r,n-r-1})$.*

7.2.4 Conclusions

When an $(n-k+1)$-out-of-n system or a coherent system having its signature vector as $s = (s_1, s_2, \ldots, s_r, 0, \ldots, 0)$ just fails at time $t (> 0)$, some components from these systems still remain alive. These unfailed components may be used for life tests or other purposes. Motivated by this, this section introduces the concepts of conditional spacings and normalized conditional ones based on an $(n-k+1)$-out-of-n system or a coherent system. First, the marginal survival function and their joint survival one of conditional spacings are obtained. Then, the hazard rate order: $D^{t,*}_{i-k,n-k-1} \leq_{hr} D^{t,*}_{i+1-k,n-k-1}$ and $D^{t,*}_{i-k,n-k-1} \leq_{hr} D^{t,*}_{i-k,n-k-2}$ are conducted under the $(n-k+1)$-out-of-n system setting. Also, other stochastic behaviors of the conditional spacings are gained when the parent distribution of component lifetimes is DFR. Moreover, the hazard rate order is strengthened to the likelihood ratio order given the component lifetimes are independent and identically distributed as Exponential. And for two different $(n-k+1)$-out-of-n systems having two different sets of components, we get the likelihood ratio order between corresponding spacings given that there exists the likelihood ratio order between their component lifetimes. And some statistical properties and the dependence of conditional spacings are investigated under the $(n-k+1)$-out-of-n system setting. In the end, the reliability function of conditional spacings based on a coherent system can be represented as a mixture of the conditional spacings based on an $(n-k+1)$-out-of-n system with conditional signature as mixture coefficients and the reliability function of conditional spacings based on a coherent system can be represented as a mixture of the conditional spacings based on an $(n-k+1)$-out-of-n system with conditional signature as mixture coefficients, and the usual stochastic order between conditional spacings is get under the coherent system setting.

7.3 Conditional Spacings from Heterogeneous Exponential Random Variables

The conditional spacings are investigated based on a series system with n heterogeneous Exponential component lifetimes. First the survival function and their joint distribution of conditional spacings are obtained. Then, we show that the survival function of first spacing is Schur convex in the hazard rates of component lifetimes. Moreover, some stochastic comparisons between conditional spacings are conducted. In the end, the case of conditional sample range is researched.

7.3.1 Introductions and Preliminaries

Exponential distributions play an important role in many fields such as reliability theory and life testing etc. Due to the memoryless of the Exponential distributions a lot of work have been done in the literature on stochastic comparisons of spacings from independent but non-identically distributed exponential variables, including Pledger and Khaledi and Kochar (2001), Kochar and Ma (1999), and Kochar and Rojo (1996) etc.

In some applications there exists a mixture of experiments, a (random) change of scale in location may take place the random variables are still independent but no longer identically distributed. Therefore, this part focus on the conditional spacings based on a series system with independent Exponential distribution having unequal scale parameters. At first, the survival function and their joint distribution of conditional spacings are obtained based on heterogeneous Exponential random variables. Then, the first spacing is proved to be Schur convex in hazard rates. Moreover, we obtain that $D^t_{i,n-1} \leq_{st} D^t_{i+1,n-1}$, and $D^t_{1,n-1} \leq_{hr} D^t_{1,n-2}$. In the end, we consider the case of conditional sample range and give an upper bound and low one for the distribution of sample range in the sense of usual stochastic order.

7.3.2 Stochastic Results of Conditional Spacings

In this section, by utilizing the Markov property of order statistics, reliability functions of conditional spacings are first derived when the component lifetimes are independent but non-identically distributed as exponential. Next, we prove that the first spacing is Schur convex in hazard rates. Finally, some stochastic comparisons of conditional spacings are discussed.

Let X_i $(i = 1, 2, \ldots, n)$ be independent random variables with hazard rate λ_i (> 0). And let us denote $s = \lambda_1 + \lambda_2 + \cdots + \lambda_n$, $S = \{X_1, \ldots, X_n\}$, $S|i = S - \{X_i\}$, $S|i, j = S - \{X_i, X_j\}$, and so on. Next, for $i = 1, 2, \ldots, n-1$ and fixed $t > 0$, let $D^t_{i,n-1} = \left(X_{i+1,n} - X_{i,n} \mid X_{1,n} = t\right)$ and $D^{t,*}_{i,n-1} = (n-i)\left(X_{i+1,n} - X_{i,n} \mid X_{1,n} = t\right)$ be the corresponding conditional spacings and conditional normalized spacings. We first derive their reliability functions.

For given $X_{1,n} = t$, we first observe that the conditional joint density function of $(X_{2,n}, \ldots, X_{n,n})$ as

$$
\begin{aligned}
&\frac{f_{X_{1,n},X_{2,n},\ldots,X_{n,n}}(t, x_{2,n}, \ldots, x_{n,n})}{f_{X_{1,n}}(t)}\\
&= \frac{\sum_P f_{i_1}(t) f_{i_2}(x_{2,n}) \cdots f_{i_n}(x_{n,n})}{se^{-st}}, \quad t < x_{2,n} < \cdots < x_{n,n}\\
&= \frac{1}{s}\sum_P \lambda_{i_1} e^{-(\lambda_{i_1}-s)t} \lambda_{i_2} e^{-\lambda_{i_2} x_{2,n}} \cdots \lambda_{i_n} e^{-\lambda_{i_n} x_{n,n}}, \quad s - \lambda_{i_1} = \lambda_{i_2} + \lambda_{i_3} + \cdots + \lambda_{i_n}\\
&= \sum_P \frac{\lambda_{i_1}}{s} \lambda_{i_2} e^{-\lambda_{i_2}(x_{2,n}-t)} \cdots \lambda_{i_n} e^{-\lambda_{i_n}(x_{n,n}-t)}, \quad t < x_{2,n} < \cdots < x_{n,n},
\end{aligned}
$$

where $s = \sum_{i=1}^n \lambda_i$ and $\sum_P$ is summation over all $n!$ permutations of $\{1, 2, \ldots, n\}$.

We thus see that, given $X_{1,n} = t$, $(X_{2,n} - t, \ldots, X_{n,n} - t)$ are distributed as order statistics from $n-1$ exponential variables having parameters $\{\lambda_1, \ldots, \lambda_n\} \backslash \{\lambda_{i_1}\}$ with probability $\frac{\lambda_{i_1}}{s}$. Hence, we have the conditional density of $X_{2,n} - t$ given $X_{1,n} = t$, is readily seen to be $(s - \lambda_{i_1})e^{-(s-\lambda_{i_1})x}$ with probability $\frac{\lambda_{i_1}}{s}$, i.e., for $x > 0$,

$$P(D^t_{1,n-1} > x) = \sum_{i=1}^{n} \frac{\lambda_i}{s} e^{-(s-\lambda_i)x}. \tag{7.3}$$

Remark From (7.3), the reliability function of the first conditional spacing is a mixture of Exponential distribution $e^{-(s-\lambda_i)x}$ with corresponding weight λ_i/s for $i = 1, \ldots, n$.

Recall that the vector $\boldsymbol{y}$ is said to majorize the vector $\boldsymbol{x}$ (written as $\boldsymbol{x} \overset{m}{\leqslant} \boldsymbol{y}$) if $\sum_{i=1}^{j} y_{(i)} \leqslant \sum_{i=1}^{j} x_{(i)}$ for $j \in \{1, \ldots, n-1\}$, and $\sum_{i=1}^{n} y_{(i)} = \sum_{i=1}^{n} x_{(i)}$, where $\{x_{(1)} \leqslant \cdots \leqslant x_{(n)}\}$ and $\{y_{(1)} \leqslant \cdots \leqslant y_{(n)}\}$ denote the increasing arrangement of the components of the vector $\boldsymbol{x} = (x_1, x_2, \ldots, x_n)$ and $\boldsymbol{y} = (y_1, y_2, \ldots, y_n)$, respectively. A real-valued function ϕ defined on a set $\mathcal{A} \subset \mathbb{R}^n$ is said to be Schur-convex (Schur-concave) on $\mathcal{A}$ if $\boldsymbol{x} \overset{m}{\leqslant} \boldsymbol{y} \Rightarrow \phi(\boldsymbol{x}) \leqslant (\geqslant) \phi(\boldsymbol{y})$. For more details about these definitions, please see Robertson and Wright (1982).

The following theorem shows the Schur convex property of the first spacing $D^t_{1,n-1}$.

Theorem 7.3.1 For fixed $t > 0$, the survival function of $D^t_{1,n-1}$ is Schur convex in $(\lambda_1, \lambda_2, \ldots, \lambda_n)$.

Proof At each fixed point $x > 0$, $\lambda_i e^{\lambda_i x}$ is convex in λ_i, for each $i = 1, 2, \ldots, n$. Hence, from Marshall and Olkin (2010), the survival function $\sum_{i=1}^{n} \frac{\lambda_i}{s} e^{(\lambda_i - s)x} = \frac{e^{-sx}}{s} \sum_{i=1}^{n} \lambda_i e^{\lambda_i x}$ is Schur convex in $(\lambda_1, \lambda_2, \ldots, \lambda_n)$ for each x. Thus the proof of the result is completed.

For fixed time t and any $x \geq 0$, we obtain the reliability function of the second conditional spacing $(X_{3,n} - X_{2,n} > x \mid X_{1,n} = t)$ as follows.

$$\begin{aligned}
P(D^t_{2,n-1} > x) &= P(X_{3,n} - X_{2,n} > x \mid X_{1,n} = t) \\
&= \sum_{i=1}^{n} P(X_{3,n} - X_{2,n} > x, X_{1,n} = X_i \mid X_{1,n} = t) \\
&= \sum_{i=1}^{n} P(X_{3,n} - X_{2,n} > x \mid X_{1,n} = X_i = t) P(X_{1,n} = X_i \mid X_{1,n} = t).
\end{aligned}$$

It is noted that, for fixed $t > 0$ and any $x > 0$,

$$\begin{aligned}
&P(X_{3,n} - X_{2,n} > x \mid X_{1,n} = X_i = t) \\
= &\sum_{j \neq i, j=1}^{n} P(X_{3,n} - X_{2,n} > x, X_{2,n} = X_j \mid X_{1,n} = X_i = t) \\
= &\sum_{j \neq i, j=1}^{n} \int_t^{+\infty} P(X_{3,n} - y > x \mid X_{2,n} = X_j = y, X_{1,n} = X_i = t) \\
&\quad \times P(X_{2,n} = X_j = y \mid X_{1,n} = X_i = t) dy
\end{aligned}$$

$$= \sum_{j\neq i, j=1}^{n} \int_{t}^{+\infty} e^{-(s-\lambda_i-\lambda_j)x} \lambda_j e^{-(s-\lambda_i)(y-t)} dy$$

$$= \sum_{j\neq i, j=1}^{n} \frac{\lambda_j}{s-\lambda_i} e^{-(s-\lambda_i-\lambda_j)x}. \tag{7.4}$$

Thus, we have

$$P(D^t_{2,n-1} > x) = \sum_{i=1}^{n} \sum_{j\neq i, j=1}^{n} \frac{\lambda_i \lambda_j}{s(s-\lambda_i)} e^{-(s-\lambda_i-\lambda_j)x} \tag{7.5}$$

$$= \sum_{i=1}^{n} P(\min S = X_i) \sum_{j\neq i, j=1}^{n} P(\min S|i = X_j) P(\min S|i, j > x).$$

Remark Also, from (7.4), given a series system lifetime equals component with lifetime X_i and have failed at time t, the reliability of the second conditional spacing does not depend on this component lifetime; And obviously, $\frac{1}{s}\sum_{i=1}^{n}\sum_{j\neq i,j=1}^{n}\frac{\lambda_i\lambda_j}{s-\lambda_i} = 1$, hence, by (7.5), the reliability function of the second conditional spacing is also a mixture of Exponential distribution.

Furthermore, for fixed time t and any $x > 0$, the reliability function of the third conditional spacing $D^t_{3,n-1}$ is given as follows.

$$P(D^t_{3,n-1} > x) = P(X_{4,n} - X_{3,n} > x \mid X_{1,n} = t)$$

$$= \sum_{i=1}^{n} P(X_{4,n} - X_{3,n} > x, X_{1,n} = X_i \mid X_{1,n} = t)$$

$$= \sum_{i=1}^{n} P(X_{4,n} - X_{3,n} > x \mid X_{1,n} = X_i = t) P(X_{1,n} = X_i \mid X_{1,n} = t).$$

It is noted that, by the Markov property of order statistics, for fixed $t > 0$,

$$P(X_{4,n} - X_{3,n} > x \mid X_{1,n} = X_i = t)$$

$$= \sum_{j\neq i, j=1}^{n} P(X_{4,n} - X_{3,n} > x, X_{2,n} = X_j \mid X_{1,n} = X_i = t)$$

$$= \sum_{j\neq i, j=1}^{n} \int_{t}^{+\infty} P(X_{4,n} - X_{3,n} > x, X_{2,n} = X_j = y \mid X_{1,n} = X_i = t) dy$$

$$= \sum_{j\neq i, j=1}^{n} \int_{t}^{+\infty} P(X_{4,n} - X_{3,n} > x \mid X_{2,n} = X_j = y, X_{1,n} = X_i = t)$$

$$\times P(X_{2,n} = X_j = y \mid X_{1,n} = X_i = t) dy$$

$$= \sum_{j\neq i, j=1}^{n} \int_{t}^{+\infty} \frac{e^{-(s-\lambda_i-\lambda_j)x}}{s-\lambda_i-\lambda_j} \sum_{k\neq i, j, k=1}^{n} \lambda_k e^{\lambda_k x} \lambda_j e^{-(s-\lambda_i)(y-t)} dy$$

$$= \sum_{j\neq i, j=1}^{n} \sum_{k\neq i, j, k=1}^{n} \frac{\lambda_j}{s-\lambda_i} \frac{e^{-(s-\lambda_j-\lambda_i)x}}{s-\lambda_j-\lambda_i} \lambda_k e^{\lambda_k x},$$

where the fourth equality follows from the following facts

$$\begin{aligned}
&P(X_{4,n} - X_{3,n} > x \mid X_{2,n} = X_j = y, X_{1,n} = X_i = t) \\
&= \sum_{k\neq i\neq j, k=1}^{n} P(X_{4,n} - X_{3,n} > x, X_{3,n} = X_k \mid X_{2,n} = X_j = y, X_{1,n} = X_i = t) \\
&= \sum_{k\neq i\neq j, k=1}^{n} \int_{y}^{+\infty} P(X_{4,n} - X_{3,n} > x, X_{3,n} = X_k = z \mid X_{2,n} = X_j = y, X_{1,n} = X_i = t) dz \\
&= \sum_{k\neq i\neq j, k=1}^{n} \int_{y}^{+\infty} P(X_{4,n} - X_{3,n} > x \mid X_{3,n} = X_k = z, X_{2,n} = X_j = y, X_{1,n} = X_i = t) \\
&\quad \times P(X_{3,n} = X_k = z \mid X_{2,n} = X_j = y, X_{1,n} = X_i = t) dz \\
&= \sum_{k\neq i\neq j, k=1}^{n} \int_{y}^{+\infty} e^{-(s-\lambda_k-\lambda_j-\lambda_i)x} \lambda_k e^{-(s-\lambda_j-\lambda_i)(z-y)} dz \\
&= \sum_{k\neq i\neq j, k=1}^{n} \frac{e^{-(s-\lambda_j-\lambda_i)x}}{s-\lambda_j-\lambda_i} \lambda_k e^{\lambda_k x},
\end{aligned}$$

and

$$\begin{aligned}
P(X_{2,n} = X_j = y \mid X_{1,n} = X_i = t) &= \frac{P(X_{2,n} = X_j = y, X_{1,n} = X_i = t)}{P(X_{1,n} = X_i = t)} \\
&= \lambda_j e^{-(s-\lambda_i)(y-t)}.
\end{aligned}$$

Combining the above facts, at last, we have

$$\begin{aligned}
P(D_{3,n-1}^t > x) &= \sum_{i=1}^{n} \sum_{\substack{j\neq i \\ j=1}}^{n} \sum_{\substack{k\neq i\neq j \\ k=1}}^{n} \frac{\lambda_i \lambda_j \lambda_k}{s(s-\lambda_i)(s-\lambda_j-\lambda_i)} e^{-(s-\lambda_i-\lambda_j-\lambda_k)x} \\
&= \sum_{i=1}^{n} P(\min S = X_i) \sum_{j\neq i, j=1}^{n} P(\min S|i = X_j) P(\min S|i, j > x) \\
&\quad \sum_{k\neq j\neq i, j=1}^{n} P(\min S|i, j = X_k) P(\min S|i, j, k > x).
\end{aligned}$$

Let $\lambda_{i_0} = 0$, then, according to the above expressions, we can conclude that, for fixed time t and any $x \geq 0$, the reliability function of the $(m-1)$-th $(m = 2, \ldots, n)$ conditional spacing $D_{m-1,n}^t$ is given as

$$\begin{aligned}
&P(D^t_{m-1,n-1} > x) \\
&= P(X_{m,n} - X_{m-1,n} > x \mid X_{1,n} = t) \\
&= \sum_{i_1=1}^{n} \sum_{\substack{i_2 \neq i_1 \\ i_2=1}}^{n} \cdots \sum_{\substack{i_{m-1} \neq i_{m-2} \neq \cdots \neq i_1 \\ i_{m-1}=1}}^{n} \frac{\prod_{k=1}^{m-1} \lambda_{i_k}}{\prod_{k=0}^{m-2} (s - \sum_{l=0}^{k} \lambda_{i_l})} e^{-(s - \sum_{k=1}^{m-1} \lambda_{i_k})x} \qquad (7.6) \\
&= \sum_{i_1=1}^{n} P(\min S = X_{i_1}) \\
&\quad \sum_{\substack{i_2 \neq i_1 \\ i_2=1}}^{n} P(\min S | i_1 = X_{i_2}) \cdots \\
&\quad \sum_{\substack{i_{m-2} \neq i_{m-3} \neq \cdots \neq i_1 \\ i_{m-2}=1}}^{n} P(\min S | i_1, \ldots, i_{m-3} = X_{i_{m-2}}) \\
&\quad \sum_{\substack{i_{m-1} \neq i_{m-2} \neq \cdots \neq i_1 \\ i_{m-1}=1}}^{n} P(\min S | i_1, \ldots, i_{m-2} = X_{i_{m-1}}) P(\min S | i_1, \ldots, i_{m-1} > x),
\end{aligned}$$

where $\{i_1, \ldots, i_{m-1}\}$ is any permutation of subsets of size $m-1$ from $\{1, \ldots, n\}$, and $\min S|i_1 = S - \{X_{i_1}\}$,..., $\min S|i_1, \ldots, i_{m-3} = S - \{X_{i_1}, \ldots, X_{i_{m-3}}\}$, etc., that is, $\lambda_{i_1}, \ldots, \lambda_{i_{m-1}}$ are from $\{\lambda_1, \ldots, \lambda_n\}$ and not identical.

As shown in the above reliability function of the second conditional spacing, we can obtain that, for $i \in \{2, \ldots, n-1\}$, fixed $t > 0$ and any $x \geq 0$, the survival function of the normalized conditional spacing $D^{t,*}_{m-1,n-1}$ is

$$P(D^{t,*}_{m-1,n-1} > x) = \sum_{i_1=1}^{n} \sum_{\substack{i_2 \neq i_1 \\ i_2=1}}^{n} \cdots \sum_{\substack{i_{m-1} \neq i_{m-2} \neq \cdots \neq i_1 \\ i_{m-1}=1}}^{n} \frac{\prod_{k=1}^{m-1} \lambda_{i_k}}{\prod_{k=0}^{m-2} (s - \sum_{l=0}^{k} \lambda_{i_l})} e^{-(s - \sum_{k=1}^{m-1} \lambda_{i_k})x/(n-m)}.$$

Kochar and Kirmani (1995) obtained $D_{i,n} \leq_{hr} D_{i+1,n}$ for $i = 1, \ldots, n-1$, given that $X_1, \ldots, X_n$ are a random sample of size n from a DFR distribution. However, it is difficult to extend this result to the conditional spacings. Fortunately, we can construct usual stochastic order between $D^t_{i,n-1}$ and $D^t_{i+1,n-1}$ in the following theorem.

Theorem 7.3.2 *Let X_i be independent Exponential distribution with hazard rate $\lambda_i (i = 1, 2, \ldots, n)$. Set $s = \lambda_1 + \cdots + \lambda_n$. Then, for fixed $t > 0$ and $m = 3, \ldots, n$, $D^t_{m-2,n-1} \leq_{st} D^t_{m-1,n-1}$.*

Proof According to (7.6), for fixed $t > 0$, $x \geq 0$, and $m = 3, \ldots, n$,

$$\begin{aligned}
P(D^t_{m-2,n-1} > x) &= P(X_{m-1,n} - X_{m-2,n} > x \mid X_{1,n} = t) \\
&= \sum_{i_1=1}^{n} \sum_{\substack{i_2 \neq i_1 \\ i_2=1}}^{n} \cdots \sum_{\substack{i_{m-2} \neq i_{m-3} \neq \cdots \neq i_1 \\ i_{m-2}=1}}^{n} \frac{\prod_{k=1}^{m-2} \lambda_{i_k}}{\prod_{k=0}^{m-3} (s - \sum_{l=0}^{k} \lambda_{i_l})} e^{-(s - \sum_{k=1}^{m-2} \lambda_{i_k})x}.
\end{aligned}$$

It is noted that, for $x \geq 0$,

$$
\begin{aligned}
&P(D^t_{m-1,n-1} > x) - P(D^t_{m-2,n-1} > x) \\
&= \sum_{i_1=1}^{n} \sum_{\substack{i_2 \neq i_1 \\ i_2=1}}^{n} \cdots \sum_{\substack{i_{m-2} \neq i_{m-3} \neq \cdots \neq i_1 \\ i_{m-2}=1}}^{n} \frac{\prod_{k=1}^{m-2} \lambda_{i_k}}{\prod_{k=0}^{m-3} (s - \sum_{l=0}^{k} \lambda_{i_l})} e^{-(s - \sum_{k=1}^{m-2} \lambda_{i_k})x} \\
&\quad \times \left[\sum_{\substack{i_{m-1} \neq i_{m-2} \neq \cdots \neq i_1 \\ i_{m-1}=1}}^{n} \frac{\lambda_{i_{m-1}}}{s - \sum_{l=1}^{m-2} \lambda_{i_l}} e^{\lambda_{i_{m-1}} x} - 1 \right] \\
&= \sum_{i_1=1}^{n} \sum_{\substack{i_2 \neq i_1 \\ i_2=1}}^{n} \cdots \sum_{\substack{i_{m-2} \neq i_{m-3} \neq \cdots \neq i_1 \\ i_{m-2}=1}}^{n} \frac{\prod_{k=1}^{m-2} \lambda_{i_k}}{\prod_{k=0}^{m-2} (s - \sum_{l=0}^{k} \lambda_{i_l})} e^{-(s - \sum_{k=1}^{m-2} \lambda_{i_k})x} \\
&\quad \times \left[\sum_{\substack{i_{m-1} \neq i_{m-2} \neq \cdots \neq i_1 \\ i_{m-1}=1}}^{n} \frac{\lambda_{i_{m-1}}}{s - \sum_{l=1}^{m-2} \lambda_{i_l}} (e^{\lambda_{i_{m-1}} x} - 1) \right]
\end{aligned}
$$

is nonnegative for $x \geq 0$, and hence the result holds.

Naturally, it is expected that there should exist some relations between $D^t_{i,n-2}$ and $D^t_{i,n-1}$ for $i = 1, 2, \ldots, n-2$, however, it is difficult to construct for us. Fortunately, we can investigate the first conditional spacings some stochastic order.

Theorem 7.3.3 *Let X_i be independent Exponential distribution with hazard rate λ_i $(> 0)(i = 1, \ldots, 4)$. Then, for fixed $t > 0$, $D^t_{1,n-1} \leq_{hr} D^t_{1,n-2}$.*

Proof According to (7.6), for fixed time t and any $x > 0$, the reliability functions of $D^t_{1,3}$ and $D^t_{1,2}$ are

$$
P(D^t_{1,n-1} > x) = P(X_{2,n} - X_{1,n} > x \mid X_{1,n} = t) = \frac{e^{-(\sum_{i=1}^{n} \lambda_i)x}}{\sum_{i=1}^{n} \lambda_i} \times \sum_{i=1}^{n} \lambda_i e^{\lambda_i x},
$$

$$
P(D^t_{1,n-2} > x) = P(X_{2,n-1} - X_{1,n-1} > x \mid X_{1,n-1} = t) = \frac{e^{-(\sum_{i=1}^{n-1} \lambda_i)x}}{\sum_{i=1}^{n-1} \lambda_i} \times \sum_{i=1}^{n-1} \lambda_i e^{\lambda_i x},
$$

respectively. In order to get the result, it is sufficient to prove

$$
g(x) = \frac{e^{-(\sum_{i=1}^{n-1} \lambda_i)x} \times \sum_{i=1}^{n-1} \lambda_i e^{\lambda_i x}}{e^{-(\sum_{i=1}^{n} \lambda_i)x} \times \sum_{i=1}^{n} \lambda_i e^{\lambda_i x}} = \frac{\sum_{i=1}^{n-1} \lambda_i e^{(\lambda_i + \lambda_n)x}}{\sum_{i=1}^{n} \lambda_i e^{\lambda_i x}}
$$

is increasing in $x > 0$. Let $g'(x)$ is the derivative of $g(x)$, then, by some computations, we obtain

$$g'(x) \overset{sgn}{=} e^{\lambda_n x} \sum_{i=1}^{n-1} \lambda_i(\lambda_i + \lambda_n)e^{\lambda_i x} \times \sum_{i=1}^{n} \lambda_i e^{\lambda_i x} - \sum_{i=1}^{n} \lambda_i^2 e^{\lambda_i x} \times \sum_{i=1}^{n-1} \lambda_i e^{\lambda_i x} e^{\lambda_n x}$$
$$= \lambda_n e^{\lambda_n x} \sum_{i=1}^{n-1} \lambda_i^2 (e^{2\lambda_i x} + e^{(\lambda_i+\lambda_n)x}) + 2\lambda_n e^{\lambda_n x} \sum_{i=1}^{n-1} \sum_{j=1, j\neq i}^{n-1} \lambda_i \lambda_j e^{(\lambda_i+\lambda_j)x}$$

is nonnegative for $x > 0$. Therefore, the result holds.

Kochar and Korwar (1996) proved that $D_{1,n}$ is independent of $(D_{2,n}, \ldots, D_{n,n})$ under the condition that $X_1, \ldots, X_n$ have Exponential distribution with survival function $\bar{F}_i(t) = exp(-\lambda_i t)$, $t > 0$, for $i = 1, \ldots, n$. We want to know whether or not this is true for conditional spacings, that is, is $D^t_{1,n-1}$ independent of $(D^t_{2,n-1}, \ldots, D^t_{n-1,n-1})$? To answer this question, in what follows, let's look at the joint distribution of conditional spacings. For simplicity, we just consider the case of the first three conditional spacings. For fixed $t > 0$, and any $x \geq 0$, $y \geq 0$, $z \geq 0$,

$$P(D^t_{1,n-1} > x, D^t_{2,n-1} > y, D^t_{3,n-1} > z)$$
$$= P(X_{2,n} - X_{1,n} > x, X_{3,n} - X_{2,n} > y, X_{4,n} - X_{3,n} > z \mid X_{1,n} = t)$$
$$= \sum_{i=1}^{n} P(X_{2,n} - t > x, X_{3,n} - X_{2,n} > y, X_{4,n} - X_{3,n} > z, X_{1,n} = X_i \mid X_{1,n} = t)$$
$$= \sum_{i=1}^{n} \frac{\lambda_i}{s} P(X_{2,n} - t > x, X_{3,n} - X_{2,n} > y, X_{4,n} - X_{3,n} > z \mid X_{1,n} = X_i = t)$$
$$= \sum_{i=1}^{n} \sum_{j\neq i, j=1}^{n} \frac{\lambda_i}{s} P(X_{2,n} - t > x, X_{3,n} - X_{2,n} > y, X_{4,n} - X_{3,n} > z, X_{2,n} = X_j \mid X_{1,n} = X_i = t)$$
$$= \sum_{i=1}^{n} \sum_{j\neq i, j=1}^{n} \int_t^{+\infty} P(X_{2,n} - t > x, X_{3,n} - X_{2,n} > y, X_{4,n} - X_{3,n} > z, X_{2,n} = X_j = h \mid X_{1,n} = X_i = t) \frac{\lambda_i}{s} dh.$$

Noting that, for fixed $t > 0$,

$$\int_t^{+\infty} P(X_{2,n} - t > x, X_{3,n} - X_{2,n} > y, X_{4,n} - X_{3,n} > z, X_{2,n} = X_j = h \mid X_{1,n} = X_i = t)dh$$
$$= \int_{t+x}^{+\infty} P(X_{3,n} - h > y, X_{4,n} - X_{3,n} > z \mid X_{2,n} = X_j = h, X_{1,n} = X_i = t) \times P(X_{2,n} = X_j = h \mid X_{1,n} = X_i = t)dh$$
$$= \int_{t+x}^{+\infty} \sum_{m\neq i\neq j, m=1}^{n} P(X_{3,n} - h > y, X_{4,n} - X_{3,n} > z, X_{3,n} = X_m \mid X_{2,n} = X_j = h, X_{1,n} = X_i = t) \times P(X_{2,n} = X_j = h \mid X_{1,n} = X_i = t)dh$$
$$= \int_{t+x}^{+\infty} \sum_{m\neq i\neq j, m=1}^{n} P(X_{3,n} - h > y, X_{4,n} - X_{3,n} > z \mid X_{3,n} = X_m, X_{2,n} = X_j = h, X_{1,n} = X_i = t) \times P(X_{3,n} = X_m \mid X_{2,n} = X_j = h, X_{1,n} = X_i = t) \times P(X_{2,n} = X_j = h \mid X_{1,n} = X_i = t)dh$$

$$= \int_{t+x}^{+\infty} \sum_{m \neq i \neq j, m=1}^{n} \frac{\lambda_m}{(s - \lambda_j - \lambda_i)} e^{-(s-\lambda_m-\lambda_j-\lambda_i)z} e^{-(s-\lambda_j-\lambda_i)y} \lambda_j e^{-(s-\lambda_i)(h-t)} dh$$

$$= \sum_{m \neq i \neq j, m=1}^{n} \frac{\lambda_j \lambda_m}{(s - \lambda_i)(s - \lambda_j - \lambda_i)} e^{-(s-\lambda_m-\lambda_j-\lambda_i)z} e^{-(s-\lambda_j-\lambda_i)y} e^{-(s-\lambda_i)x},$$

where the fourth equality follows from the following fact that

$$\begin{aligned}
&P(X_{3,n} - h > y, X_{4,n} - X_{3,n} > z \mid X_{3,n} = X_m, X_{2,n} = X_j = h, X_{1,n} = X_i = t) \\
&\times P(X_{3,n} = X_m \mid X_{2,n} = X_j = h, X_{1,n} = X_i = t) \\
&= \int_h^{+\infty} P(X_{3,n} - h > y, X_{4,n} - X_{3,n} > z \mid X_{3,n} = X_m = u, X_{2,n} = X_j = h, X_{1,n} = X_i = t) \\
&\quad \times P(X_{3,n} = X_m = u \mid X_{2,n} = X_j = h, X_{1,n} = X_i = t) du \\
&= \int_{h+y}^{+\infty} P(X_{4,n} - u > z \mid X_{3,n} = X_m = u, X_{2,n} = X_j = h, X_{1,n} = X_i = t) \\
&\quad \times P(X_{3,n} = X_m = u \mid X_{2,n} = X_j = h, X_{1,n} = X_i = t) du \\
&= \int_{h+y}^{+\infty} e^{-(s-\lambda_m-\lambda_j-\lambda_i)z} \lambda_m e^{-(s-\lambda_j-\lambda_i)(u-h)} du \\
&= \frac{\lambda_m}{(s - \lambda_j - \lambda_i)} e^{-(s-\lambda_m-\lambda_j-\lambda_i)z} e^{-(s-\lambda_j-\lambda_i)y}.
\end{aligned}$$

Therefore, for fixed $t > 0$, and any $x > 0$, $y > 0$, $z > 0$, we have

$$\begin{aligned}
&P(D_{1,n-1}^t > x, D_{2,n-1}^t > y, D_{3,n-1}^t > z) \\
&= \sum_{i=1}^{n} \sum_{j \neq i, j=1}^{n} \sum_{m \neq i \neq j, m=1}^{n} \frac{\lambda_i \lambda_j \lambda_m}{s(s - \lambda_i)(s - \lambda_j - \lambda_i)} e^{-(s-\lambda_m-\lambda_j-\lambda_i)z} e^{-(s-\lambda_j-\lambda_i)y} e^{-(s-\lambda_i)x}.
\end{aligned} \tag{7.7}$$

Thus, for fixed $t > 0$, and any $y \geq 0$, $z \geq 0$, it holds

$$\begin{aligned}
&P(D_{2,n-1}^t > y, D_{3,n-1}^t > z) \\
&= \sum_{i=1}^{n} \sum_{j \neq i, j=1}^{n} \sum_{m \neq i \neq j, m=1}^{n} \frac{\lambda_i \lambda_j \lambda_m}{s(s - \lambda_i)(s - \lambda_j - \lambda_i)} e^{-(s-\lambda_m-\lambda_j-\lambda_i)z} e^{-(s-\lambda_j-\lambda_i)y}.
\end{aligned} \tag{7.8}$$

According to (7.3), (7.7) and (7.8), the first spacing $D_{1,n-1}^t$ is not independent of $(D_{2,n-1}^t, D_{3,n-1}^t)$, and hence $(D_{2,n-1}^t, \ldots, D_{n-1,n-1}^t)$. Also, it is easy to find the coefficients of the joint reliability function of $(D_{1,n-1}^t, D_{2,n-1}^t, D_{3,n-1}^t)$ and those of $(D_{2,n-1}^t, D_{3,n-1}^t)$ are the same as the ones of $D_{3,n-1}^t$, that is, the coefficients of the joint reliability function of $(D_{1,n-1}^t, D_{2,n-1}^t, D_{3,n-1}^t)$ and those of $(D_{2,n-1}^t, D_{3,n-1}^t)$ are identical and depend only on $D_{3,n-1}^t$. By this property, we can directly give the joint reliability function of the whole spacing $(D_{1,n-1}^t, D_{2,n-1}^t, \ldots, D_{n-1,n-1}^t)$ as follows

$$
\begin{aligned}
&P(D^t_{1,n-1} > x_1, D^t_{2,n-1} > x_2, \ldots, D^t_{n-1,n-1} > x_{n-1}) \\
&= \sum_{i_1=1}^{n} \sum_{\substack{i_2 \neq i_1 \\ i_2=1}}^{n} \cdots \sum_{\substack{i_{n-1} \neq i_{m-2} \neq \cdots \neq i_1 \\ i_{n-1}=1}}^{n} \frac{\prod_{k=1}^{n-1} \lambda_{i_k}}{\prod_{k=0}^{n-2}(s - \sum_{l=0}^{k} \lambda_{i_l})} \prod_{k=1}^{n-1} e^{-(s-\sum_{l=1}^{k} \lambda_{i_l})x_k}
\end{aligned}
$$

for any $x_1 \geq 0, x_2 \geq 0, \ldots, x_{n-1} \geq 0$.

In what follows, we consider conditional sample range $(X_{n,n} - X_{1,n} \mid X_{1,n} = t)$ (denoted by $R^t_{\lambda_i}$) under the condition that $X_1, \ldots, X_n$ are independent and have Exponential distribution with corresponding hazard rates $\lambda_1, \ldots, \lambda_n$. Its distribution function is

$$
\begin{aligned}
F_{R^t_{\lambda_i}}(x) &= P(X_{n,n} - X_{1,n} \leq x \mid X_{1,n} = t) \\
&= \sum_{i=1}^{n} P(X_{n,n} - X_{1,n} \leq x, X_{1,n} = X_i \mid X_{1,n} = t) \\
&= \sum_{i=1}^{n} P(X_{n,n} - X_{1,n} \leq x \mid X_{1,n} = X_i = t) P(X_{1,n} = X_i \mid X_{1,n} = t) \\
&= \sum_{i=1}^{n} \prod_{\substack{j \neq i \\ j=1}}^{n} \frac{\lambda_i}{s}(1 - e^{-\lambda_j x}) \\
&= \frac{\prod_{j=1}^{n}(1 - e^{-\lambda_j x})}{s} \sum_{i=1}^{n} \frac{\lambda_i}{1 - e^{-\lambda_i x}},
\end{aligned}
$$

for fixed time t and any $x > 0$. If component lifetimes are independent and identically distributed as Exponential with hazard rate λ, then the distribution function of conditional sample range $(X_{n,n} - X_{1,n} \mid X_{1,n} = t)$ (denoted by R^t_{λ}) is given by

$$
\begin{aligned}
P(R^t_{\lambda} \leq x) &= P(X_{n,n} - X_{1,n} \leq x \mid X_{1,n} = t) \\
&= \sum_{i=1}^{n} \prod_{\substack{j \neq i \\ j=1}}^{n} \frac{\lambda_i}{s}(1 - e^{-\lambda x}).
\end{aligned}
$$

It is not difficult to get the following result and hence the proof is omitted here. It in fact gives an upper bound and low one for $R^t_{\lambda_i}$.

Theorem 7.3.4 *Let X_i $(i = 1, 2, \ldots, n)$ be independent Exponential distribution with hazard rate λ_i, Y_i $(i = 1, 2, \ldots, n)$ be independent Exponential distribution with the same hazard rate λ. Let $\lambda_{max} = \max\{\lambda_1, \ldots, \lambda_n\}$, $\lambda_{min} = \min\{\lambda_1, \ldots, \lambda_n\}$. Set $s = \lambda_1 + \cdots + \lambda_n$. Then, for fixed $t > 0$,*

$$
R^t_{\lambda_{min}} \leq_{st} R^t_{\lambda_i} \leq_{st} R^t_{\lambda_{max}}.
$$

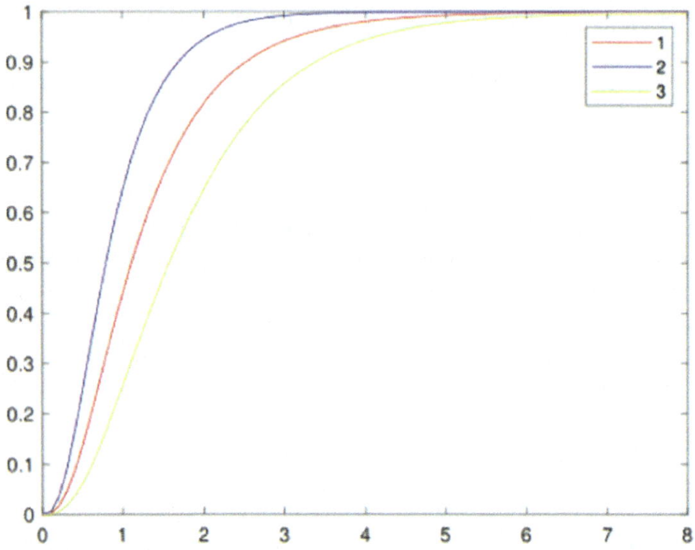

Fig. 7.4 The distribution function of $R^t_{\lambda_i}$, $R^t_{\lambda_{max}}$ and $R^t_{\lambda_{min}}$

For example, let X_1, X_2, X_3, X_4 be independent Exponential distribution with corresponding hazard rates $\lambda_1 = 1$, $\lambda_1 = 1.4$, $\lambda_1 = 1.8$, $\lambda_1 = 2$. Obviously, $\lambda_{max} = 2$, $\lambda_{min} = 1$. For fixed $t > 0$, the distribution functions $R^t_{\lambda_{min}}$, $R^t_{\lambda_i}$ and $R^t_{\lambda_{max}}$ are plotted in Fig. 7.4, where the red curve, the blue one and the orange one represents the distribution function of $R^t_{\lambda_i}$, $R^t_{\lambda_{max}}$ and $R^t_{\lambda_{min}}$, respectively.

7.3.3 Conclusion

This section focus on the conditional spacings based on a series system with independent Exponential distribution having unequal scale parameters. By utilizing the Markov property of order statistics, the survival function and their joint distribution of conditional spacings are obtained when component lifetimes are Exponential random variables. Then, the first spacing is proved to be Schur convex about hazard rates. Also, we obtain that $D^t_{i,n-1} \leq_{st} D^t_{i+1,n-1}$ and $D^t_{1,n-1} \leq_{hr} D^t_{1,n-2}$, that is, $D^t_{i,n-1}$ is stochastically increasing in i in the sense of usual stochastic order. We conjecture that the strong order between $D^t_{i,n-1}$ and $D^t_{i+1,n-1}$ should exist, which may be an open problem. Finally, we consider the case of conditional sample range and give an upper bound and low one for the distribution of sample range in the sense of usual stochastic order.

7.4 On Generalized Conditional Spacings of k-Out-of-n Systems

This section develops the concepts of generalized conditional spacings and generalized normal conditional spacings based on used k-out-of-n systems, which incorporate the factors of system operating time and number of faulty components. First, the survival function of the generalized conditional spacings are obtained. The stochastic and hazard rate order relationships between different generalized conditional spacings and generalized normal conditional spacings are constructed when the parent distributions are of the $DFR(IFR)$ and ILR classes, respectively.

7.4.1 Introduction

Given a k-out-of-n system consisting of n independent and identically distributed components, for $i \in \{1, 2, \ldots, n\}$, X_i represents the lifetime of the ith component, and $X_{i:n}$ be the ith smallest order statistic of $X_1, X_2, \ldots, X_n$. To preserve the state of the data at time t of the used k-out-of-n system, we need to define the statistic $S_n(t)$, which represents the number of observations in the sample $\{X_1, \ldots, X_n\}$ that do not exceed the running time t of k-out-of-n system. Clearly $S_n(t) \leq n - k + 1$. For $s + 1 \leq i < j \leq n$, let

$$D^t_{i,j:n|s} \equiv \left(X_{j:n} - X_{i:n} \mid S_n(t) = s\right), \quad D^{*,t}_{i,j:n|s} \equiv (n-i)\left(X_{j:n} - X_{i:n} \mid S_n(t) = s\right) \tag{7.9}$$

be the corresponding generalized conditional spacing and generalized normal conditional spacing based on the residual life of $n - s$ components in the used k-out-of-n system that are still alive at time t. In particular, when $j = i + 1$, shorthand $D^t_{i,j:n|s}$, $D^{*,t}_{i,j:n|s}$ for $D^t_{i:n|s}$, $D^{*,t}_{i:n|s}$, i.e.,

$$D^t_{i:n|s} \equiv (X_{i+1:n} - X_{i:n} \mid S_n(t) = s), \quad D^{*,t}_{i:n|s} \equiv (n-i)(X_{i+1:n} - X_{i:n} \mid S_n(t) = s) \tag{7.10}$$

be the corresponding conditional spacing and normalized conditional spacing on the used k-out-of-n system. If we assume that n components are placed on test at time 0, then $D^t_{i:n|s}$ may be regarded as the conditional differences between consecutive observations, $D^{*,t}_{i:n|s}$ represents the total test time observed between $X_{i+1:n}$ and $X_{i:n}$ under the condition that s components have already failed.

7.4.2 Stochastic Properties of Generalized Conditional Spacings

We start by an agreement on some symbols that will be used in the lemmas and theorems. Let $X_1, \ldots, X_n$ be nonnegative independent random variables with the identical distribution as X, and $X_{1:n} \leq X_{2:n} \leq \cdots X_{n:n}$ denote the corresponding order statistics. Let $Y_{i:n-s}$ be the ith order statistics of independent random variables $Y_1, \ldots, Y_{n-s}$, which have the same distribution as $Y \equiv (X - t|X > t)$. Denote the distribution functions of random variables X and Y as F and F_t^U, density functions f and f_t^U, and survival functions $\bar{F}$ and $\bar{F}_t^U$, respectively. It is easy to check that $F_t^U(\cdot) = \frac{F(t+\cdot)-F(t)}{\bar{F}(t)}$.

For the concise proof of theorems, we introduce some lemmas.

Lemma 7.4.1 (Lillo et al. 2001) *Let $X_1, \ldots, X_n$ be independent and identically distributed random variables, then $X_{i:n} \leq_{lr} X_{p:m}$ whenever $p \geq i$ and $p - i \geq m - n$.*

Remark (Lillo et al. 2001) If X_1 strengthens to ILR, then the result of the Lemma 7.4.1 can be strengthened to $X_{i:n} \leq_{lr\uparrow} X_{p:m}$. But using DLR instead of ILR, the result $X_{i:n} \leq_{lr\downarrow} X_{p:m}$ does not necessarily hold.

Let $B_{\alpha,\beta}(\cdot)$ denote the cumulative distribution function of a beta distribution with parameters $\alpha \geq 0$ and $\beta \geq 0$ with probability density function

$$b_{\alpha,\beta}(\mu) = \frac{\Gamma(\alpha+\beta)}{\Gamma(\alpha)\Gamma(\beta)}\mu^{\alpha-1}(1-\mu)^{\beta-1}, 0 \leq \mu \leq 1.$$

Then the cumulative distribution function and survival function of the ith order statistic $X_{i:n}$ can be written as $Pr(X_{i:n} \leq x) = B_{i,n-i+1}(F(x))$ and $Pr(X_{i:n} \geq x) = B_{n-i+1,i}(\bar{F}(x))$, respectively.

Lemma 7.4.2 (Nanda and Shaked 2001)

(a) If $X \leq_{hr} Y$, then $X_{i:n} \leq_{hr} Y_{p:m}$ whenever $p \geq i$ and $p - i \geq m - n$.
(b) If $F \leq_{hr} G$, then $B_{\alpha,\beta}(\bar{F}) \leq_{hr} B_{p,q}(\bar{G})$ for all non-negative integers α, β, p, q such that $\alpha \geq p$ and $\beta \leq q$.

Remark Replacing all $\leq_{hr}$ in Lemma 7.4.2 with $\leq_{lr\uparrow}$, Lemma 7.4.2 still holds by Lillo et al. (2001).

Lemma 7.4.3 (Hu and Wei 2001) *If F is $DFR(IFR)$, then $F_{y_1}^U \leq_{hr} (\geq_{hr}) F_{y_2}^U$ for $y_1 \leq y_2$.*

Lemma 7.4.4 (Shanthikumar and Yao 1991) *Let Z_1 and Z_2 be independent random variables. Then $Z_1 \leq_{lr} Z_2$ if and only if $E[g(Z_2, Z_1)] \geq E[g(Z_1, Z_2)]$ for all $g \in \mathcal{G}_{lr}$, where*

$$\mathcal{G}_{lr} = \{g | g(x, y) \geq g(y, x) \text{ whenever } x \geq y\}.$$

Given a k-out-of-n system consisting of n independent and identically distributed components, for $i \in \{1, 2, \ldots, n\}$, X_i represents the lifetime of the ith component. Below we give the survival functions of $D^t_{i,j:n|s}$ and $D^{*,t}_{i:n|s}$ introduced by Eq. (7.9).

Theorem 7.4.5 *For $s + 1 \leq i < j \leq n$, any $x > 0$, the survival function of $D^t_{i,j:n|s}$ is*

$$Pr(D^t_{i,j:n|s} > x) = E\left[B_{n-j+1,j-i}(\bar{F}^U_{t+Y_{i-s:n-s}}(x))\right]$$

where $E(\cdot)$ is the expectation of a random variable.

Proof For $k + 1 \leq i < j \leq n$, any $x > 0$,

$$\begin{aligned}
\bar{F}_{D^t_{i,j:n|s}}(x) &= Pr(D^t_{i,j:n|s} > x) \\
&= Pr\left(X_{j:n} - X_{i:n} > x \mid S_n(t) = s\right) \\
&= Pr\left(Y_{j-s:n-s} - Y_{i-s:n-s} > x\right) \\
&= \int_0^{+\infty} Pr\left(Y_{j-s:n-s} - Y_{i-s:n-s} > x \mid Y_{i-s:n-s} = y\right) dF_{Y_{i-s:n-s}}(y).
\end{aligned}$$

Recall that the conditional distribution of $Y_{j-s:n-s} - Y_{i-s:n-s}$ given $Y_{i-s:n-s} = y$ is the same as the unconditional distribution of the $(j-i)$th order statistic of a random sample of size $n - i$ from the distribution $F^U_{t+y}(\cdot)$.

$$\begin{aligned}
\bar{F}_{D^t_{i,j:n|s}}(x) &= \int_0^{+\infty} B_{n-j+1,j-i}(\bar{F}^U_{t+y}(x))dF_{Y_{i-s:n-s}}(y), \ \forall x \geq 0. \\
&= E\left[B_{n-j+1,j-i}(\bar{F}^U_{t+Y_{i-s:n-s}}(x))\right].
\end{aligned}$$

Remark From Theorem 7.4.5 it is easy to obtain the survival function of generalized normal conditional spacing $D^{*,t}_{i:n|s}$ is $Pr(D^{*,t}_{i:n|s} > x) = E\left[B_{n-j+1,j-i}\left(\bar{F}^U_{t+Y_{i-s:n-s}}\left(\frac{x}{n-i}\right)\right)\right]$.

In the following, we will perform some stochastic comparisons among generalized conditional spacings.

Theorem 7.4.6 *Let $X_1, \ldots, X_n$ be independent and identically distributed component lifetimes of a k-out-of-n system with X_1 being a DFR distribution, for $\forall\, t_1 \leq t_2$, $p - h \geq i - s$ and $q - j \geq p - i \geq m - n$. Then*

$$D^{t_1}_{i,j:n|s} \leq_{st} D^{t_2}_{p,q:m|h}.$$

Proof As shown in the above theorem, we can obtain that

$$\bar{F}_{D^{t_1}_{i,j:n|s}}(x) = \int_0^{+\infty} B_{n-j+1,j-i}(\bar{F}^U_{t_1+y}(x))dF_{Y_{i-s:n-s}}(y), \ \forall x \geq 0.$$

$$\bar{F}_{D^{t_2}_{p,q:m|h}}(x) = \int_0^{+\infty} B_{m-q+1,q-p}(\bar{F}^U_{t_2+y}(x))dF_{Y_{p-h:m-h}}(y), \ \forall x \geq 0.$$

Since the parameters satisfy $q-j \geq p-i \geq m-n$, by Lemma 7.4.2 (b),

$$B_{n-j+1,j-i}(\bar{F}^{U}_{t_2+y}(\cdot)) \leq_{hr} B_{m-q+1,q-p}(\bar{F}^{U}_{t_2+y}(\cdot)).$$

Since the hazard rate order implies the usual stochastic order, we have

$$\bar{F}_{D^{t_2}_{p,q:m|h}}(x) \geq \int_0^{+\infty} B_{n-j+1,j-i}(\bar{F}^{U}_{t_2+y}(x))dF_{Y_{p-h:m-h}}(y), \ \forall x \geq 0.$$

By Lemmas 7.4.3 and 7.4.2 (b), the function $B_{n-j+1,j-i}(\bar{F}^{U}_{y}(x))$ is increasing in y for each x. Hence,

$$\bar{F}_{D^{t_2}_{p,q:m|h}}(x) \geq \int_0^{+\infty} B_{n-j+1,j-i}(\bar{F}^{U}_{t_1+y}(x))dF_{Y_{i-s:n-s}}(y) = \bar{F}_{D^{t_1}_{i,j:n|s}}(x), \ \forall x \geq 0,$$

since Lemma 7.4.1 implies $Y_{i-s:n-s} \leq_{lr} Y_{p-h:m-h}$ and hence $Y_{i-s:n-s} \leq_{st} Y_{p-h:m-h}$ for $\forall\, t_1 \leq t_2$, $p-h \geq i-s$, and $p-i \geq m-n$. We get the required result.

Corollary 7.4.7 *The corner labels in Theorem 7.4.6 have many constraints, but the conclusion holds when the conditions take an equal sign. For ease of application, we list some special cases.*

(1) If $p \geq i$, then $D^t_{i,i+m:n|s} \leq_{st} D^t_{p,p+m:n|s}$.(Conditional m-spacings)
(2) If $q \geq j$, then $D^t_{i,j:n|s} \leq_{st} D^t_{i,q:n|s}$.
(3) If $n \geq m$, then $D^t_{i,j:n|s} \leq_{st} D^t_{i,j:m|s}$.
(4) If $s \geq h$, then $D^t_{i,j:n|s} \leq_{st} D^t_{i,j:n|h}$.
(5) If $t_1 \leq t_2$, then $D^{t_1}_{i,j:n|s} \leq_{st} D^{t_2}_{i,j:n|s}$.

Example 7.4.8 Suppose a k-out-of-n system with n components whose lifetimes are IID follows the first Pareto distribution, i.e., $f(x) = \frac{1}{x^2}$, $(x \geq 1)$, then X_1 is clearly DFR. From Theorem 7.4.5, we can obtain generalized conditional spacings and survival functions, respectively.

$$D^t_{3,6:6|1} = (X_{6:6} - X_{3:6} \mid S_6(t) = 1), \ \bar{F}_{D^t_{3,6:6|1}}(x) = \int_0^{+\infty} \frac{20t^4 y}{(t+y)^6} \cdot \left[1 - \frac{x^3}{(t+x+y)^3}\right] dy,$$

$$D^t_{3,5:5|2} = (X_{5:5} - X_{3:5} \mid S_5(t) = 2), \ \bar{F}_{D^t_{3,5:5|2}}(x) = \int_0^{+\infty} \frac{3t^3}{(t+y)^3} \cdot \frac{t+2x+y}{(t+x+y)^2} dy,$$

$$D^t_{2,5:6|1} = (X_{5:6} - X_{2:6} \mid S_6(t) = 1), \ \bar{F}_{D^t_{2,5:6|1}}(x) = \int_0^{+\infty} \frac{5t^5}{(t+y)^6} \cdot \left[1 - \frac{x^3(4t+x+4y)}{(t+x+y)^4}\right] dy,$$

$$D^t_{3,5:6|1} = (X_{5:6} - X_{3:6} \mid S_6(t) = 1), \ \bar{F}_{D^t_{3,5:6|1}}(x) = \int_0^{+\infty} \frac{20t^4 y(t+3x+y)}{(t+x+y)^3(t+y)^4} dy,$$

$$D^t_{3,5:6|2} = (X_{5:6} - X_{3:6} \mid S_6(t) = 2), \ \bar{F}_{D^t_{3,5:6|2}}(x) = \int_0^{+\infty} \frac{4t^4(t+3x+y)}{(t+x+y)^3(t+y)^3} dy.$$

Suppose $t = 1$, We can obtain the curves $g_1(x) = \bar{F}_{D^1_{3,6:6|1}}(x)$, $g_2(x) = \bar{F}_{D^1_{3,5:5|2}}(x)$, $g_3(x) = \bar{F}_{D^1_{2,5:6|1}}(x)$, $g_4(x) = \bar{F}_{D^1_{3,5:6|1}}(x)$ and $g_5(x) = \bar{F}_{D^1_{3,5:6|2}}(x)$ as shown in Fig. 7.5. From the positional relationships of the function curves, we can determine conclusions corresponding to Corollary 7.4.7: (1)$D^1_{2,5:6|1} \leq_{st} D^1_{3,6:6|1}$, (2) $D^1_{3,5:6|1} \leq_{st} D^1_{3,6:6|1}$, (3) $D^1_{3,5:6|2} \leq_{st} D^1_{3,5:5|2}$, (4)$D^1_{3,5:6|2} \leq_{st} D^1_{3,5:6|1}$.
For $t = 2$, we can obtain the curves $g_2^{t=2}(x) = \bar{F}_{D^2_{3,5:5|2}}(x)$ and $g_5^{t=2}(x) = \bar{F}_{D^2_{3,5:6|2}}(x)$ as shown in Fig. 7.6. Based on the positional relationship of the function curves, we can determine that the conclusion corresponding to Corollary 7.4.7: (5) $D^1_{3,5:5|2} \leq_{st} D^2_{3,5:5|2}$ and $D^1_{3,5:6|2} \leq_{st} D^2_{3,5:6|2}$.

Theorem 7.4.9 *Let $X_1, \ldots, X_n$ be independent and identically distributed component lifetimes of a k-out-of-n system with X_1 being a $DFR(IFR)$ distribution, for $s \geq h$, then*

$$D^t_{i,j:n|s} \leq_{hr} (\geq_{hr}) D^t_{i,j:n|h}.$$

Proof By Theorem 7.4.5, the survival functions of $D^t_{i,j:n|s}$ and $D^t_{i,j:n|h}$ can be obtained as follows,

$$\bar{F}_{D^t_{i,j:n|s}}(x) = E\left[B_{n-j+1,j-i}(\bar{F}^U_{t+Y_{i-s:n-s}}(x))\right],$$
$$\bar{F}_{D^t_{i,j:n|h}}(x) = E\left[B_{n-j+1,j-i}(\bar{F}^U_{t+Y_{i-h:n-h}}(x))\right].$$

The conclusion $D^t_{i,j:n|s} \leq_{hr} D^t_{i,j:n|h}$ to be proved is equivalent to proving

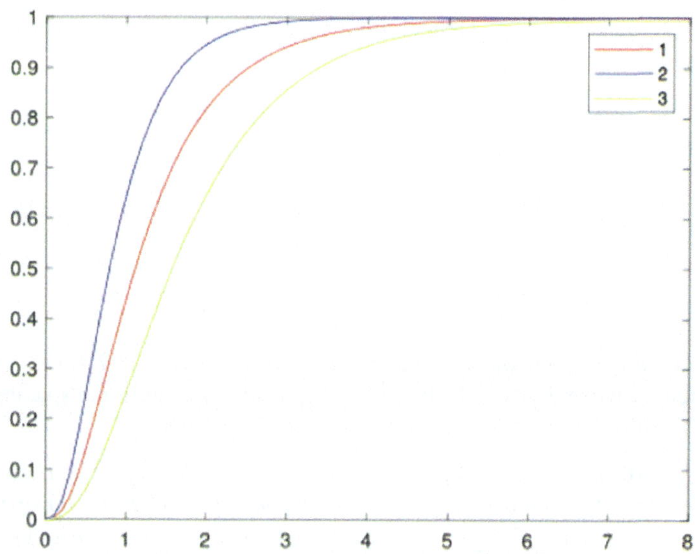

Fig. 7.5 Survival functions of generalized conditional spacings

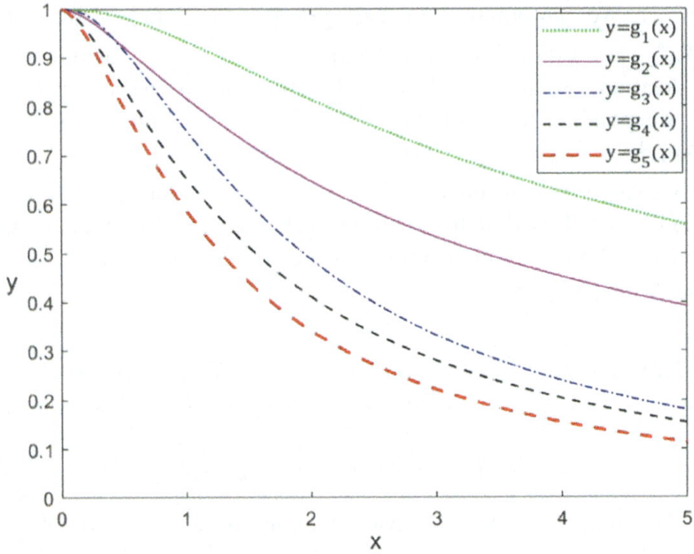

Fig. 7.6 Survival functions of generalized conditional spacings

$$\frac{E\left[B_{n-j+1,j-i}(\bar{F}^U_{t+Y_{i-s:n-s}}(x_1))\right]}{E\left[B_{n-j+1,j-i}(\bar{F}^U_{t+Y_{i-h:n-h}}(x_1))\right]} \geq \frac{E\left[B_{n-j+1,j-i}(\bar{F}^U_{t+Y_{i-s:n-s}}(x_2))\right]}{E\left[B_{n-j+1,j-i}(\bar{F}^U_{t+Y_{i-h:n-h}}(x_2))\right]}, \; for \; 0 < x_1 \leq x_2. \tag{7.11}$$

Since F is DFR, by Lemmas 7.4.3 and 7.4.2(b), one obtains $B_{n-j+1,j-i}(\bar{F}^U_{y_1}) \leq_{hr} B_{n-j+1,j-i}(\bar{F}^U_{y_2})$, for $y_1 \leq y_2$. Hence, $g(y_2, y_1) = B_{n-j+1,j-i}(\bar{F}^U_{y_1}(x_1)) \cdot B_{n-j+1,j-i}(\bar{F}^U_{y_2}(x_2)) \in \mathcal{G}_{lr}$ for $x_1 \leq x_2$.

By Lemma 7.4.1 $Y_{i-s:n-s} \leq_{lr} Y_{i-h:n-h}$ for $s \geq h$. So applying Lemma 7.4.4 yields that, for $x_1 \leq x_2$,

$$E\left[B_{n-j+1,j-i}(\bar{F}^U_{t+Y_{i-s:n-s}}(x_1))\right] \cdot E\left[B_{n-j+1,j-i}(\bar{F}^U_{t+Y_{i-h:n-h}}(x_2))\right]$$

$$\geq E\left[B_{n-j+1,j-i}(\bar{F}^U_{t+Y_{i-h:n-h}}(x_1))\right] \cdot E\left[B_{n-j+1,j-i}(\bar{F}^U_{t+Y_{i-s:n-s}}(x_2))\right].$$

That is, inequality (7.11) holds, so that the proof of the Theorem 7.4.9 is completed.

In particular, when $j = i + 1$, the following result on conditional spacings holds. The proof is similar to that of Theorem 7.4.9 so it is omitted.

Theorem 7.4.10 *Let $X_1, \ldots, X_n$ be independent and identically distributed component lifetimes of a k-out-of-n system with X_1 being a $DFR(IFR)$ distribution. Then $D^t_{i:n|s} \leq_{hr} (\geq_{hr}) D^t_{(i+1):(n+1)|s}$ for fixed $i \in \{s+1, s+2, \ldots, n-1\}$.*

If the conditional IFR in Theorem 7.4.9 is enhanced to ILR, the following stronger conclusion can be obtained.

Theorem 7.4.11 *Let $X_1, \ldots, X_n$ be independent and identically distributed component lifetimes of a k-out-of-n system with X_1 being a ILR distribution, for $\forall\, t_1 \le t_2$, $s \ge h$, then*

$$D^{t_1}_{i,j:n|s} \ge_{hr} D^{t_2}_{i,j:n|h}.$$

Proof Using the conclusions of the above remarks, Theorem 7.4.11 can be proved using a similar proof procedure to that of Theorem 7.4.9.

Remark If DLR is used in Theorem 7.4.11 instead of ILR, it is still open whether the inverse partial order conclusion holds. Under the conditions of Theorem 7.4.11, it is easy to obtain

$$D^{t_1}_{i,j:n|s} \ge_{hr} D^{t_2}_{i,j:n|h} \quad \Longleftrightarrow \quad \begin{array}{ccccc} & & D^{t_2}_{i,j:n|s} & & \\ & \overset{\ge_{hr}}{\nearrow}\,(1) & & (3)\,\overset{\ge_{hr}}{\searrow} & \\ D^{t_1}_{i,j:n|s} & & & & D^{t_2}_{i,j:n|h}. \\ & \underset{\ge_{hr}}{\searrow}\,(2) & & (4)\,\underset{\ge_{hr}}{\nearrow} & \\ & & D^{t_1}_{i,j:n|h} & & \end{array}$$

Example 7.4.12 Suppose a series system has six components whose lifetimes are IID follows the Gamma distribution, i.e., $f(x) = xe^{-x}$, $(x \ge 0)$, then X is clearly ILR. By definition $D^t_{i,j:n|s} \equiv \left(X_{j:n} - X_{i:n} \mid S_n(t) = s\right)$ (refer to the Eq. 7.9), we have

$$D^t_{3,5:6|2} = (X_{5:6} - X_{3:6} \mid S_6(t) = 2)\,,\ D^t_{3,5:6|1} = (X_{4:4} - X_{2:4} \mid S_6(t) = 1)\,.$$

From Theorem 7.4.5 we can calculate survival functions

$$\bar{F}_{D^t_{3,5:6|2}}(x) = \int_0^{+\infty} \frac{4(1+t+x+y)(1+t+y)(t+y)}{(1+t)^4 e^{2x+4y}} \cdot \left[3 - \frac{2(1+t+x+y)}{(1+t+y)e^x}\right] dy,$$

$$\bar{F}_{D^t_{3,5:6|1}}(x) = \int_0^{+\infty} \frac{20(1+t+x+y)^2(1+t+y)(t+y)}{(1+t)^4 e^{2x+4y}} \left[3 - \frac{2(1+t+x+y)}{(1+t+y)e^x}\right]\left[1 - \frac{1+t+y}{(1+t)e^y}\right] dy.$$

To determine the hazard rate relationship of $D^1_{3,5:6|2}$, $D^2_{3,5:6|2}$, $D^1_{3,5:6|1}$ and $D^2_{3,5:6|1}$, let

$$g_6(x) = \frac{\bar{F}_{D^2_{3,5:6|2}}(x)}{\bar{F}_{D^1_{3,5:6|2}}(x)},\ g_7(x) = \frac{\bar{F}_{D^1_{3,5:6|1}}(x)}{\bar{F}_{D^1_{3,5:6|2}}(x)},\ g_8(x) = \frac{\bar{F}_{D^2_{3,5:6|1}}(x)}{\bar{F}_{D^2_{3,5:6|2}}(x)},\ g_9(x) = \frac{\bar{F}_{D^2_{3,5:6|1}}(x)}{\bar{F}_{D^1_{3,5:6|1}}(x)},$$

and draw their curves as shown in Fig. 7.7. From the monotonicity of curves in Fig. 7.7, which corresponds to the conclusion of the last Remark,

$$(1)\ D^1_{3,5:6|2} \ge_{hr} D^2_{3,5:6|2},\ (2)\ D^1_{3,5:6|2} \ge_{hr} D^1_{3,5:6|1},\ (3)\ D^2_{3,5:6|2} \ge_{hr} D^2_{3,5:6|1},\ (4)\ D^1_{3,5:6|1} \ge_{hr} D^2_{3,5:6|1}.$$

When $j = i + 1$, we have the following hazard rate order relation for between generalized normal condition spacings.

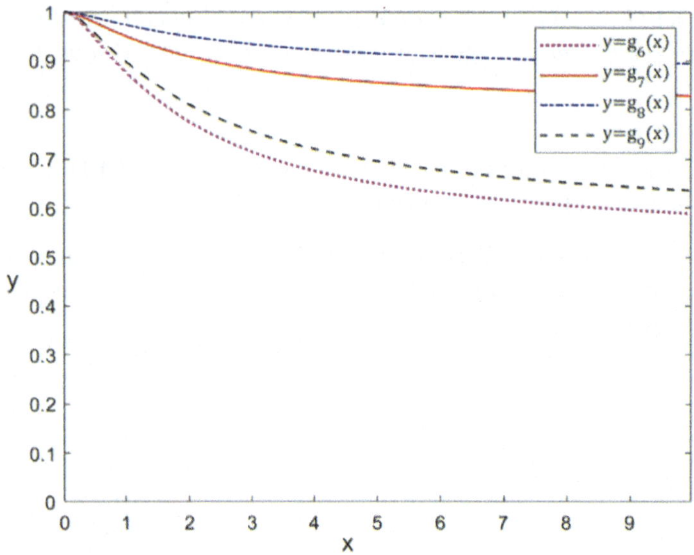

Fig. 7.7 The curves of functions $g_i(x)$

Theorem 7.4.13 *Let $X_1, \ldots, X_n$ be independent and identically distributed component lifetimes of a k-out-of-n system with X_1 being a DFR distribution. Then, for fixed $t > 0$,*

$$D^{*,t}_{i:n|s} \leq_{hr} D^{*,t}_{(i+1):n|s}, \; i \in \{s+1, s+2, \ldots, n-2\}.$$

Proof For $j = i + 1$, we further simplify the survival function of $D^t_{i:n|s}$.

$$\begin{aligned}
\bar{F}_{D^t_{i:n|s}}(x) &= Pr\left(X_{i+1:n} - X_{i:n} > x \mid S_n(t) = s\right) \qquad (7.12)\\
&= \int_0^{+\infty} Pr\left(Y_{i-s+1:n-s} - Y_{i-s:n-s} > x \mid Y_{i-s:n-s} = y\right) dF_{Y_{i-s:n-s}}(y)\\
&= \binom{n-s}{i-s} \int_0^{+\infty} [\bar{F}^U_t(y+x)]^{n-i}(i-s)[F^U_t(y)]^{i-s-1} f^U_t(y)dy\\
&= \binom{n-s}{i-s} \int_0^{+\infty} [\bar{F}^U_t(y+x)]^{n-i} dF_{Y_{i-s,i-s}}(y)\\
&= \binom{n-s}{i-s} E[\bar{F}^U_t(Y_{i-s,i-s}+x)]^{n-i}.
\end{aligned}$$

The survival functions of $D^{*,t}_{i:n|s}$ and $D^{*,t}_{(i+1):n|s}$ are as

$$\bar{F}_{D^{*,t}_{i:n|s}}(x) = \binom{n-s}{i-s} E\left[\bar{F}^U_t\left(Y_{i-s,i-s} + \frac{x}{n-i}\right)\right]^{n-i},$$

$$\bar{F}_{D^{*,t}_{(i+1):n|s}}(x) = \binom{n-s}{i-s+1} E\left[\bar{F}^U_t\left(Y_{i-s+1,i-s+1} + \frac{x}{n-i-1}\right)\right]^{n-i-1}.$$

The conclusion $D^{*,t}_{i:n|s} \leq_{hr} D^{*,t}_{(i+1):n|s}$ to be proved is equivalent to proving

$$\frac{E\left[\bar{F}_t^U\left(Y_{i-s,i-s}+\frac{x_1}{n-i}\right)\right]^{n-i}}{E\left[\bar{F}_t^U\left(Y_{i-s+1,i-s+1}+\frac{x_1}{n-i-1}\right)\right]^{n-i-1}} \geq \frac{E\left[\bar{F}_t^U\left(Y_{i-s,i-s}+\frac{x_2}{n-i}\right)\right]^{n-i}}{E\left[\bar{F}_t^U\left(Y_{i-s+1,i-s+1}+\frac{x_2}{n-i-1}\right)\right]^{n-i-1}},\ for\ 0<x_1\leq x_2. \tag{7.13}$$

Let $M(x,y)=[\bar{F}_t^U(y+\frac{x}{n-i-1})]^{n-i-1}$, $N(x,y)=[\bar{F}_t^U(y+\frac{x}{n-i})]^{n-i}$, for $0<x_1\leq x_2$, $0<y_1\leq y_2$, $0<x$, $0<y$, then

(a) $\frac{M(x,y_2)}{N(x,y_1)}$ is increasing in x.
(b) $\frac{N(x,y_2)}{N(x,y_1)}$ is increasing in x.
(c) $\frac{M(x,y)}{N(x,y)}$ is increasing in y.

Next the conclusion (a) is verified.

$$ln\left(\frac{M(x,y_2)}{N(x,y_1)}\right)=(n-i-1)ln\bar{F}_t^U\left(\frac{x}{n-i-1}+y_2\right)-(n-i)ln\bar{F}_t^U\left(\frac{x}{n-i}+y_1\right).$$

Calculate the partial derivatives of the above equation with respect to x, and then get the right-hand side of the following equation to be non-negative, based on the DFR property of X_1.

$$\frac{\partial}{\partial x}ln\left(\frac{M(x,y_2)}{N(x,y_1)}\right)=-\frac{f_t^U(\frac{x}{n-i-1}+y_2)}{\bar{F}_t^U(\frac{x}{n-i-1}+y_2)}+\frac{f_t^U(\frac{x}{n-i}+y_1)}{\bar{F}_t^U(\frac{x}{n-i}+y_1)}\geq 0.$$

Thus, for $0<y_1\leq y_2$, $\frac{M(x,y_2)}{N(x,y_1)}$ is increasing in x. Conclusions (b) and (c) are verified in a similar way, So they are omitted.

From (a) and (c), $\frac{M(x_2,y_2)}{N(x_2,y_2)}-\frac{M(x_1,y_1)}{N(x_1,y_1)}\geq\frac{M(x_1,y_2)}{N(x_1,y_2)}-\frac{M(x_2,y_1)}{N(x_2,y_1)}$ and $\frac{M(x_2,y_2)}{N(x_2,y_2)}-\frac{M(x_1,y_1)}{N(x_1,y_1)}\geq 0$.

From (b), $N(x_2,y_2)\cdot N(x_1,y_1)\geq N(x_1,y_2)\cdot N(x_2,y_1)$. Consequently,

$$M(x_2,y_2)N(x_1,y_1)-M(x_1,y_1)N(x_2,y_2)\geq M(x_1,y_2)N(x_2,y_1)-M(x_2,y_1)N(x_1,y_2). \tag{7.14}$$

By the fact that the random variables $Y_{i-s,i-s}$ and $Y_{i-s+1,i-s+1}$ are independent, inequality (7.14) and $Y_{i-s,i-s}\leq_{lr}Y_{i-s+1,i-s+1}$, it follows that

$$\begin{aligned}
&E\left[N(x_2,Y_{i-s,i-s})M(x_1,Y_{i-s+1,i-s+1})\right]-E\left[N(x_1,Y_{i-s,i-s})M(x_2,Y_{i-s+1,i-s+1})\right]\\
&=\int\int_{y_1\leq y_2}[N(x_2,y_1)M(x_1,y_2)-N(x_1,y_1)M(x_2,y_2)]h(y_1,y_2)dy_1dy_2\\
&\quad+\int\int_{y_1>y_2}[N(x_2,y_1)M(x_1,y_2)-N(x_1,y_1)M(x_2,y_2)]h(y_1,y_2)dy_1dy_2\\
&=\int\int_{y_1\leq y_2}[N(x_2,y_1)M(x_1,y_2)-N(x_1,y_1)M(x_2,y_2)]h(y_1,y_2)dy_1dy_2
\end{aligned}$$

$$+\int\int_{y_1<y_2}[N(x_2,y_2)M(x_1,y_1)-N(x_1,y_2)M(x_2,y_1)]h(y_2,y_1)dy_2dy_1$$

$$(Note\ that\ here\ N(x_2,y_1)M(x_1,y_2)-N(x_1,y_1)M(x_2,y_2)\leq 0)$$

$$\leq\int\int_{y_1\leq y_2}[N(x_2,y_1)M(x_1,y_2)-N(x_1,y_1)M(x_2,y_2)$$

$$+N(x_2,y_2)M(x_1,y_1)-N(x_1,y_2)M(x_2,y_1)]h(y_2,y_1)dy_1dy_2\leq 0,$$

where $h(\cdot,\cdot)$ denotes the joint density function of $Y_{i-s,i-s}$ and $Y_{i-s+1,i-s+1}$. This implies that inequality (7.13) holds, so the proof of Theorem 7.4.13 is complete.

Corollary 7.4.14 Under the assumptions of Theorem 7.4.13 we further have

$$D^{*,t}_{i:(n+1)|s}\leq_{hr}D^{*,t}_{i:n|s}\leq_{hr}D^{*,t}_{(i+1):(n+1)|s},\ i\in\{s+1,s+2,\ldots,n-1\}.$$

Proof Similar to the calculation of Eq. (7.12), the survival functions of $D^{*,t}_{i:n|s}$, $D^{*,t}_{i:(n+1)|s}$ and $D^{*,t}_{(i+1):(n+1)|s}$ can be obtained as follows, respectively

$$\bar{F}_{D^{*,t}_{i:n|s}}(x)=\binom{n-s}{i-s}E\left[\bar{F}_t^U\left(Y_{i-s,i-s}+\frac{x}{n-i}\right)\right]^{n-i},$$

$$\bar{F}_{D^{*,t}_{i:(n+1)|s}}(x)=\binom{n-s+1}{i-s}E\left[\bar{F}_t^U\left(Y_{i-s,i-s}+\frac{x}{n-i+1}\right)\right]^{n-i+1},$$

$$\bar{F}_{D^{*,t}_{(i+1):(n+1)|s}}(x)=\binom{n-s+1}{i-s+1}E\left[\bar{F}_t^U\left(Y_{i-s+1,i-s+1}+\frac{x}{n-i}\right)\right]^{n-i}.$$

The conclusion $D^{*,t}_{i:(n+1)|s}\leq_{hr}D^{*,t}_{i:n|s}$ to be proved is equivalent to proving

$$\frac{E[\bar{F}_t^U(Y_{i-s,i-s}+\frac{x_1}{n-i+1})]^{n-i+1}}{E[\bar{F}_t^U(Y_{i-s,i-s}+\frac{x_1}{n-i})]^{n-i}}\geq\frac{E[\bar{F}_t^U(Y_{i-s,i-s}+\frac{x_2}{n-i+1})]^{n-i+1}}{E[\bar{F}_t^U(Y_{i-s,i-s}+\frac{x_2}{n-i})]^{n-i}},\ for\ 0<x_1\leq x_2.\tag{7.15}$$

The conclusion $D^{*,t}_{i:n|s}\leq_{hr}D^{*,t}_{(i+1):(n+1)|s}$ to be proved is equivalent to proving

$$\frac{E[\bar{F}_t^U(Y_{i-s,i-s}+\frac{x_1}{n-i})]^{n-i}}{E[\bar{F}_t^U(Y_{i-s+1,i-s+1}+\frac{x_1}{n-i})]^{n-i}}\geq\frac{E[\bar{F}_t^U(Y_{i-s,i-s}+\frac{x_2}{n-i})]^{n-i}}{E[\bar{F}_t^U(Y_{i-s+1,i-s+1}+\frac{x_2}{n-i})]^{n-i}},\ for\ 0<x_1\leq x_2.\tag{7.16}$$

By a process similar to that used to prove inequality (7.13), it follows that inequalities (7.15) and (7.16) hold. Thus the proof is complete.

Since X_1 is DFR if and only if $Y=(X_1-t|X_1>t)$ is DFR, the following results are easily obtained according to Kochar and Kirmani (1995). Hence their proofs are omitted.

Theorem 7.4.15 *Let $X_1, \ldots, X_n$ be independent and identically distributed component lifetimes of a k-out-of-n system with X_1 being a DFR distribution. Then*

(1) $D^t_{i:n|s} \leq_{disp} D^t_{i+1:n|s}$ *for* $i \in \{s+1, \ldots, n-2\}$*;*
(2) $Var(D^t_{i:n|s}) \leq Var(D^t_{i+1:n|s})$ *for* $i \in \{s+1, \ldots, n-2\}$*;*
(3) $D^t_{i:n+1|s} \leq_{disp} D^t_{i:n|s}$ *for* $i \in \{s+1, \ldots, n-1\}$*;*
(4) $Var(D^t_{i:n+1|s}) \leq Var(D^t_{i:n|s})$ *for* $i \in \{s+1, \ldots, n-1\}$,

where $Var(\cdot)$ is the variance of a random variable.

Conclusions

The number of components that fail and the cumulative operating time of the system are closely related to the remaining life of the used k-out-of-n system. Inspired by this, the section presents the concepts of generalized conditional spacing and generalized normal conditional spacing based on the used k-out-of-n systems, which incorporate the factors of system operating time and number of faulty components. First, survival functions of generalized conditional spacings are obtained. Then, the stochastic order $D^{t_1}_{i,j:n|s} \leq_{st} D^{t_2}_{p,q:m|h} (\forall t_1 \leq t_2, p-h \geq i-s, \text{and}\, q-j \geq p-i \geq m-n)$ and the hazard rate order $D^t_{i,j:n|s} \leq_{hr} D^t_{i,j:n|h} (s \geq h)$ are obtained when the parent distribution of component lifetimes is DFR. In particular, when $j = i+1$, the hazard rate ordinal relations $D^t_{i:n|s} \leq_{hr} D^t_{(i+1):(n+1)|s}$ and $D^{*,t}_{i:n|s} \leq_{hr} D^{*,t}_{(i+1):n|s}$ are obtained. Moreover, $D^{t_1}_{i,j:n|s} \geq_{hr} D^{t_2}_{i,j:n|h} (\forall t_1 \leq t_2, s \geq h)$ is obtained when the parent distribution family is strengthened to ILR. Finally it is worth noting the replacement of the DFR in Theorems 7.4.6 and 7.4.13 by the IFR, and it is still open whether the reverse partial order of the corresponding conclusion holds.

References

Afify WM (2010) On estimation of the Exponential Pareto Distribution under different sample schemes. Stat Methodol 8:393–402

Al-Kadim KA, Boshi MA (2013) Exponential Pareto distribution. Math Theory Model 3:135–146

Amari SV, McLaughlinand L, Pham H (2006) Costeffective condition-based maintenance using Markov decision processes. In: Proceedings of annual reliability and maintainability symposium, pp 464–469

Arnold BC, Balakrishnan N, Nagaraja HN (1992) A first course in order statistics. Wiley, New York

Asadi M (2004) On the mean residual life and the mean past life of the k-out-of-n systems. In: Proceedings of the 7th Iranian statistical conference, invited papers, vol 17. Allameh Tabatabaie University, Tehran, pp 55–77

Asadi M (2006) On the mean past lifetime of components of a parallel system. J Stat Plan Inference 136:1197–1206

Asadi M, Bairamov I (2005) A note on the mean residual life function of a parallel system. Commun Stat Theory Methods 34:475–484

Asadi M, Bayramoglu I (2006) The mean residual life function of a k-out-of-n structure at the system level. IEEE Trans Reliab 2:314–318

Aven T (1990) Availability formulae for standby systems of similar units that are preventively maintained. IEEE Trans Reliab 39:603–606

Bagdonavicius V, Nikulin M (2001) Accelerated life models: modeling and statistical analysis. Florida, Chapman & Hall/CRC

Bairamov I, Ahsanullah M, Akhundov I (2002) A residual life function of a system having parallel or series structures. J Stat Theory Appl 1:119–132

Bairamov I, Arnold BC (2008) On the residual life lengths of the remaining components in an $n-k+1$ out of n systems. Stat Probab Lett 78:945–952

Balakrishnan N (2007) Permanents, order statistics, outliers, and robustness. Revista Matematica Complutense 20:7–107

Balakrishnan N, Asadi M (2012) A proposed measure of residual life of live components of a coherent system. IEEE Trans Reliab 61:41–49

Balakrishnan N, Rao CR (1998) Handbook of statistics 16-order statistics. Theory and Methods Elsevier, New York

Bapat RB, Beg MI (1989) Order statistics for nonidentically distributed variables and permanents. Sankhya: Indian J Stat 51:79–93

Barlow RE, Proschan F (1966) Inequalities for linear combinations of order statistics from restricted families. Ann Math Stat 37:1574–1592

Z. Zhang, *Conditional Order Statistics and Conditional Spacings*,
https://doi.org/10.1007/978-981-96-7151-9

Barlow RE, Proschan F (1975) Statistical theory of reliability and life testing. Holt, Rinehart and Winston, New York
Belzunce F, Franco M, Ruiz JM (1999) On aging properties based on the residual life of k-out-of-n systems. Probab Eng Inf Sci 13:193–199
Belzunce F, Franco M, Ruiz JM (2001) On partial orderings between coherent systems with different structures. Probab Eng Inf Sci 15:273–293
Bhattacharya D, Samaniego FJ (2008) On the optimal allocation of components within coherent systems. Stat Probab Lett 78:938–943
Birnbaum ZW, Saunders SC (1958) A statistical model for life-length of materials. J Am Stat Assoc 53:151–160
Birolini A (1994) Quality and reliability of technical systems. Springer, New York
Block H, Savits T, Singh H (1998) The reversed hazard rate function. Probab Eng Inf Sci 15:95–102
Blyth CR (1972) Some probability paradoxes in choice from among random alternatives. Am Stat Assoc 67:366–373
Boland PJ (1997) A reliability comparison of basic systems using hazard rate functions. Appl Stoch Model Data Anal 13:377–384
Boland PJ (2001) Signature of indirect majority systems. J Appl Probab 38:597–603
Boland PJ, El-Neweihi E, Proschan F (1988) Active redundancy allocation in coherent systems. Probab Eng Inf Sci 2:343–353
Boland PJ, El-Neweihi E, Proschan F (1991) Redundancy importance and allocation of spares in coherent systems. J Stat Plan Inference 29:55–66
Boland PJ, El-Neweihi E, Proschan F (1992) Stochastic order for redundancy allocations in series and parallel systems. Adv Appl Probab 24:161–171
Boland PJ, Proschan F, Tong YL (1989) Allocations arrangement of components via pairwise rearrangements. Nav Res Logist 36:807–815
Boland PJ, Samaniego F (2003) The signature of a coherent system and its applications in reliability. In: Soyer R, Mazzuchi T, Singpurwalla N (eds) Mathematical reliability: an expository perspective. International series in operations research and management science, vol 67. Kluwer, pp 1–29
Boland PJ, Singh H, Cukic B (2004) The stochastic precedence ordering with applications in sampling and testing. J Appl Probab 41:73–82
Bueno VC (2005) Minimal standby redundancy allocation in a k-out-of-n: F system of dependent components. Eur J Oper Res 165:786–793
Bueno VC, Carmo M (2007) Active redundancy allocation for a k-out-of-n: F system of dependent components. Eur J Oper Res 2:1041–1051
Burkschat M (2009) Multivariate dependence of spacings of generalized order statistics. J Multivar Anal 100:1093–1106
Cao J, Wang Y (1991) The NBUC and NWUC classes of life distributions. J Appl Probab 31:473–479
Carlson RL, Kardomateas GA (1996) An introduction to fatigue in metals and composites. Chapman and Hall, New York
Cha JH, Mi J, Yun WY (2008) Modelling a general standby system and evaluation of its performance. Appl Stoch Model Bus Ind 24:159–169
Chandra NK, Roy D (2001) Some results on the reversed hazard rate. Probab Eng Inf Sci 12:69–70
Chen Y (1994) Classes of life distributions and renewal counting process. J Appl Probab 31:1110–1115
Cochran WG (1952) The χ^2 test of goodness of fit. Ann Math Stat 23:315–345
Daniels HE (1945) The statistical theory of the strength bundles of threads. Proc R Soc Math Phys Eng Sci 183:405–435
Darling DA (1957) The Kolmogorov-Smirnov, Cramér-von Mises tests. Ann Math Stat 28:823–838
David HA, Nagaraja HN (2003) Order statistics. Wiley, New Jersey
Deshpande JV, Kochar SC, Singh H (1991) Aspects of positive aging. J Appl Probab 28:473–479

Ebeling CE (1997) An introduction to reliability and maintainability engineering. McGraw-Hill, Boston

Efron B (1982) The Jacknife, the bootstrap and other re-sampling plans CBMS-NSF regional conference series in applied mathematics society, vol 38. SIAM Philadelphia

El-Neweihi E, Proschan F, Sethuraman J (1986) Optimal allocation of components in parallel-series and series-parallel systems. J Appl Probab 23:770–777

El-Neweihi E, Sethuraman J (1993) Optimal allocation under partial ordering of lifetimes of components. J Appl Probab 25:914–925

Epstein B (1960) Estimation from life test data. Technometrics 2:447–454

Eryilmaz S (2013) On residual lifetime of coherent systems after the r-th failure. Stat Pap 54:243–250

Es A, Mtb C (2019) Stochastic comparisons on conditional residual lifetime and inactivity time of coherent systems with exchangeable components. Stat Probab Lett 145:327–337

Esary JD, Marshall AW (1970) Coherent life functions. SIAM J Appl Math 18:810–814

Finkelstein MS (2004) Mortality in varying environment. In: Working paper of the max planck institute for demographic research, vol 29. Germany, Rostock, pp 145–158

Finkelstein MS (2007) On statistical and information-based virtual age of degrading system. Reliab Eng Syst Saf 92:676–681

Finkelstein MS (2009) Virtual age of non-repairable objects. Reliab Eng Syst Saf 94:666–669

Finkelstein MS, Hazra NK (2016) on stochastic comparisons for load-sharing series and parallel systems. Probab Eng Inf Sci 31:311–329

Finkelstein MS, Finkelstein M (2008) Failure rate modelling for reliability and risk. Springer, London, UK

Finkelstein MS, Vaupel JW (2006) The relative tail of longevity and the mean remaining lifetime. Demogr Res 14(6):111–138

Gasemyr J, Natvig B (1998) The posterior distribution of the parameters of component lifetimes based on autopsy data in a shock model. Scand J Stat 25:271–292

Gasemyr J, Natvig B (2001) Bayesian inference based on partial monitoring of components with applications to preventive system maintenance. Nav Res Logist 48:551–557

Goliforushani S, Asadi M (2011) Stochastic ordering among inactivity times of coherent systems. Sankhya 73:241–262

Goliforushani S, Asadi M, Balakrishnan N (2012) On the residual and inactivity times of the components of used coherent systems. J Appl Probab 49:385–404

Gross AJ, Clark VA, Liu V (1971) Estimation of survival parameters when one of two organs must function for survival. Biometrics 68:369–377

Guess F, Proschan F (1988) Mean residual life: theory and applications. In: Krishnaiha PR, Rao CR (eds) Handbook of statistics, vol 7. North-Holland, Amsterdam, pp 215–224

Gupta RC, Gupta PL, Gupta RD (1998) Modeling failure time data by Lehman alternatives. Commun Stat-Theory Methods 27:887–904

Gurler S, Bairamov I (2009) Parallel and k-out-of-n: G system with nonidentical components and their mean residual life functions. Appl Math Model 33:1116–1125

Hall P (1988) Theoretical comparison of bootstrap confidence intervals. Ann Stat 16:327–953

Hollander M, Proschar F (1984) Nonparametric concepts and methods in reliability. Handbook of statistics Nonparametric methods 34:613–655

Hu T, Jin W, Khaledi BE (2007a) Ordering conditional distributions of generalized order statistics. Probab Eng Inf Sci 21:401–417

Hu T, Li X, Xu M, Zhuang W (2007b) Some new results on ordering conditional distributions of generalized order statistics. Statistics 21:401–417

Hu T, Wang F, Zhu Z (2006) Stochastic comparisons and dependence of spacings from two samples of exponential random variables. Commun Stat-Theory Methods 35:979–988

Hu T, Wei Y (2001) Stochastic comparisons of spacings from restricted families of distributions. Stat Probab Lett 53:91–99

Hu T, Zhuang W (2005) Stochastic properties of p-spacings of generalized order statistics. Probab Eng Inf Sci 19:257–276
Hu T, Zhuang W (2006) Stochastic comparisons of m-spacings. J Stat Plan Inference 136:33–42
Jasinski K, Navarro J, Rychlik T (2009) Bounds on variances of lifetimes of coherent and mixed systems. J Appl Probab 46:894–908
Jelinski Z, Moranda P (1972) Software reliability research. Stat Comput Perform Eval 465–484
Jewell NP, Kalbfleisch JD (1996) Marker processes in survival analysis. Lifetime Data Anal 2:15–19
Johnson NL, Kotz S, Balakrishnan N (1994) Continuous univariate distributions
Kapur KC, Lamberson LR (1977) Reliability in engineering design
Karlin S (1968) Total positivity. Stanford University
Karlin S, Rinott F (1980) Classes of orderings of measures and related correlation inequalities: multivariate reverse rule distributions. J Multivarite Ann 10:499–516
Kececioglu D (1991) Reliability engineering handbook. Prentice Hall, Englewood Cliffs NJ
Kelkin Nama M, Asadi M (2014) Stochastic properties of components in a used coherent system. Methodol Comput Appl Probab 16:675–691
Kelkin Nama M, Asadi M, Zhang Z (2013) On the residual life lengths of the remaining components in a coherent system. Metrika 76:979–996
Keseling C (1999) Conditional distributions of generalized order statistics and some characterizations. Metrika 49:27–40
Khaledi B, Kochar SC (2000) Dependence among spacings. Probab Eng InfAl Sci 14:461–472
Khaledi B, Kochar SC (1999) Stochastic ordering between distributions and their sample spacings. Stat Probab Lett 44:161–166
Khaledi B, Kochar SC (2001) Stochastic properties of spacings in a single-outlier exponential model. Probab Eng Inf Sci 15:401–408
Khaledi B, Shaked M (2007) Ordering conditional lifetimes of coherent systems. J Stat Plan Inference 137:1173–1184
Kijima M (1989) Some results for repairable systems with general repair. Adv Appl Probab 26:89–102
Kim SH, David HA (1990) On the dependence structure of order statistics and concomitants of order statistics. J Stat Plan Inference 24:363–368
Kim H, Kvam PH (2004) Reliability estimation based on system data with an unknown load share rule. Lifetime Data Anal 10:83–94
Kochar SC (1999) On stochastic ordering between distributions and their sample spacings. Stat Probab Lett 42:345–352
Kochar SC, Kirmani S (1995) Some results on normalized spacings from restricted families of distributions. J Stat Plan Inference 46:47–57
Kochar SC, Korwar R (1996) Stochastic orders for spacings of heterogeneous Exponential random variables. J Multivar Anal 57:69–83
Kochar SC, Ma C (1999) Dispersive ordering of convolutions of exponential random variables. Stat Probab Lett 43:321–324
Kochar SC, Mukerjee H, Samaniego FJ (1999) The signature of a coherent system and its application to comparisons among systems. Nav Res Logist 46:507–523
Kochar SC, Rojo J (1996) Some new results on stochastic comparisons of spacings from heterogeneous Exponential distributions. J Multivar Anal 59:272–281
Kochar SC, Xu M (2011) Stochastic comparisons of spacings from heterogeneous samples. In: Wells M, SenGupta A (eds) Advances in directional and linear statistics. Springer, Physica-Verlag HD, pp 113–130
Kotz S, Shanbhag DN (1980) Some new approaches to probability distributions. Adv Appl Probab 12:903–921
Kuo W, Prasad VR (2000) An annotated overview of system reliability optimization. IEEE Trans Reliab 49:176–187
Kuo W, Prasad VR, Tillman F, Hwang L (2001) Optimization system design. Cambridge University Press

Kvam PH, Day D (2001) The multivariate Polya distribution in combat modeling. Nav Res Logist 48:1–17

Kvam PH, Pena EA (2005) Estimating load-sharing properties in a dynamic reliability system. J Am Stat Assoc 100:262–272

Langberg NA, Leon RV, Proschan F (1980) Characterizations of nonparametric classes of life distributions. Ann Probab 8:1163–1170

Lillo R, Nanda A, Shaked M (2001) Preservation of some likelihood ratio stochastic orders by order statistics. Stat Probab Lett 51:111–119

Li T, Zhang Z (2024) Generalized conditional spacings and their stochastic properties. AIMS Math 9(23909):C23923

Li X, Chen J (2004) Aging properties of the residual life length of k-out-of-n systems with independent but non-identical components. Appl Stoch Model Bus Ind 20:143–153

Li X, Hu X (2008) Some new stochastic comparisons for redundancy allocations in series and parallel systems. Stat Probab Lett

Li X, Lu J (2003) Stochastic comparisons on residual life and inactivity time of series and parallel systems. Probab Eng Inf Sci 17:267–275

Li X, Wu YD, Zhang Z (2013) On allocation of general standby redundancy to series and parallel systems. Commun Stat-Theory Methods 42:4056–4069

Li X, Zhang Z (2008) Some stochastic comparisons of conditional coherent systems. Appl Stoch Model Bus Ind 24:541–549

Li X, Zhang Z, Wu YD (2009) Some new results involving general standby systems. Appl Stoch Model Bus Ind 25:632–642

Li X, Zhao P (2006) Some aging properties of the residual life of k-out-of-n system. IEEE Trans Reliab 55:535–541

Li X, Zhao P (2008) Stochastic comparison on general inactivity time and general residual life of k-out-of-n systems. Commun Stat-Simul Comput 37:1005–1019. Li and Lu

Li X, Zuo MJ (2002) On the behaviour of some new ageing properties based upon the residual life of k-out-of-n systems. J Appl Probab 39:426–433

Li X, Zuo MJ (2004) Stochastic comparison of residual life and inactivity time at a random time. Stoch Model 2:229–235

Lin HH, Chen KH, Wang RT (1993) A multivariate exponential shared-load model. IEEE Trans Reliab 42:165–171

Mahmoud MAE, Yhiea NM, El-Said SM (2016) Estimation of parameters for the exponentiated Pareto distribution based on progressively type-II right censored data. J Egyptian Math Soc 24:421–436

Marshall AW, Olkin I, Arnold B (2010) Inequalities: theory of majorization and its applications, 2nd edn. Springer, New York

Marshall AW, Proschan F (1972) Classes of applicable in replacement, with renewal theory applications. Proceedings of the sixth Berkeley symposium on mathematical statistics and probability I:395–415

McCool J (2006) Testing for dependency of failure times in life testing. Technometrics 48:41–48

Meng FC (1996) More on optimal allocation of components in coherent systems. J Appl Probab 33:548–556

Meilijson I (1981) Estimation of the lifetime distribution of the parts from autopsy statistics of the machine. J Appl Probab 18:829–838

Misra N, Meulen EC (2003) On stochastic properties of m-spacings. J Stat Plan Inference 115:683–697

Mi J, Shaked M (2002) Stochastic dominance of random variables implies the dominance of their order statistics. J Indian Stat Assoc 40:161–168

Misra N, Gupta N, Dutt Dhariyal I (2008) Stochastic properties of residual life and inactivity time at a random time. Stoch Model 24:89–102

Müller A, Stoyan D (2002) Comparison methods for stochastic models and risks. Wlley, Westsusse

Nanda AK, Jain K, Singh H (1998) Preservation of some partial orderings under the formation of coherent systems. Stat Probab Lett 39:123–131

Nanda A, Shaked M (2001) The hazard rate and the reversed hazard rate orders, with applications to order statistics. Ann Inst Stat Math 53:853–864

Navarro J, Balakrishnan N, Samaniego FJ (2008a) Mixture representations of residual lifetimes of used systems. J Appl Probab 45:1097–1112

Navarro J, Balakrishnan N, Samaniego FJ (2011) Signature-based representations for the reliability of systems with heterogeneous components. J Appl Probab 48:856–867

Navarro J, Eryilmaz S (2007) Mean residual lifetimes of consecutive-k-out-of-n systems. J Appl Probab 44:82–98

Navarro J, Hernandez PJ (2008) Mean residual life functions of finite mixtures, order statistics and coherent systems. Metrika 67:277–298

Navarro J, Rubio R (2010) Comparisons of coherent systems using stochastic precedence 19:469–486

Navarro J, Ruiz JM, Sandoval CJ (2005) A note on comparisons among coherent systems with dependent components using signatures. Stat Probab Lett 72:179–185

Navarro J, Ruiz JM, Sandoval CJ (2007) Properties of coherent systems with dependent components. Commun Stat-Simul Comput 36:175–191

Navarro J, Rychlik T (2007) Reliability and expectation bounds for coherent systems with exchangeable components. J Multivar Anal 98(1):102–113

Navarro J, Samaniego FJ, Balakrishnan N, Bhattacharya D (2008b) On the application and extension of system signatures in engineering reliability. Nav Res Logist 55:313–327

Navarro J, Samaniego FJ, Balakrishnan N (2010) The joint signature of coherent systems with shared components. J Appl Probab 47:235–253

Navarro J, Shaked M (2006) Hazard rate ordering of order statistics and systems. J Appl Probab 43:391–408

Navarro J, Spizzichino F (2010) Comparisons of series and parallel systems with components sharing the same copula. Appl Stoch Model Bus Ind 26:775–791

Nelson W (1982) Applied Life data analysis. Wiley, New York

Nelson W (1990) Accelerated testing: statistical models, test plans, and data analysis. New York

Oanceaa B, Andrei T, Pirjol D (2017) Income inequality in Romania: the exponential-Pareto distribution. XXPhys A 469:486–498

Ortega JM, Rheinbold CR (1970) Iterative solution of nonlinear equation in several variables. Academic Press, New York

Park C (2010) Parameter estimation for reliability of load sharing systems. IIE Trans 42:753–765

Parvardeh A, Balakrishnan N (2013) Conditional residual lifetimes of coherent systems. Stat Probab Lett 83:2664–2672

Pledger P, Proschan F (1971) Comparisons of order statistics and of spacings from heterogeneous distributions. In: Rustagi JS (ed) Optimizing Methods in Statistics. Academic Press, New York, pp 89–113

Poursaeed MH (2010) A note on the mean past and the mean residual life of a (n-k+1)-out-of-n system under multi monitoring June. Stat Pap 51:409–419

Poursaeed MH, Nematollahi AR (2008) On the mean past and the mean residual life under double monitoring. Commun Stat-Theory Methods 37:1119–1133

Prasad VR, Kuo W, Kim KM (1999) Optimal allocation of identical, multi functional spares in a series system. IEEE Trans Reliab 2:118–126

Rausand M, Høland A (2004) System reliability theory: models, statistical methods, and applications (2nd edn). New Jersey

Ravichandran N (1990) Stochastic Methods in reliability theory. New York

Righter R, Shankthikumar JG (1992) Extremal properties of the FIFO discipline in queueing networks. J Appl Probab 29:967–978

Robertson T, Wright FT (1982) On measuring the conformity of a parameter set to a trend, with applications. Ann Stat 4:1234–1245

Romera R, Valdés JE, Zequeira RI (2004) Active-redundancy allocation in systems. IEEE Trans Reliab 53:313–318
Rosen BW (1964) Tensile failure of fibrous composites. Am Inst Aeronaut Astronaut 2:1985–1991
Ross SM (1996) Stochastic process, 2nd edn. Wiley, New York
Samadi P, Rezaei M, Chahkandi M (2017) On the residual lifetime of coherent systems with heterogeneous components. Metrika 80:1–14
Samaniego F (1985) On closure of the IFR class under formation of coherent systems. IEEE Trans Reliab 34:69–72
Samaniego FJ (2007) System signatures and their applications in engineering reliability. New York
Samaniego FJ, Balakrishnan N, Navarro J (2009) Dynamic signatures and their use in comparing the reliability of new and used systems. Nav Res Logist 56:577–591
Satyanarayana A, Prabhakar A (1978) A new topological formula and rapid algorithm for reliability analysis of complex networks. IEEE Trans Reliab 27:82–100
Scheuer EM (1988) Reliability of a m-out-of-n system when component failure induces higher failure rates in survivors. IEEE Trans Reliab 37:73–74
Shaked M, Shanthikumar JG (1987) Characterization of some first passage times using log-concavity and log-convexity as aging notions. Probab Eng Inf Sci 1:279–291
Shaked M, Shanthikumar JG (1992) Optimal allocation of resources to nodes of parallel and series systems. Adv Appl Probab 24:894–914
Shaked M, Shanthikumar JG (1994) Stochastic orders and their applications. Academic Press, San Diego
Shaked M, Shanthikumar JG (1998) Two variability orders. Probab Eng Inf Sci 12:1–23
Shaked M, Shanthikumar JG (2007) Stochastic orders. Springer, New York
Shaked M, Suarez-Llorens A (2003) On the comparison of reliability experiments based on the convolution order. J Am Stat Assoc 98:693–702
Shaked M, Tong YL (1984) Stochastic ordering of spacings from dependent random variables. Inequalities Stat Probab IMS Lect Notes-Monogr Ser 5:141–149
Shanthikumar GJ, Yao D (1991) Bivariate characterization of some stochastic order relations. Adv Appl Probab 23:642–659
Shao J, Lamberson LR (1991) Modeling a shared-load k-out-of-n: G system. IEEE Trans Reliab 40:205–209
She J, Pecht MG (1992) Reliability of a k-out-of-n warm-standby system. IEEE Trans Reliab 41:72–75
Shechner Z (1984) A load sharing model: the linear break down rule. Nav Res Logist 31:137–144
Singh H, Misra N (1994) On redundancy allocations in systems. J Appl Probab 31:1004–1014
Singh B, Sharma KK, Kumar A (2008) A classical and Bayesian estimation of a k components load sharing parallel system. Comput Stat Data Anal 52:5175–5185
Singh B, Sharma KK, Kumar A (2009) Inferences on the dynamic system model with discrete failure time distribution. Stat Methods Appl 18:521–542
Singh B, Gupta PP (2012) Load-sharing system model and its application to the real data set. Math Comput Simul 82:1615–1629
Singpurwalla ND (1995) Survival in dynamic environments. Stat Sci 10:86–103
Stoyan D (1983) Comparison methods for queues and other stochastic models. New York
Tang YH, Zhang J (2008) New model for load-sharing k-out-of-n: G system with different components. J Syst Eng Electron 19:748–751
Tavangar M, Asadi M (2010) A study on the mean past lifetime of the components of $n-k+1$-out-of-n system at the system level. Metrika 12:59–73
Torrado N, Lillo RE (2013) Likelihood ratio order of spacings from two heterogeneous samples. J Multivar Anal 114:338–348
Torrado N, Lillo RE (2018) Likelihood ratio comparisons among spacings related to both one or two samples. Statistics 49:831–841
Torrado N, Lillo RE, Wiper MP (2010) On the conjecture of Kochar and Korwar. J Multivar Anal 5:1274–1283

Valdés JE, Zequeira RI (2003) On the optimal allocation of an active redundancy in a two-component series system. Stat Probab Lett 63:325–332

Vanderperre EJ (2004) Reliability analysis of a renewable multiple cold standby system. Oper Res 32:288–292

Vapnik V, Navarro J, Rubio R (2010) Comparisons of coherent systems using stochastic precedence. TEST 19:469–486

Wang KS, Huang JJ, Tsai Y, TandHsu FS (2000) Study of loading policies for unequal strength shared-load system. Reliab Eng Syst Saf 67:119–128

Wang YT, Morris RJT (1985) Load sharing in distributed systems. IEEE Trans Comput 34:204–217

Xie H, Hu T (2007) Ordering conditional distributions of spacings of generalized order statistics. J Iran Stat Soc 6:155–171

Xiong WJ, Li DQ, Zhang Z (2020) The number of failed components in a conditional coherent operating system. J Math Res Appl 40:101–110

Xiong WJ, Zhang Z (2020) The number of surviving components of failed coherent systems. Chinese J Appl Probab Stat 36:151–161

Xu M, Li X (2008) Some further results on the winner's rent in the second-price business auction. Sankhya 70:124–133

Yue D, Cao J (2000) Some results on the residual life at random time. Acta Math Appl Sin 16:436–443

Yun WY, Cha JH (2010) A stochastic model for a general load-sharing system under overload condition. Appl Stoch Model Bus Ind 26:624–638

Zhang Z (2010a) Ordering conditional general coherent systems with exchangeable components. J Stat Plan Inference 140:454–460

Zhang Z (2010b) Mixture representations of inactivity times of conditional coherent systems and their applications. J Appl Probab 47:876–885

Zhang Z (2011) Ordering new conditional residual lifetimes of k-out-of-n systems. Commun Stat-Theory Methods 40:1591–1600

Zhang Z, Balakrishnan N (2015) Representations for reliability functions of conditional coherent systems with INID components and ordered properties. J Syst Eng Electron 26:1316–1324

Zhang Z, Balakrishnan N (2016) Representations of the inactivity time for coherent system with heterogeneous components and some ordered properties. Metrika 79:113–126

Zhang Z, Balakrishnan N (2017) Stochastic properties and parameter estimation for a general load-sharing system. Commun Stat-Theory Mothods 46:747–760

Zhang Z, Balakrishnan N, Zhao ZY (2024) On conditional spacings and their properties under coherent system setting. Statistics 58(194):C208

Zhang Z, Chang Y (2011) MRL and MPL functions of parallel systems with INID components under double monitoring. Commun Stat-Simul Comput 40:1514–1523

Zhang Z, Li X (2010) Some new results on stochastic orders and aging properties of coherent systems. IEEE Trans Reliab 59:718–724

Zhang Z, Meeker WQ (2014) The residual lifetime of surviving components from failed coherent systems. IEEE Trans Reliab 63:534–542

Zhang Z, Meeker WQ (2013) Mixture representations of reliability in coherent systems and preservation results under double monitoring. Commun Stat-Theory Methods 42:385–397

Zhang Z, Yang YH (2010) Ordered properties on the residual life and inactivity time of (n-k+1)-out-of-n systems under double monitoring. Stat Probab Lett 80:711–717

Zhang Z, Yang YH, Balakrishnan N (2024) On conditional spacings from heterogeneous exponential random variables. Commun Stat-Simul Comput 53:5816–5827

Zhang Z, Yang YH, Li DQ (2022) Estimation of parameters for load-sharing parallel systems under exponential pareto distribution. Proc Inst Mech Eng Part O-J Risk Reliab 236:248–255

Zhang Z, Yang YH, Ji XJ (2020) On stochastic behaviors of load-sharing parallel systems. Commun Stat-Theory Methods 49:5392–5405

Zhang Z, Yang L, Yang YH (2022) The conditional spacings and their stochastic properties. Stat Probab Lett 186:109466

Subject Index

Z. Zhang, *Conditional Order Statistics and Conditional Spacings*,
https://doi.org/10.1007/978-981-96-7151-9

The manufacturer's authorised representative in the EU is Springer Nature Customer Service Centre GmbH, Europaplatz 3, 69115 Heidelberg, Germany. If you have any concerns regarding our products, please contact ProductSafety@springernature.com

Printed and bound by CPI Group (UK) Ltd, Croydon, CR0 4YY
30/09/2026
02225303-0008